UNITEXT

La Matematica per il 3+2

Volume 175

Editor-in-Chief

Alfio Quarteroni, Politecnico di Milano, Milan, Italy
  École Polytechnique Fédérale de Lausanne (EPFL), Genova, Switzerland

Series Editors

Luigi Ambrosio, Scuola Normale Superiore, Pisa, Italy

Paolo Biscari, Politecnico di Milano, Milan, Italy

Ciro Ciliberto, Università di Roma "Tor Vergata", Rome, Italy

Camillo De Lellis, Institute for Advanced Study, Princeton, USA

Lorenzo Rosasco, DIBRIS, Università degli Studi di Genova, Genova, Italy
  Center for Brains Mind and Machines, Massachusetts Institute of Technology,
Cambridge, Massachusetts, USA
  Istituto Italiano di Tecnologia, Genova, Italy

The **UNITEXT - La Matematica per il 3+2** series is designed for undergraduate and graduate academic courses, and also includes books addressed to PhD students in mathematics, presented at a sufficiently general and advanced level so that the student or scholar interested in a more specific theme would get the necessary background to explore it.

Originally released in Italian, the series now publishes textbooks in English addressed to students in mathematics worldwide.

Some of the most successful books in the series have evolved through several editions, adapting to the evolution of teaching curricula.

Submissions must include at least 3 sample chapters, a table of contents, and a preface outlining the aims and scope of the book, how the book fits in with the current literature, and which courses the book is suitable for.

For any further information, please contact the Editor at Springer:
francesca.bonadei@springer.com

THE SERIES IS INDEXED IN SCOPUS

Andrea Carpignani · Massimo Pappalardo

Theory and Methods of Optimisation

 Springer

Andrea Carpignani [iD]
Department of Mathematics
The Radcliffe School
Wolverton, Milton Keynes, England

Massimo Pappalardo [iD]
Department of Computer Science
University of Pisa
Pisa, Italy

ISSN 2038-5714   ISSN 2532-3318 (electronic)
UNITEXT
ISSN 2038-5722   ISSN 2038-5757 (electronic)
La Matematica per il 3+2
ISBN 978-3-032-01513-6   ISBN 978-3-032-01514-3 (eBook)
https://doi.org/10.1007/978-3-032-01514-3

Cover illustration: "Depiction of a snowflake" by Lucy Carpignani (Bletchley, 2025)

This Springer imprint is published by the registered company Springer Nature Switzerland AG
The registered company address is: Gewerbestrasse 11, 6330 Cham, Switzerland

If disposing of this product, please recycle the paper.

To Lucy
and to Paolo Acquistapace
A friend and a constant source of inspiration

Andrea

To Dina and Martina
my wife and my daughter
who make it all worthwhile

Massimo

Preface

This book grew from a shared intellectual journey, one that began with a collaborative exchange of ideas of linear programming and gradually expanded into a deeper, more systematic study of optimisation between the two authors. What united us from the outset was not only a mutual interest in the subject but more importantly our complementary perspectives: one honed through decades of teaching mathematics, with an emphasis on clarity, accessibility, and pedagogy, the other shaped by a long academic career in Operational Research, blending theoretical innovation with practical applications.

A.C. is an expert educator with over twenty years of experience in distilling complex mathematical ideas into teachable insights; M.P. has spent forty years advancing the field of optimisation through research, while guiding generations of students of Engineering, Mathematics and Informatics. Our partnership, rooted in a common academic heritage at the University of Pisa, reflects the enduring legacy of an institution that prizes both the purity of mathematical theories and the applied fields of engineering and science, where advanced mathematical tools are used as tools to generate progress in science and technology. Though deeply appreciative of abstract theories, we find particular fascination in the interplay between rigorous mathematics and real-world problem-solving, where elegant proofs meet impactful models.

This dual perspective shapes the book's approach. Our book is indeed designed to be both mathematically precise and pedagogically deliberate, favouring simple but effective examples and developing not only the fundamental tools of optimisation, but also the mindset necessary to advance in the study of this subject. Every theorem is presented with complete proofs, the methods chosen have been selected to give an overview of the most classical and basic ones, showing how that specific method is derived, what their limitations are and what the range of its applicability is. In transitioning from the first part and the second part, more rigorous and abstract to the third part, which is the most concrete and algorithmic, we have taken special care to ensure that reader understands the importance of the mathematical tools developed and approaches the more applied problems with awareness, providing examples and insights that illuminate the underlying principles without diminishing their formal rigour.

There are many people that we feel obliged to thank for their advice and support. We are grateful to our colleagues, students, and mentors who have influenced our thinking over the years particularly those at Pisa, where our shared journey began.

Specifically, A.C. wishes to take a moment to thank his mentors, colleagues and lifelong friends, who all contributed, through many years of patience and perseverance, to maintain alive his unconditional love for mathematics: Francesca Acquistapace, Paolo Acquistapace, Fabrizio Broglia, Alessandro Iacopetti, Brendan Lowry, Antonio Tarsia. A particular thank however goes to Lucy Carpignani, without whom A.C. would have never been able to spend so much time doing maths.

Historical introduction

The study of optimisation has deep historical roots, originating from the search of maxima and minima in geometry and physics. The earliest contributions to optimisation can be traced back to the work of LEONHARD EULER (1707–1783) and JOSEPH-LOUIS LAGRANGE (1736–1813) on *calculus of variations*, that constitutes an example of optimisation problem. These pioneers laid the mathematical foundations for solving problems involving continuous quantities, such as determining the shape of a curve that minimises energy or maximises area. Lagrange, in particular, may be considered the father of optimisation, having introduced the multiplier method to address constrained optimisation problems, providing a powerful theoretical framework that remains influential in modern optimisation theory.

The transition from these classical problems to more general formulations of optimisation was driven by the needs of applied sciences and economics. The growth of industrial and economic systems in the 19th and early 20th centuries demanded mathematical tools to efficiently allocate resources. These practical demands led to the formalisation of constrained optimisation problems in finite-dimensional spaces. One of the milestones in this transition was the development of convex analysis, which provided a rigorous mathematical framework to study problems with convex constraints and objective functions. The theory of convex sets and convex functions, as developed by mathematicians such as HERMANN MINKOWSKI (1864–1909), introduced geometric concepts that were instrumental in the analysis of feasible regions and optimal solutions.

The origins of linear programming are often attributed to the pioneering work of the American mathematician GEORGE B. DANTZIG (1914–2005). During World War II, while serving as a civilian with the US Air Force, Dantzig formulated the general linear programming framework and developed the simplex algorithm. This powerful method allowed for efficient optimisation of resource allocation, helping to minimise costs for the military. Due to the war,

Dantzig's contributions were initially classified, but after declassification, he was able to publish the simplex algorithm in 1947, marking a milestone in the field.

Although Dantzig is generally considered the father of linear programming, its roots extend further back. The earliest known work dates to JEAN-BAPTISTE JOSEPH FOURIER (1768–1830), who proposed a method for solving linear programming problems as early as 1827. Despite this early contribution, linear programming remained largely neglected until the 1930s, when the Soviet mathematician LEONID V. KANTOROVICH (1912–1986) explored its practical applications. Kantorovich's innovative work, which included using linear programming to address issues in economics and resource management, was initially overlooked in the USSR but eventually gained recognition, earning him the Nobel Prize in Economic Sciences in 1975.

The study of optimisation continued to evolve beside linear programming. Inspired by Lagrange's multiplier method to address equality constrained problems, HAROLD W. KUHN (1925–2014), ALBERT W. TUCKER (1905–1995), and WILLIAM KARUSH (1917–1997) introduced a set of conditions which are now called the Karush-Kuhn-Tucker (KKT) conditions to address inequality constrained problems.

In the modern era, mathematical programming has become the backbone of machine learning and artificial intelligence. At its core, machine learning involves solving optimisation problems, where the goal is to minimise a loss function that quantifies the error between predictions made by a model and actual observations. Deep learning, a subset of machine learning, has pushed these challenges further by introducing highly non-convex and high-dimensional optimisation problems. For instance, neural networks, which are central to deep learning, require iterative optimisation of their parameters to fit data and generalise to unseen examples.

The optimisation techniques used in deep learning rely heavily on gradient-based methods. These methods exploit the structure of the optimisation problem, often leveraging the differentiability of the loss function and computational efficiency of back-propagation to calculate gradients. Moreover, advanced techniques such as adaptive learning rates, momentum methods, and regularisation are employed to improve convergence and generalisation in practice.

Optimisation's influence extends beyond traditional machine learning and deep learning frameworks. It drives advances in combinatorial optimisation, integer programming, and network optimisation, which are crucial for solving large-scale problems such as transportation networks, resource allocation, and vehicle routing. Classical applications, including engineering design, economic modelling, and physics simulations, continue to leverage the power of optimisation. The field's adaptability underscores its enduring importance as both a theoretical discipline and a practical toolkit for solving complex, real-world challenges.

Audience and background

This book originates from the courses of 'Operations Research' for student in Engineering and of 'Theory and Methods of Optimisation' for graduate students in Mathematics, that one of the authors held at the University of Pisa in the last years. The book is primarily intended for graduate students in mathematics who wish to delve into the rigorous study of optimisation techniques, but it is mostly self contained and for this reason it may be used with success also by undergraduate students who want to delve immediately into the world of applied mathematics. Furthermore, the book serves both as an introduction and as a comprehensive resource for those looking to apply mathematical principles to real-world problems, and to young researchers in pure and applied mathematics who wish to apply the techniques of mathematical optimisation to applied mathematical problems, or who wish to learn the theoretical foundation of the methods utilised in optimisation.

A solid foundation in mathematical analysis, particularly regarding functions of one and several variables, familiarity with matrix algebra, basics of linear algebra, and a basic understanding of the topology of Euclidean spaces, are expected. However, key definitions and results in both differential calculus and topology are recalled within the text for the benefit of the reader, ensuring that the material remains largely self-contained and can be accessed from the earlier stages of undergraduates studies.

Each chapter builds on the preceding material, ensuring a cohesive and logical progression from foundational theory to advanced methods and each method is described in details, showing the pseudo-code for the algorithms, and rigorously proving the convergence of the method. Together, the chapters provide a rigorous yet accessible treatment of optimisation theory and methods.

All chapters are enriched by numerous numerical examples that show in detail the structures, the problems and the methods discussed, giving to the reader a model to follow to apply the same concepts to different problems. Every lemma, proposition, theorem and corollary in the book come with a proof, no matter how easy it might be, and all proofs are discussed in detail to underpin the logical and mathematical background where the theory sets, giving to the book a duplex nature of a theoretical and applied mathematical monograph.

Comments on the content of individual chapters

The book is divided into three parts, reflecting the natural progression from the theoretical foundations to practical algorithms to solve a mathematical programming problem: a comprehensive introduction to convex analysis, the theory of linear and non-linear programming, and finally the development of some fundamental and traditional methods of linear and non-linear programming, such as the simplex method, and gradient descent methods.

The first part starts with the introduction of the fundamental concepts of convex analysis, forming the theoretical background for the rest of the book and it is divided into two chapters. Chapter I explores the theory of convex sets, their fundamental geometric and topological properties, such as the separation between convex sets, and then specialises in the study of cones and then on the detailed study of polyhedra. This chapter includes in particular key results on the structure of polyhedra, and contains an elementary proof of Minkowski-Weyl Theorem, which yields the general representation of a polyhedral set in $\mathbb{R}^n$. The chapter ends with the proof of a theorem of the alternative due to THEODORE S. MOTZKIN (1908–1970) that is going to be pivotal in the development of the main tools of non-linear programming.

We then move on focusing on convex functions. This is the content of Chapter II, where we examine the definition of convex functions. After discussing the fundamental properties of a convex function, we shall delve into the general notion of sub-gradient, which generalises the one of gradient of a differentiable function in the more general case of a convex function. Finally, we shall focus on the special case of differentiable and twice-differentiable functions, showing how the properties of convex functions take an even more natural form in these special cases.

The second part of the book transitions from the general theory to the framework of a mathematical programming problem. In Chapter III we start by providing a broad introduction to the general notion of an optimisation problem in finite dimension spaces, also called a mathematical programming problem. In this chapter we shall also explore in detail the notion of 'equivalent transformation', which is, roughly speaking, the tool that enables to change the 'shape' of an optimisation problem without affecting the optimal solutions, and maintaining a simple relationship with between the original optimisation problem and the new reshaped problem. This chapter also includes an introductory analysis of optimisation problems for convex and concave functions, and terminates with the introduction of the concept of coercive functions, and its links to the existence of global optimal solutions.

In Chapter IV we discuss in detail the one of linear programming problem, starting by showing how practical problems may be modelled as linear programs. Once the general notion of linear program has been introduced, we shall show that to every linear program we can associate another linear program, called its 'dual', which helps to decide whether a feasible solution is optimal, and prove the Complementary Slackness Theorem, which gives a condition to establish whether a feasible solution is optimal. Furthermore, utilising the theoretical knowledge of polyhedra developed in Chapter I, we shall then focus on the feasible region of a linear programming problem and prove the Fundamental Theorem of Linear Programming. The algebraic treatment of basic and non-basic feasible solutions is also explored, showing the numerical method to calculate basic and non-basic feasible solutions.

Next we move onto addressing optimisation problems involving non-linear functions. This is done in Chapter V, where we introduce concepts such as feasible, descent, and tangent directions, and explore the notion of regular regions. After showing the basic relationships between these sets of directions, we delve into the research of the solution of a non-linear programming problem proving the famous KKT conditions, which generalise the Lagrange's multipliers method. The chapter then terminates with a discussion on second-order conditions for optimality.

Chapter VI extends and generalises the notion of duality to non-linear programming. Further, it examines the relationship between Lagrangian duality and saddle points, and terminates showing how duality in linear programming is indeed a particular case of that of non-linear programming and some examples of application of duality in these special cases, where numerical calculations can be carried out.

The third part of the book focuses on the so called 'methods', which means the algorithms for solving optimisation problems, and in particular we shall discuss the simplex method to solve linear programming problems, and, for non-linear programs, we shall focus on two classes of algorithms: those that solve unconstrained problems, and those that solve constrained problems. For each of these we shall give the most common and widely used algorithms, we shall prove their convergence, and we shall describe their limitations.

In particular, in Chapter VII the general notion of multi-function is discussed, and this notion is then utilised to define the general concept of algorithm. Having defined in a mathematical form what an optimisation algorithm is, the rest of the chapter deals with the general problem of convergence of an iterative algorithm and the main result proved in this chapter shall be a theorem due to W. Zangwill, giving a general condition for the convergence of an algorithm. This theorem shall then be used to prove the convergence of several algorithms.

The details of the simplex algorithm are thoroughly described in Chapter VIII where we state the simplex method both for problems written in 'primal standard form' and those written in 'dual standard form', showing in each case that the algorithm is correct and terminates in a finite number of steps. To this end, we shall start the chapter by reviewing the famous and useful Sherman-Morrison formula, and then we shall delve in more detail into the link between the algebra of a linear program and the geometry of the feasible region. We shall also delve into the problem of choosing the starting vertex, discussing in particular the so called 'two phased method' which consists in creating an auxiliary linear programming problem whose solution offers a starting vertex for the original problem. This method shall be developed both for a linear program in primal standard form and in dual standard form.

Moving onto the methods for non-linear programming, in Chapter IX we describe the most traditional, but very commonly used, algorithms for solving unconstrained non-linear optimisation problems, starting from the methods involving the research of the stationary points utilising suitable algorithms to find the zeros of a real function, such as bisection and Newton-Raphson's algorithm, and using general theoretical criteria to ascertain when the search of a stationary point leads in particular to an optimal solution. In the same chapter we shall also analyse the widely used algorithm of gradient descent. The chapter distinguishes between exact and inexact gradient descent methods, providing a comprehensive overview of this class of problems. In this chapter we also delve into more complex algorithms such as conjugate gradient method and the sub-gradient method, that can be used, in lieu of gradient descent, when the objective function is not differentiable. Finally, Chapter X gives a snapshot of the most classical methods for solving constrained optimisation problems. In particular, we are going to discuss Frank-Wolfe algorithm, the projected gradient algorithm, and penalty methods, emphasising their applications and their limitations, and giving an idea of how to possibly overcome some of the limitations of these methods.

The first part is quite complete and gives a deep overview of convex analysis in a finite-dimensional space. The second part offers the most traditional and important concepts of theory of optimisation in finite dimensional spaces and is sufficiently complete to enable the reader to understand how to handle, from a theoretical perspective, any non-linear programming problem. The third part, on the other hand, consciously only discusses a selection of methods of optimisation. The reader should be aware that out there in the wild there are a myriad of other methods that have been developed with different advantages and disadvantages. In this presentation, however, the authors have focused more on developing the methodology that the student should adopt when facing a new method. The reader who is interested in the whole variety of methods, once grasped this general methodology, will find it easy to approach more advanced textbooks that focus on methods.

Notations and conventions used throughout the book

A final note as regard to the notations and conventions in this book. For the most part, we have tried to follow the traditional notations and, where possible, the well established customs. The words 'greater than' and 'smaller than' are always meant in a weak sense, i.e. admitting the possibility of the two quantities to be equal. Coherently, a positive number may be zero, and a negative number may be zero as well. In other words, in this book zero is both a positive and a negative number. When the strict inequality is intended, we use the phrases 'strictly positive' and 'strictly negative'.

Any vector is always intended to be a column vector, even when this vector is selected as the row of a matrix. No vector, in this book, is ever a row-vector. Furthermore, we have decided to use the symbol $\langle u, v \rangle$ to denote the scalar product of two vectors in $\mathbb{R}^n$.

Another consistent convention we have used is to indicate the index of a sequence of vectors with the superscript, and the indices of real numbers (scalars) with the subscript. This should help the reader not to confuse the superscript with a power, for real numbers and to distinguish a sequence of vectors from the coordinates of one vector (the former being a superscript index, and the latter a subscript index).

With the intent to make references throughout the book more readable, all definitions, propositions, theorems, corollary, formulae, etc, are labelled with a single increasing number, preceded by the number corresponding to the section, within that chapter, where the reference belongs and wrapped into round brackets. So, the reference $(p.q)$ is the qth label in in section p of the current chapter. This label is always on the left-hand side of the page, to make the search even faster. Chapters are labelled with capital Roman numbers, and a reference to a label in a different chapter is made by prepending the Roman number relative to the chapter to the label. So, for example $r(p.q)$ represents the qth label, in section p of Chapter r.

Contents

III   Methods of Optimisation

Part I

Convex Analysis

Chapter I
Convex Sets

Convexity is an extremely important notion in the study of optimisation problems. Indeed, convex sets, polyhedra, separation and disjoint convex sets are often used to tackle optimisation problems. This chapter is devoted to give a summary of the most important geometric and algebraic properties of convex sets and to show the proof of the most important tools that are going to be used in the analysis of optimisation problems.

1. Convex sets and convex hulls

In this section, we shall introduce the basic definitions regarding convex sets.

(1.1) Definition. A set E in $\mathbb{R}^n$ is said to be **convex** if the **line segment** joining two points of the set also belongs to the set. In other words, if x^1 and x^2 are two points in E, then the point $\lambda x^1 + (1 - \lambda)x^2$ must also belong to E for each λ in $[0, 1]$. Sometimes we shall use the symbol $[x^1, x^2]$ to denote the line segment joining the points x^1 and x^2.

Figure 1.1 shows a convex set, (a), and a non-convex set, (b). Notice how in (b) the line joining the points x^1 and x^2 does not lie entirely in the set.

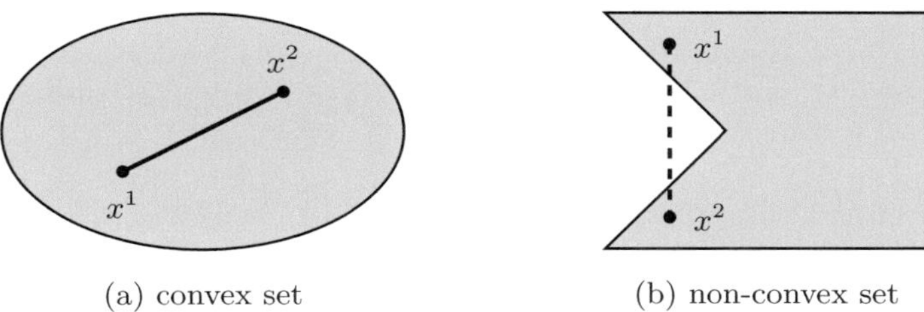

(a) convex set   (b) non-convex set

Figure 1.1. The figure shows a convex set, (a), and a non-convex set, (b). Note how in (b) the line joining the points x^1 and x^2 does not lie entirely in the set.

© The Author(s), under exclusive license to Springer Nature Switzerland AG 2026
A. Carpignani and M. Pappalardo, *Theory and Methods of Optimisation*,
UNITEXT 175, https://doi.org/10.1007/978-3-032-01514-3_1

The following are all examples of convex sets.

(1.2) Example. If α is a vector in $\mathbb{R}^n$ and β a real number, the set

$$E = \left\{ x \in \mathbb{R}^n : \langle \alpha, x \rangle = \beta \right\}$$

is called a **hyper-plane** in $\mathbb{R}^n$. This is a convex set. Indeed, if x^1 and x^2 are in E, then $\langle \alpha, x^1 \rangle = \beta$ and $\langle \alpha, x^2 \rangle = \beta$. If $x = \lambda x^1 + (1 - \lambda)x^2$, then

$$
\begin{aligned}
\langle \alpha, x \rangle &= \langle \alpha, \lambda x^1 + (1 - \lambda)x^2 \rangle \\
&= \lambda \langle \alpha, x^1 \rangle + (1 - \lambda)\langle \alpha, x^2 \rangle \\
&= \lambda \beta + (1 - \lambda)\beta \\
&= \beta.
\end{aligned}
$$

Notice that, if $\bar{x}$ is a point in E, we have $\langle \alpha, \bar{x} \rangle = \beta$, so the equation $\langle \alpha, x \rangle = \beta$ can be rewritten in the form $\langle \alpha, x \rangle = \langle \alpha, \bar{x} \rangle$, and using the linearity of the scalar product, we can also rewrite the definition of E equivalently in the form $E = \{x \in \mathbb{R}^n : \langle \alpha, x - \bar{x} \rangle = 0\}$. In other words, the vector α is perpendicular to all vectors of the form $(x - \bar{x})$ for $x \in E$, i.e. it is perpendicular to the surface of the hyper-plane E.

(1.3) Example. If α is a vector in $\mathbb{R}^n$ and β is a real number, the set

$$E = \left\{ x \in \mathbb{R}^n : \langle \alpha, x \rangle \leq \beta \right\}$$

is called a **half-space** in $\mathbb{R}^n$. It is not difficult to show, repeating the same proof and following the same steps as the ones demonstrated in Example (1.2), that this is indeed a convex set, and that, if $\bar{x}$ is a point in E, the set E can always be written in the form $E = \{x \in \mathbb{R}^n : \langle \alpha, x - \bar{x} \rangle \leq 0\}$. Finally, if $H = \{x \in \mathbb{R}^n : \langle \alpha, x - \bar{x} \rangle = 0\}$ is a hyper-plane, this splits the space into two half-spaces, namely

$$H^+ = \left\{ x \in \mathbb{R}^n : \langle \alpha, x - \bar{x} \rangle \geq 0 \right\} \quad \text{and} \quad H^- = \left\{ x \in \mathbb{R}^n : \langle \alpha, x - \bar{x} \rangle \leq 0 \right\}.$$

(1.4) Lemma. *The intersection of an arbitrary non-empty family of convex sets is itself a convex set.*

Proof. Let $\mathcal{H}$ be an arbitrary non-empty family of convex sets and denote by E the intersection of all the sets in this family. Let x^1 and x^2 be two points in E, and let λ be a real number in $[0, 1]$. If X is any set of the sets belonging to $\mathcal{H}$, from the fact that $E \subseteq X$ it follows that x^1 and x^2 in particular belong to X, hence the point $x = \lambda x^1 + (1 - \lambda)x^2$ also belongs to X, because X is by hypothesis a convex set. Since x belongs to all X in $\mathcal{H}$, and since E is the intersection of such X in $\mathcal{H}$, it follows that x belongs to E as well, thus E is convex. $\qquad \square$

(1.5) Definition. If H is a non-empty set in $\mathbb{R}^n$ the **convex hull** of H is denoted by $\mathrm{conv}(H)$ and it is the smallest (with respect to inclusion) convex set that contains H. In other words, the convex hull of H is a convex set that contains H and it is included in every other convex set that contains H.

(1.6) Definition. If $x^1, \ldots, x^p$ are points in $\mathbb{R}^n$ and $\lambda_1, \ldots, \lambda_p$ are positive real numbers with $\sum_{k=1}^p \lambda_k = 1$, the weighted average $\sum_{k=1}^p \lambda_k x^k$ is called a **convex combination** of the points $x^1, \ldots, x^p$. If, in addition, $\lambda_k \neq 1$ for all $k = 1, \ldots, p$, we shall say that the convex combination is **proper**.

(1.7) Lemma. *The convex hull of a non-empty set H coincides with the collection of all convex combinations of points of H. In other words, x in $\mathrm{conv}(H)$ if and only if x can be represented in the form $x = \sum_{k=1}^p \lambda_k x^k$, where $\sum_{k=1}^p \lambda_k = 1$, $\lambda_k \geq 0$ and $x^k \in H$ for all $k = 1, \ldots, p$. In particular, a set E is convex if and only if it coincides with its convex hull.*

Proof. Let H be a non-empty set, and denote by E the collection of all convex combinations of elements of H. This set is convex by definition and contains H, so it contains the convex hull $\mathrm{conv}(H)$ as well. On the other hand, since $\mathrm{conv}(H)$ contains H, it also contains all convex combinations of elements of H, whence E is included in $\mathrm{conv}(H)$. This proves that the two sets must be equal. $\square$

(1.8) Example. In the Euclidean plane $\mathbb{R}^2$, consider the points $x^1 = (1,1)$, $x^2 = (3,1)$, and $x^3 = (2,3)$. It is not difficult to show that the point $x = (2,2)$ is a convex combination of the points x^1, x^2 and x^3. In fact,

$$x = \tfrac{1}{4}x^1 + \tfrac{1}{4}x^2 + \tfrac{1}{2}x^3.$$

It is possible to show that the convex hull of the points x^1, x^2 and x^3 is the triangle with vertices in these three points, as shown in Figure 1.2.

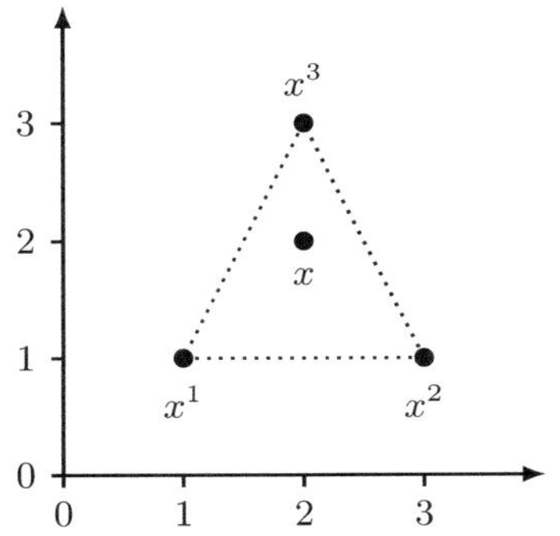
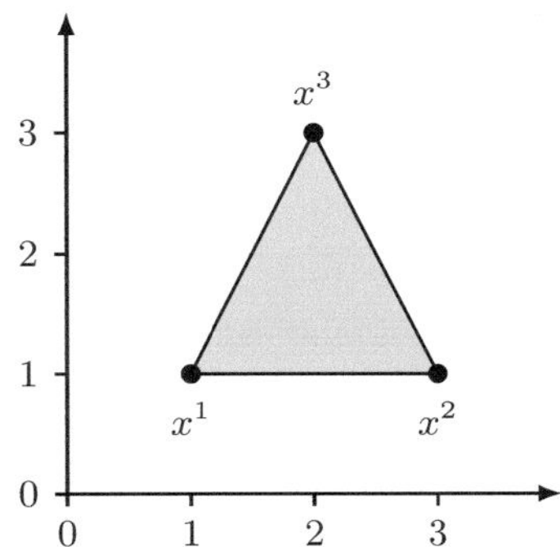

Figure 1.2. On the left, the point x is convex combination of the points x^1, x^2, x^3, hence lies inside the triangle formed by these three points. On the right, the shaded area shows the convex hull of the points x^1, x^2 and x^3.

The previous remark may erroneously induce to think that all elements of a convex set are convex combination of other elements. A careful look at Example (1.8), however, shows that the points x^1, x^2 and x^3 cannot be written as convex combination of the other points in the set. At least not in a proper way. This justifies the following definition.

(1.9) Definition. Let E be a convex set in $\mathbb{R}^n$. If there exists a point x in E that is not a proper convex combination of other points in E, this point is said to be an **extreme** point of E. The set of all extreme points of E is denoted by $\text{ext}(E)$.

2. Topological properties of convex sets

Convex sets have a wealth of nice topological properties that are particularly intuitive. Before delving into at least the simplest of them, however, we shall recall, for convenience and with the intent to fix the terminology and notation, the most important topological terms. It is expected, however, that the reader already has some understanding of basic topology.

(2.1) Definition. Given a point x in $\mathbb{R}^n$ and a number $r > 0$, we shall denote by $B(x, r)$ the set $\{y \in \mathbb{R}^n : \|x - y\| < r\}$ of all points in $\mathbb{R}^n$ whose distance from x is smaller than r, and we shall call it the **open ball** with centre x and radius r. A set V shall be called a **neighbourhood** of a point x if it contains an open ball centred in x. In other words, V is a neighbourhood of x if there exists a number $\varepsilon > 0$, such that $B(x, \varepsilon) \subseteq V$. The set of all neighbourhoods of x shall be denoted by $N(x)$.

(2.2) Definition. Let E be a subset in $\mathbb{R}^n$. A point x is said to be in the **closure** of E, which is denoted by $\overline{E}$, if for every neighbourhood $V \in N(x)$, the intersection $V \cap E$ is not empty. If a set E coincides with its closure, this set is said to be **closed**. A point x is said to be in the **interior** of E, which is denoted by $\text{int}(E)$, if there exists a neighbourhood $V \in N(x)$ such that $V \subseteq E$. If a set E coincides with its interior, the set is said to be **open**. Finally, a point x is said to be in the **boundary** of E, which is denoted by ∂E, if $x \in \overline{E} \setminus \text{int}(E)$.

The first topological property of convex sets is the following.

(2.3) Proposition. *Let E be a convex set in $\mathbb{R}^n$. Then $\overline{E}$ is also convex.*

Proof. Let x^1 and x^2 be two points in $\overline{E}$ and a real number λ with $\lambda \in [0, 1]$. We wish to prove that $\lambda x^1 + (1 - \lambda)x^2 \in \overline{E}$. To this end, let us consider two sequences $\{u^k\}$ and $\{v^k\}$ in E, the first convergent towards x^1, and the second convergent towards x^2, and let us take the sequence $\{w^k\}$ defined, for every index k, by $w^k = \lambda u^k + (1 - \lambda)v^k$. Since E is a convex set, one has $w^k \in E$, for every k. On the other hand, the sequence $\{w^k\}$, of elements of E, converges towards the point $\lambda x^1 + (1 - \lambda)x^2$, which thus must lie in $\overline{E}$. $\square$

Some very intuitive topological properties of a convex set are a consequence of the following theorem, whose proof is actually a bit tedious, but whose result is indeed very intuitive and it is going to be extremely rewarding.

(2.4) Theorem. *Let E be a convex set in $\mathbb{R}^n$, and assume that the interior of E is non-empty. Let $x^1 \in \overline{E}$ and $x^2 \in \mathrm{int}(E)$. Then*

$$\lambda x^1 + (1 - \lambda)x^2 \in \mathrm{int}(E)$$

for each $\lambda \in (0, 1)$.

Proof. Let x^1 be a point in the closure $\overline{E}$ of E, x^2 a point in the interior $\mathrm{int}(E)$ of E, and set $x = \lambda x^1 + (1 - \lambda)x^2$, with $\lambda \in (0, 1)$. By definition of interior, there exists a positive real number ε such that $B(x^2, \varepsilon) \subseteq E$. Take a number $0 < \delta < (1 - \lambda)\varepsilon$ and set $\varrho = \lambda^{-1}[(1 - \lambda)\varepsilon - \delta]$. Note that δ has been carefully chosen in such a way that $\varrho > 0$. To complete the proof, it suffices to prove that $B(x, \delta) \subseteq E$. To this end, take $w \in B(x, \delta)$ and observe that, since x^1 lies on the closure of E, there exists a point $u \in B(x^1, \varrho) \cap E$. In other words, there exists a point $u \in E$ such that $\|x^1 - u\| < \varrho$. Now, let us set $v = (w - \lambda u)/(1 - \lambda)$. Then,

$$\begin{aligned}
\|x^2 - v\| &= \left\| \frac{x - \lambda x^1}{1 - \lambda} - \frac{w - \lambda u}{1 - \lambda} \right\| \\
&\leq \frac{1}{1 - \lambda}\left(\|x - w\| + \lambda\|x^1 - u\| \right) \\
&< \frac{1}{1 - \lambda}\left(\delta + \lambda\varrho \right) \\
&= \varepsilon
\end{aligned}$$

where, in the first identity, we have rearranged $x = \lambda x^1 + (1 - \lambda)x^2$ to make x^2 the subject, and we have substituted it into the norm, the strict inequality follows from the fact that $w \in B(x, \delta)$, and $u \in B(x^1, \varrho)$, and the last identity follows from the definition of ϱ. This means that $v \in B(x^2, \varepsilon)$. In summary, $u \in E$, $v \in E$ and $w = \lambda u + (1 - \lambda)v \in E$. It is thus shown that every point $w \in B(x, \delta)$ lies in E, and the theorem is proved. $\square$

(2.5) Corollary. *In the same hypotheses as in Theorem (2.4), the interior of E is also convex.*

Proof. To prove the first statement, observe that, if x^1 and x^2 are in $\mathrm{int}(E)$, then in particular x^1 lies in the closure of E. Hence, Theorem (2.4) ensures that $\lambda x^1 + (1 - \lambda)x^2 \in \mathrm{int}(E)$, for every $\lambda \in (0, 1)$. $\square$

(2.6) Corollary. *In the same hypotheses as in Theorem (2.4), $\overline{\mathrm{int}(E)} = \overline{E}$.*

Proof. It is a general fact that $\mathrm{int}(E) \subseteq E$, hence $\overline{\mathrm{int}(E)} \subseteq \overline{E}$. It remains to prove the converse inclusion. To this end, let $x \in \overline{E}$ and $y \in \mathrm{int}(E)$. Then, by means of Theorem (2.4), we know that $\lambda x + (1 - \lambda)y \in \mathrm{int}(E)$ for every $0 < \lambda < 1$. Taking the limit as the real number $\lambda \to 1^-$ immediately gives $x \in \overline{\mathrm{int}(E)}$, and this completes the proof. $\square$

(2.7) Corollary. *In the same hypotheses as in (2.4), $\mathrm{int}(\overline{E}) = \mathrm{int}(E)$.*

Proof. It is a general fact that $E \subseteq \overline{E}$, hence $\mathrm{int}(E) \subseteq \mathrm{int}(\overline{E})$. It remains to prove the converse inclusion. To this end, let $x \in \mathrm{int}(\overline{E})$. We need to show that $x \in \mathrm{int}(E)$. Since $x \in \mathrm{int}(\overline{E})$, there exists a positive number ε such that $\|y - x\| < \varepsilon$ implies $y \in \overline{E}$. Now, let $y \neq x$ belong to $\mathrm{int}(E)$ and let $z = (1 + r)x - ry$, where $r = \varepsilon/(2\|x - y\|)$. Since $\|z - x\| = \frac{1}{2}\varepsilon$, we have $z \in \overline{E}$. On the other hand, $x = \lambda z + (1 - \lambda)y$, with $\lambda = 1/(1+r) \in (0, 1)$. Since, $z \in \overline{E}$ and $y \in \mathrm{int}(E)$, then, by means of Theorem (2.4), $x \in \mathrm{int}(E)$, and the proof is complete. $\qquad\qquad\qquad\qquad\qquad\qquad\qquad\qquad\qquad\qquad\qquad$ $\square$

3. Geometric properties of convex sets

Convex sets have important and very intuitive geometric properties. For example, if E is a convex set and z is a point that does not belong to E, we can always find a point w in E that has the minimum distance, among all points in E, from z. This is called the "closest-point theorem".

(3.1) Theorem (Closest-Point Theorem). *Let E be a non-empty, closed convex set in $\mathbb{R}^n$, and let $z \notin E$. Then there exists a unique point $w \in E$ with minimum distance from z. In other words, there exists a unique point $w \in E$ which minimises the function $x \mapsto \|z - x\|$, with $x \in E$. Furthermore, w is the minimising point if and only if $\langle z - w, x - w \rangle \leq 0$ for all $x \in E$.*

Proof. Let us first prove the *existence* of a point in E with minimum distance from z. To this end, since E is assumed to be non-empty, there exists a point y in E. Consider the closure of the ball B with centre y and radius $\|y - z\|$, the distance between y and z. Since all points in E outside of this ball have a distance from z greater than $\|y - z\|$, if such a point with minimal distance exists, it must belong to the set $E \cap B$. Now the set $E \cap B$ is closed and bounded, hence compact. Thus, by virtue of Weierstrass Theorem, the continuous function $x \mapsto \|x - z\|$ defined on $E \cap B$ must have a minimum, w. This means that $\|w - z\| \leq \|x - z\|$ for all $x \in E \cap B$. On the other hand, we already know that, if $x \in E \cap B^c$, then $\|w - z\| \leq \|y - z\| < \|x - z\|$ and therefore $\|w - z\| \leq \|x - z\|$ for all $x \in E$, i.e. w is a point with minimum distance from z.

We now need to prove the *uniqueness* of such a point. To this end, suppose that there are two points w^1 and w^2 in E such that $\|z - w^1\| = \|z - w^2\| = \min_{x \in E} \|z - x\| = \delta$. By convexity of E, then, we also have $\frac{1}{2}w^1 + \frac{1}{2}w^2 \in E$. In addition, by triangle inequality we also get

$$\begin{aligned}
\|z - (\tfrac{1}{2}w^1 + \tfrac{1}{2}w^2)\| &= \|\tfrac{1}{2}(z - w^1) + \tfrac{1}{2}(z - w^2)\| \\
&\leq \tfrac{1}{2}\|z - w^1\| + \tfrac{1}{2}\|z - w^2\| \\
&= \tfrac{1}{2}\delta + \tfrac{1}{2}\delta \\
&= \delta.
\end{aligned}$$

If strict inequality holds, this would contradict the fact that w^1 and w^2 are the points of closest distance. Therefore, equality must hold. In particular, from the relationship $\|z - (\frac{1}{2}w^1 + \frac{1}{2}w^2)\| = \delta$ it follows that

$$\|(z - w^1) + (z - w^2)\| = 2\delta = \|z - w^1\| + \|z - w^2\|.$$

Now, taking the square of this expression, expanding out, and simplifying, gives $\langle z - w^1, z - w^2 \rangle = \|z - w^1\| \cdot \|z - w^2\|$. Thus,

$$\begin{aligned}
\|w^1 - w^2\|^2 &= \|w^1 - z\|^2 + \|w^2 - z\|^2 - 2\langle w^1 - z, w^2 - z \rangle \\
&= \delta^2 + \delta^2 - 2\langle w^1 - z, w^2 - z \rangle \\
&= 2\delta^2 - 2\|w^1 - z\| \cdot \|w^2 - z\| \\
&= 2\delta^2 - 2\delta^2 \\
&= 0,
\end{aligned}$$

yielding to $w^1 = w^2$, and the uniqueness is established.

It remains to prove that $\langle z - w, z - x \rangle \leq 0$ for all $x \in E$ is both necessary and sufficient condition for w to be the point in E closest to z. To prove that the condition is *sufficient*, let $x \in E$ and observe that

$$\begin{aligned}
\|z - x\|^2 &= \|z - w + w - x\|^2 \\
&= \langle z - w + w - x, z - w + w - x \rangle \\
&= \langle z - w, z - w \rangle + 2\langle z - w, w - x \rangle + \langle w - x, w - x \rangle \\
&= \|z - w\|^2 + \|w - x\|^2 + 2\langle z - w, w - x \rangle
\end{aligned}$$

Since $\|w - x\|^2 \geq 0$ and $\langle z - w, w - x \rangle \geq 0$ by assumption, it follows that $\|z - x\|^2 \geq \|z - w\|^2$ and therefore w is the minimising point. Finally, let us prove that the condition is also *necessary*. To this end, assume that $\|z - x\|^2 \geq \|z - w\|^2$ for all $x \in E$. Let $x \in E$ and note that $w + \lambda(x - w) \in E$ for all $0 \leq \lambda \leq 1$ because E is convex. Thus,

$$\|z - w - \lambda(x - w)\|^2 \geq \|z - w\|^2.$$

Furthermore

$$\begin{aligned}
\|z - w - \lambda(x - w)\|^2 &= \langle z - w - \lambda(x - w), z - w - \lambda(x - w) \rangle \\
&= \|z - w\|^2 + \lambda^2 \|x - w\|^2 - 2\lambda\langle z - w, x - w \rangle.
\end{aligned}$$

Putting together these two expressions gives finally

$$\lambda^2 \|x - w\|^2 - 2\lambda\langle z - w, x - w \rangle \geq 0$$

for all $0 \leq \lambda \leq 1$. Dividing throughout by $\lambda > 0$ and letting $\lambda \to 0^+$, the result follows. $\qquad\square$

(3.2) Definition. In the same hypotheses as in Theorem (3.1), the unique point w with shortest distance from z is called the **projection** of z onto E, and it is denoted by $p_E(z)$. Furthermore, the function $p_E : \mathbb{R}^n \to E$ that associates to each point $z \in \mathbb{R}^n$ its projection $p_E(z)$ onto E is called the **projection map** onto E.

(3.3) Remark. Even though it is obvious by definition, for every point z, the projection map satisfies the following $\|z - p_E(z)\| = \min_{y \in E} \|z - y\|$. As a result of this, observe that, in particular, if $z \in E$, then $p_E(z) = z$. This equality holds for every point z in the closure of E as well, hence $\overline{E}$ can be described as the set $\overline{E} = \{z \in \mathbb{R}^n : p_E(z) = z\}$.

Another important property of convex sets is that, given a closed convex set E and a point z that does not belong to E, there exists a hyper-plane that separates them. The rest of this section is devoted to proving this result, which is also known as "separation theorem".

(3.4) Theorem (Separation Theorem). *Let E be a non-empty, closed convex set in $\mathbb{R}^n$, and let $z \notin E$. Then, there exist a vector $\alpha \in \mathbb{R}^n$ and a scalar β such that the hyper-plane $\langle \alpha, x \rangle = \beta$ strictly separates z from E, that is to say, such that*

$$\langle \alpha, x \rangle < \beta < \langle \alpha, z \rangle \qquad \text{for all } x \in E.$$

Proof. In the hypotheses of the theorem, by virtue of Theorem (3.1), there exists in E a point w with minimal distance from z. It suffices to prove that the hyper-plane separating z from E is given by the hyper-plane perpendicular to the segment with extremes z and w and passing through their mid-point. This idea is shown in Figure 3.1. The rest of the proof consists in checking that the convex set E does not cross the hyper-plane. Indeed, let

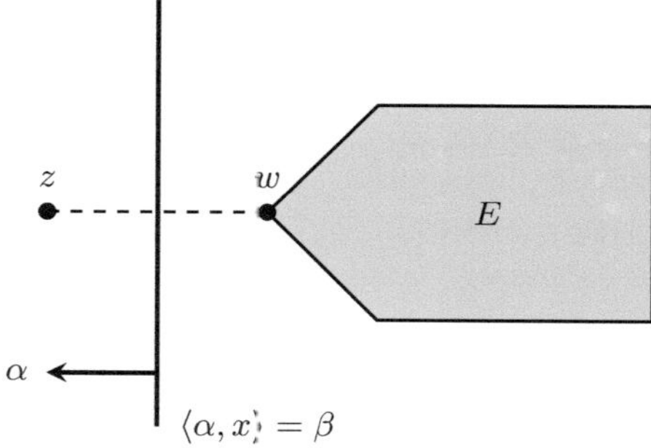

Figure 3.1. The point z is outside the convex set E. It is therefore possible to find in E a point, w, with shortest distance from z. The hyper-plane perpendicular to the segment with extremes z and w "separates" the point z from the convex set E.

us set

$$\alpha = z - w, \qquad \beta = \tfrac{1}{2}\big(\langle z, z \rangle - \langle w, w \rangle\big).$$

From these definitions,

$$\langle \alpha, z \rangle - \beta = \langle z, z \rangle - \langle w, z \rangle - \tfrac{1}{2}\langle z, z \rangle + \tfrac{1}{2}\langle w, w \rangle$$
$$= \tfrac{1}{2}\langle z, z \rangle + \tfrac{1}{2}\langle w, w \rangle - \tfrac{1}{2}\langle z, w \rangle - \tfrac{1}{2}\langle w, z \rangle$$
$$= \tfrac{1}{2}\langle z - w, z - w \rangle$$
$$= \tfrac{1}{2}\|z - w\|^2.$$

Since $z \notin E$ and $w \in E$, it must be $z \neq w$, hence $\|z - w\| > 0$ and therefore $\langle \alpha, z \rangle > \beta$. To complete the proof, we must now prove that $\langle \alpha, x \rangle < \beta$ for all $x \in E$. This can be done by contradiction. Indeed, suppose that there is a point $x \in E$ such that $\langle \alpha, x \rangle \geq \beta$. Then,

$$\langle \alpha, x \rangle - \beta = \langle z - w, x \rangle - \tfrac{1}{2}\langle z - w, z + w \rangle$$
$$= \langle z - w, x - \tfrac{1}{2}(z + w) \rangle$$
$$= \langle z - w, (x - w) - \tfrac{1}{2}(z - w) \rangle$$
$$= \langle z - w, x - w \rangle - \tfrac{1}{2}\|z - w\|^2,$$

where the first identity follows from the fact that

$$\beta = \tfrac{1}{2}\big(\langle z, z \rangle - \langle w, w \rangle\big) = \tfrac{1}{2}\langle z - w, z + w \rangle.$$

Since $\langle \alpha, x \rangle - \beta \geq 0$, this means that

$$\langle z - w, x - w \rangle \geq \tfrac{1}{2}\|z - w\|^2 > 0.$$

Now, consider a point on the segment with extremes w and x. Such a point can be written in the form $w + \lambda(x - w)$ with $0 \leq \lambda \leq 1$. In addition, since x and w both belong to E, and since E is a convex set, it follows that all $w + \lambda(x - w)$ belong to E for every $\lambda \in [0, 1]$. Consider the function

$$f(\lambda) = \|w + \lambda(x - w) - z\|^2 = \|w - z\|^2 - 2\lambda\langle z - w, x - w \rangle + \lambda^2\|x - w\|^2.$$

This is a quadratic function in λ. In particular, its derivative is

$$f'(\lambda) = 2\lambda\|x - w\|^2 - 2\langle z - w, x - w \rangle.$$

When $\lambda = 0$, we have

$$f'(0) = -2\langle z - w, x - w \rangle < 0.$$

This means that the function f is decreasing in a neighbourhood of 0, and therefore there exists a positive $\tau \in [0, 1]$ such that $f(\tau) < f(0)$, which is equivalent to

$$\|w + \tau(x - w) - z\|^2 < \|w - z\|^2$$

This means that the point $w + \tau(x - w) \in E$ has a shorter distance than w from the point z, but this is a contradiction, because, by definition, w has the shortest distance from z among all points in E. This is sufficient to prove that for all x in E, we must have $\langle \alpha, x \rangle < \beta$. $\qquad\square$

In reality, the relationship between convex sets and hyper-planes is even deeper. In fact, for a convex set we can always find hyper-planes that touch the set and leave it all in one of the two half-spaces generated by the hyper-plane. To see this in detail, let us start by giving the following definition, which ought to clarify what we mean when we say 'a hyper-plane touches a set' and 'a hyper-plane leaves the set all in one half-space'.

(3.5) Definition. Let E be a non-empty set in $\mathbb{R}^n$, and let $\bar{x} \in \partial E$ be a point in its boundary. A hyper-plane $H = \{x \in \mathbb{R}^n : \langle \alpha, x - \bar{x} \rangle = 0\}$ is called a **supporting hyper-plane** of E at $\bar{x}$ if either $E \subseteq H^+$, that is, $\langle \alpha, x - \bar{x} \rangle \geq 0$ for each $x \in E$, or else $E \subseteq H^-$, that is $\langle \alpha, x - \bar{x} \rangle \leq 0$, for each $x \in E$. If, in addition, $E \not\subseteq H$, then H is called a **proper** supporting hyper-plane of E at $\bar{x}$. Furthermore, sometimes, instead of staying that H is a supporting hyper-plane at $\bar{x}$, we shall say that the hyper-plane H **supports** E at $\bar{x}$.

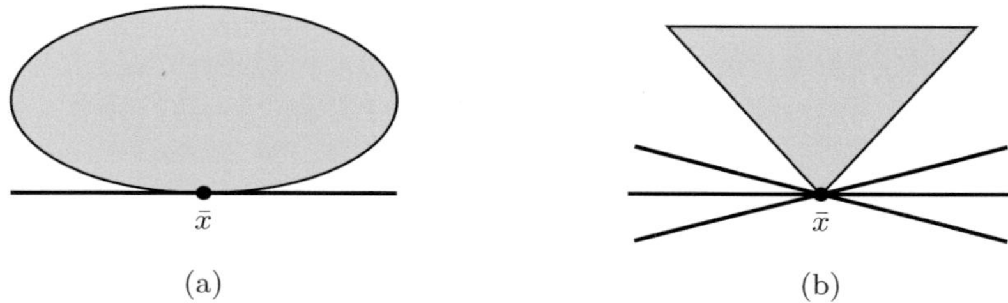

Figure 3.2. In (a) there is a unique supporting hyper-plane at the point $\bar{x}$, while in (b) there are multiple supporting hyper-planes at the point $\bar{x}$.

Of course, for a non-empty subset E of $\mathbb{R}^n$, at a given point $\bar{x}$, a supporting hyper-plane may or may not exist. Also, even when a supporting hyper-plane at a boundary point exists, it may or may not be unique. For example look at Figure 3.2: in (a) there is a unique supporting hyper-plane at $\bar{x}$, while in (b) there are multiple supporting hyper-planes at the point $\bar{x}$. However, what the sets shown in the figure have in common is that both are convex sets. Indeed, it turns out that, when the set E is convex, such a supporting hyper-plane always exists at every point $\bar{x}$ in the boundary of E. This is the content of the next theorem.

(3.6) Theorem. *Let E be a non-empty convex set in $\mathbb{R}^n$, and let $\bar{x} \in \partial E$. Then, there exists a supporting hyper-plane H at $\bar{x}$. In other words, if E is convex, then there exists a non-zero vector α such that $\langle \alpha, x - \bar{x} \rangle \leq 0$ for every $x \in \overline{E}$.*

Proof. Let $\bar{x} \in \partial E$ be a point in the boundary of E. Then, we can always find a sequence $\{z^k\}$ of points all lying outside the closure of E that approaches to $\bar{x}$, as $k \to +\infty$. Since every point z^k lies outside $\overline{E}$, by virtue of Theorem (3.4), there exist a vector α^k and a scalar β_k such that

$$\langle \alpha^k, x \rangle < \beta_k < \langle \alpha^k, z^k \rangle \qquad \text{for every } x \in \overline{E}.$$

With no loss of generality, we can always assume that each of the vectors α^k is normalised (i.e. $\|\alpha^k\| = 1$): indeed, if any of these vector did not have norm 1, we always could replace it with the same vector divided by its norm, which corresponds to dividing the previous chain of inequalities by the norm of α^k. Since the sequence $\{\alpha^k\}$ is bounded (because now each of its vectors has norm 1), by virtue of Bolzano-Weierstrass Theorem, we can extract from $\{\alpha^k\}$ a sub-sequence that converges to a vector α with norm $\|\alpha\| = 1$. Thus, restricting the index k only among those of this convergent sub-sequence, the inequality $\langle \alpha^k, x \rangle < \langle \alpha^k, z^k \rangle$ holds for every $x \in \overline{E}$. Fixing any of these points, x, and taking the limit as k approaches $+\infty$, the latter inequality becomes $\langle \alpha, x \rangle \leq \langle \alpha, \bar{x} \rangle$, and, rearranging, we finally get $\langle \alpha, x - \bar{x} \rangle \leq 0$. Since this is true for every point x in the closure of E, the theorem is thereby fully proved. $\qquad\square$

The two results proved thus far show only the relationship between a convex set and a point. In particular, they establish that if a point is external to the convex set, we can achieve a 'strict' separation, and when a point lies on the boundary of a convex set, the supporting hyper-plane gives again some sort of separation, which we we may call 'weak', because it involves only weak inequalities. The next theorem shows that we can also find a separation hyper-plane between two convex sets. Figure 3.3 shows this situation.

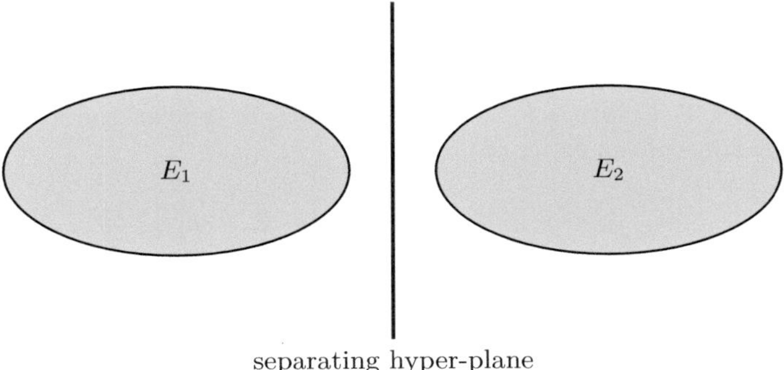

Figure 3.3. The figure shows two convex sets E_1 and E_2, with $E_1 \cap E_2$ empty, and a separating hyper-plane.

(3.7) Theorem. *Let E_1 and E_2 be two non-empty convex sets in $\mathbb{R}^n$ and suppose that $E_1 \cap E_2$ is empty. Then, there is a hyperplane that separates E_1 and E_2. In other words, there exists a vector α and a scalar β such that*

$$\langle \alpha, x \rangle \leq \beta \leq \langle \alpha, y \rangle \qquad \text{for all } x \in E_1 \text{ and } y \in E_2.$$

Proof. Consider the set $E = \{x - y : x \in E_1, y \in E_2\}$. This is a convex set, and the hypothesis that the intersection $E_1 \cap E_2$ is empty translates into the fact that $0 \notin E$. Let us start by assuming that E_1 and E_2 are closed. Then E

is also closed, and, by means of Theorem (3.4), there exists a vector α such that $\langle \alpha, z \rangle < 0$ for all $z \in E$. Since every z in E is of the form $z = x - y$, with $x \in E_1$ and $y \in E_2$, the previous inequality becomes $\langle \alpha, x \rangle < \beta < \langle \alpha, y \rangle$ as requested.

Now, if E_1 and E_2 are not necessarily closed, consider their closures, $\overline{E}_1$ and $\overline{E}_2$. If $\overline{E}_1 \cap \overline{E}_2$ remains empty, there is nothing to prove, so we may assume that the intersection $\overline{E}_1 \cap \overline{E}_2$ is not empty. Consider again the set E. Now its closure $\overline{E}$ contains 0 but since $0 \notin E$, this means 0 must lie in the boundary of E. Thus, applying now Theorem (3.6), there exists a vector α such that $\langle \alpha, z \rangle \leq 0$ for all $z \in E$. Since every z in E is of the form $z = x - y$, with $x \in E_1$ and $y \in E_2$, the previous inequality becomes $\langle \alpha, x \rangle \leq \beta \leq \langle \alpha, y \rangle$, as requested. $\qquad\qquad\square$

4. Cones and polar cones

Another important notion, linked to that of convex sets, is the one of cone.

(4.1) Definition. A non-empty set C in $\mathbb{R}^n$ is **cone** if $x \in C$ implies that $\lambda x \in C$ for all $\lambda \geq 0$. In other words, if C contains a point x distinct from the origin, then it also contains the whole ray from the origin, passing through x. In addition, if C is also convex, C is called a **convex cone**.

Observe immediately that, as consequence of previous definition, a cone always contains 0. Figure 4.1 shows an example of convex cone, (a), and an example of non-convex cone, (b).

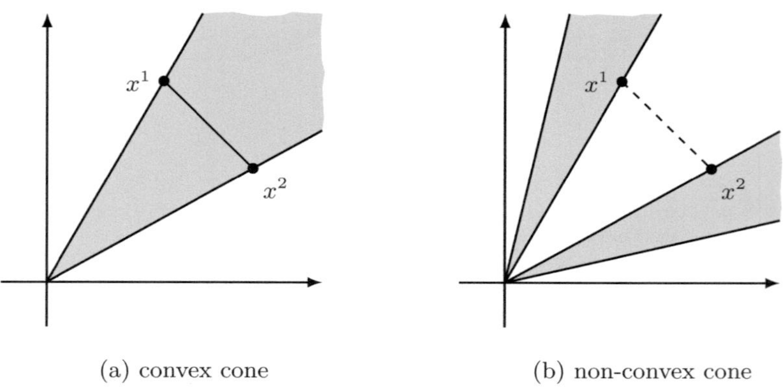

(a) convex cone   (b) non-convex cone

Figure 4.1. Figure (a) shows a convex cone: the segment joining any two points in this set is fully included in the set. Figure (b) shows a non-convex cone: in this case, even though the points x^1 and x^2 lie in the set, the segment joining them is not included in the set: in this case none of the internal points of this segment is actually included!

(4.2) Definition. Let H be a non-empty set. The smallest convex cone (with respect to inclusion) that contains H is called the **conic hull** of H, or the cone **generated** by H, and it is denoted by cone(H).

(4.3) Definition. Weighted sums of the form $\sum_{k=1}^{p} \lambda_k x^k$, where $\lambda_k \geq 0$ for all $k = 1, \ldots, p$ are called **conic combinations** of the points $x^1, \ldots, x^p$. In addition, if, at least two indexes λ_k are strictly positive, we shall say that the conic combination is **proper**.

In the same way as we have proved Lemma (1.7) we can prove the following.

(4.4) Lemma. *Let H be a non-empty set. The cone generated by H coincides with the collection of all conic combinations of points of H. In other words, $x \in$ cone(H) if and only if x can be written in the form $x = \sum_{k=1}^{p} \lambda_k x^k$, where $\lambda_k \geq 0$ and $x^k \in H$ for all $k = 1, \ldots, p$.*

(4.5) Example. Given the points $x^1 = (2, 3)$, $x^2 = (3, 1)$, it is not difficult to show that the point $x = (3, \frac{13}{6})$ is a conic combination of the points x^1 and x^2. Indeed, we have

$$x = \tfrac{1}{2} x^1 + \tfrac{2}{3} x^2.$$

The cone generated by x^1 and x^2 is shown in Figure 4.2.

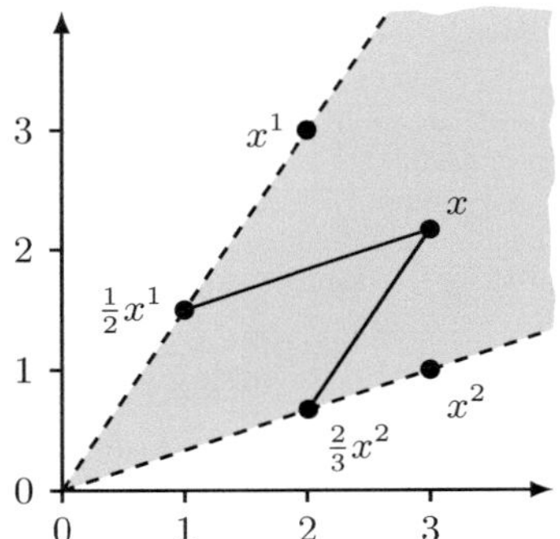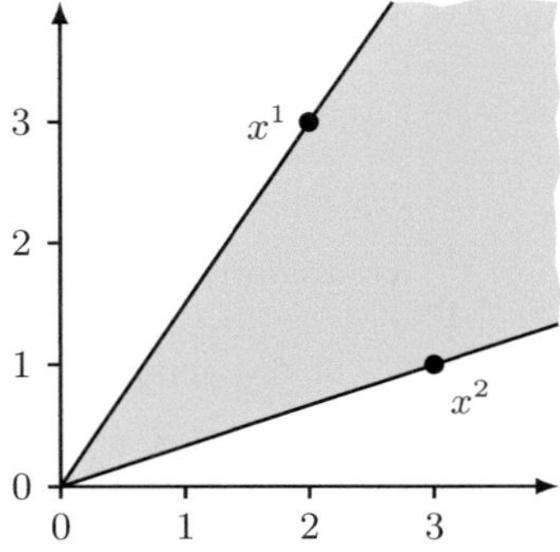

Figure 4.2. On the left it is shown the point $x = \frac{1}{2} x^1 + \frac{2}{3} x^2$, which is conic combination of the points x^1 and x^2. On the right, the shaded area represents the cone generated by the two points x^1 and x^2. Notice that the cone generated by x^1 and x^2 is, in particular, a convex set.

A special type of convex cone is the polar cone of a set, defined as follows.

(4.6) Definition. Let H be a non-empty set in $\mathbb{R}^n$. The **polar cone** of H, denoted by H^*, is given by $H^* = \{p \in \mathbb{R}^n : \langle p, x \rangle \leq 0 \text{ for all } x \in H\}$. If H is the empty set, then H^* shall be defined as the whole $\mathbb{R}^n$.

Notice that regardless of whether H is convex or not, and whether H is closed or not, H^* is always a closed convex set. Indeed, the polar cone is by definition the intersection of a family of closed half-spaces: $\{p \in \mathbb{R}^n : \langle p, x \rangle \leq 0\}$, for every $x \in H$. Since half-spaces are convex, and since intersection of closed (resp. convex) sets is closed (resp. convex), we can conclude that H^* is both closed and convex. The name 'cone' is also justified because the polar cone, H^*, of a set H is evidently a cone. Figure 4.3 shows the polar cone, C^*, of a cone C. Geometrically, the polar cone can be described as the set of all vectors that make an angle greater than $90°$ with all the vectors from the position vectors of the points in C.

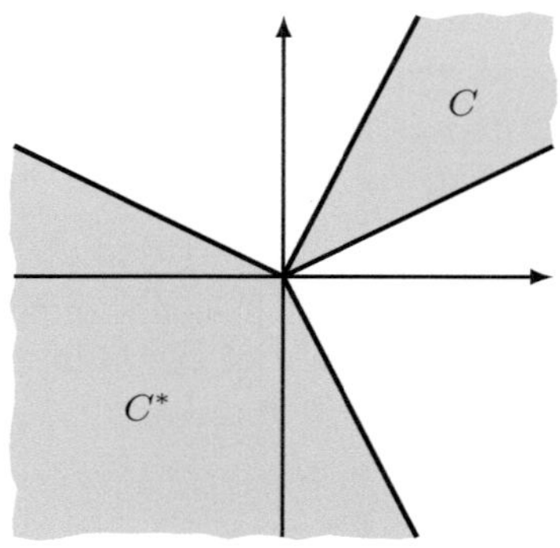

(a) the polar cone of a cone.

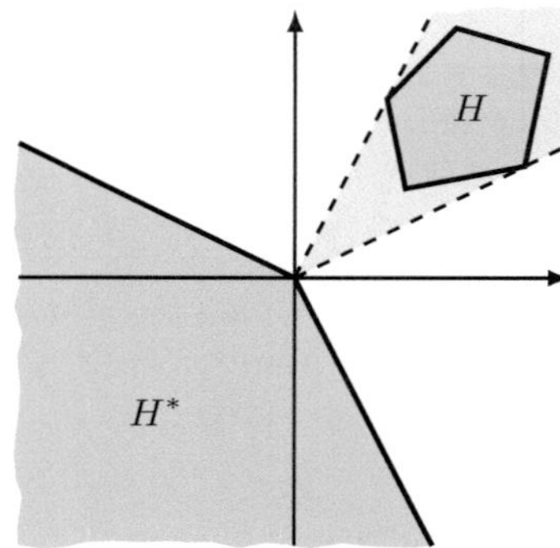

(b) the polar cone of a convex set.

Figure 4.3. Figure (a) shows a convex cone, C, and its polar cone, C^*. Notice that the boundaries of the polar cone are perpendicular to the boundaries of the original cone C. Figure (b) shows the polar cone, H^*, of a set H. The polar cone is a closed convex cone. Notice that, in this second case, the polar cone coincides with that of the cone generated by H (lighter shade of grey).

The properties of the polar cone are summarised in the following proposition.

(4.7) Proposition. *Let H and K be non-empty sets in $\mathbb{R}^n$. Then:*

(a) *If $H \subseteq K$, then $K^* \subseteq H^*$.*

(b) *$H \subseteq H^{**}$.*

(c) *$H^* = H^{***}$.*

Proof.

(a) Assume that $H \subseteq K$, and let $q \in K^*$. Then $\langle q, x \rangle \leq 0$ for all $x \in K$. Now from the fact that $H \subseteq K$, it follows that $\langle q, x \rangle \leq 0$ for all $x \in H$, hence $q \in H^*$. This proves that $K^* \subseteq H^*$.

(b) By definition

$$H^{**} = \Big\{ w \in \mathbb{R}^n : \langle w, p \rangle \leq 0 \text{ for all } p \in H^* \Big\}$$
$$= \Big\{ w \in \mathbb{R}^n : \langle p, w \rangle \leq 0 \text{ for all } p \in H^* \Big\}.$$

In particular, if $x \in H$, by definition of H^*, for all $p \in H^*$ we have $\langle p, x \rangle \le 0$ and therefore $x \in H^{**}$. This proves that $H \subseteq H^{**}$.

(c) From (b) we have $H \subseteq H^{**}$. Using (a) we get $H^{***} = (H^{**})^* \subseteq H^*$. On the other hand, by virtue of (b) again, $H^* \subseteq (H^*)^{**} = H^{***}$ which yields $H^* = H^{***}$. $\qquad\square$

The following theorem shows that for a closed convex cone, the initial geometrical intuition is even stronger, because the polar of the polar turns out to coincide with the initial cone.

(4.8) Theorem. *Let C be a non-empty closed convex cone. Then $C^{**} = C$.*

Proof. Let C be a non-empty closed convex cone. Proposition (4.7)(b) ensures that $C \subseteq C^{**}$ holds, so it suffices to prove the inclusion $C^{**} \subseteq C$. To this end, let $x \in C^{**}$, and suppose by contradiction that $x \notin C$. By virtue of Theorem (3.4), there exist a non-zero vector α and a scalar β, such that $\langle \alpha, y \rangle < \beta$ for all $y \in C$ and $\langle \alpha, x \rangle > \beta$. Since C is a closed cone, $0 \in C$, so $0 = \langle \alpha, 0 \rangle < \beta$. Hence, β is a positive number, and therefore $\langle \alpha, x \rangle > 0$. Now, let us prove that $\alpha \in C^*$. If not, $\langle \alpha, w \rangle > 0$ for some $w \in C$ necessarily different from 0, and since $\lambda w \in C$ for every $\lambda \ge 0$, we can find a suitable λ, large enough to have $\langle \alpha, \lambda w \rangle > \beta > 0$, contradicting the hypothesis that $\langle \alpha, y \rangle < \beta$ for all $y \in C$. So, $\alpha \in C^*$. Now, this is enough to complete the proof. Indeed, since $x \in C^{**} = \{ u \in \mathbb{R}^n : \langle u, v \rangle \le 0 \text{ for all } v \in C^* \}$ and $\alpha \in C^*$, this means that $\langle \alpha, x \rangle = \langle x, \alpha \rangle \le 0$, contradicting the fact that $\langle \alpha, x \rangle > 0$. $\qquad\square$

5. Polyhedra

A very important class of convex sets in $\mathbb{R}^n$ consists of polyhedra. To introduce this particularly important type of convex sets, let us first recall a convex set that we have seen in Example (1.3): a half-space is the set of all solutions of a linear inequality, $\langle a, x \rangle \le \beta$, where $\alpha \in \mathbb{R}^n$ and $\beta \in \mathbb{R}$. This is a closed set in $\mathbb{R}^n$. Before we can give the definition of polyhedron, let us introduce a notation that is going to be used throughout the sequel.

(5.1) Notation. If $u = (u_1, \ldots, u_p)$ and $v = (v_1, \ldots, v_p)$ are two vectors in $\mathbb{R}^p$, with the notation $u \le v$ we shall always mean that $u_i \le v_i$ for every $i = 1, \ldots, p$. Note that this notation induces on $\mathbb{R}^p$ a relation of partial order. Consistently with any abstract definition of partial order, we shall also write $u < v$ to mean $u \le v$ and $u \ne v$. Finally, sometimes we may wish to concisely state that $u_i < v_i$ for every $i = 1, \ldots, p$. We shall accomplish that by using the notation $u \ll v$. Clearly, for each of these notations, there will be a symmetrical one, defined in the obvious way.

(5.2) Definition. A **polyhedron** in $\mathbb{R}^n$ is the intersection of a finite number of closed half-spaces in $\mathbb{R}^n$, that is $\{x \in \mathbb{R}^n : \langle \alpha^i, x \rangle \leq \beta_i \text{ for } i = 1, \ldots, m\}$, where α^i are non-zero vectors in $\mathbb{R}^n$ and β_i are scalars, for $i = 1, \ldots, m$.

(5.3) Remark. A polyhedron in $\mathbb{R}^n$ is a closed convex set, because it is the intersection of a finite number of half-spaces which are closed convex sets. From the fact that an equation can be represented by two inequalities, a polyhedron can be represented by a finite number of inequalities and/or equalities. For example, the following are typical examples of polyhedra

$$P = \left\{ x \in \mathbb{R}^n : Ax \leq b \right\}$$

$$P = \left\{ x \in \mathbb{R}^n : Ax = b, \, x \geq 0 \right\}$$

$$P = \left\{ x \in \mathbb{R}^n : Ax \leq b, \, x \geq 0 \right\}$$

where $A \in \mathfrak{M}(m, n, \mathbb{R})$ is a $m \times n$ matrix, and $b \in \mathbb{R}^m$ is an m-vector.

(5.4) Definition. Given a polyhedron P in $\mathbb{R}^n$, an extreme point of P is called a **vertex** of P. In other words, a vertex of P is a point that cannot be written as a proper convex combination of other points of P. The set of all vertices of a polyhedron P is denoted by $\mathrm{vert}(P)$.

(5.5) Example. The set

$$P_1 = \left\{ (x_1, x_2) \in \mathbb{R}^2 : 1 \leq x_1 \leq 4, \, 1 \leq x_2 \leq 3 \right\}$$

is a bounded polyhedron. It has four vertices, namely the points $(1, 1)$, $(1, 3)$, $(4, 1)$, and $(4, 3)$.

The set
$$P_2 = \left\{ (x_1, x_2) \in \mathbb{R}^2 : x_1 \geq 1, \, x_2 \geq 1, \, x_1 + x_2 \geq 3 \right\}$$

is an unbounded polyhedron. It have two vertices, namely $(1, 2)$ and $(2, 1)$.

The set
$$P_3 = \left\{ (x_1, x_2) \in \mathbb{R}^2 : 1 \leq x_2 \leq 3 \right\}$$

is an unbounded polyhedron. It has no vertices.

(5.6) Example. In $\mathbb{R}^n$, the set

$$\Delta = \left\{ x \in \mathbb{R}^n : x \geq 0, \, \sum_{i=1}^{n} x_i = 1 \right\}$$

is called a **simplex**. It is not difficult to show that this polyhedron is the convex hull of the n elements of the canonical basis in $\mathbb{R}^n$.

(5.7) Definition. A subset of $\mathbb{R}^n$ that is at the same time a polyhedron and a cone shall be called a **polyhedral cone**.

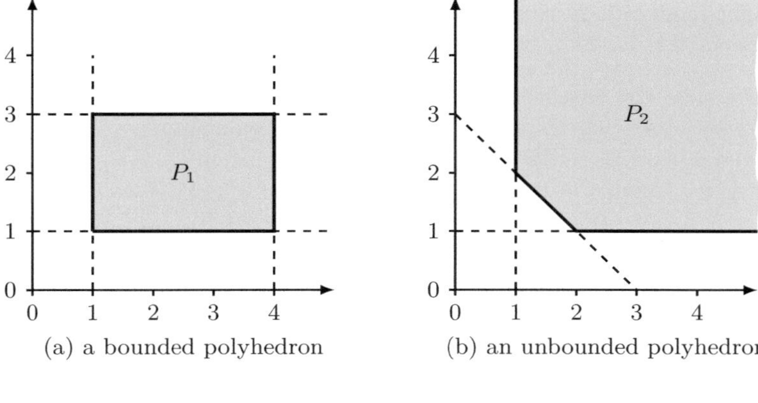

(a) a bounded polyhedron   (b) an unbounded polyhedron

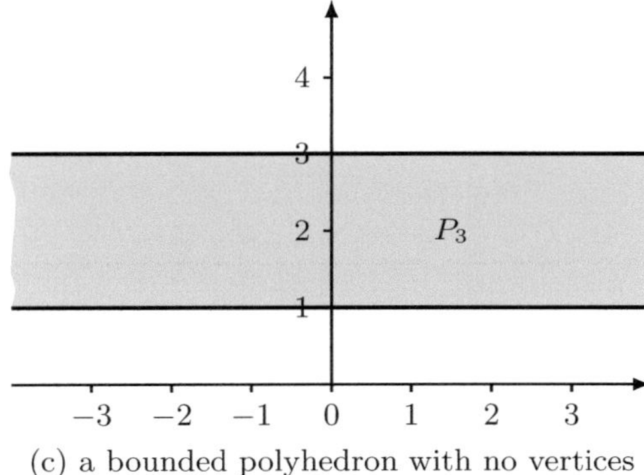

(c) a bounded polyhedron with no vertices

Figure 5.1. With reference to Example (5.5), Figure (a) shows the bounded polyhedron, P_1. Figure (b) shows the unbounded polyhedron, P_2. Figure (c) shows the unbounded polyhedron, P_3. Notice how P_1 and P_2 both have vertices, but P_3 does not have vertices.

The following proposition shows a characterisation of polyhedral cones as the set of solutions of a homogeneous system of linear inequalities.

(5.8) Proposition. *If P is a polyhedral cone in $\mathbb{R}^n$, then there exists a matrix Q with n columns, such that*

$$P = \left\{ x \in \mathbb{R}^n : Qx \leq 0 \right\}.$$

Proof. From the fact that P is by hypothesis a polyhedron, we know that it can always be written in the form $\{ x \in \mathbb{R}^n : Qx \leq q \}$, where Q is a $m \times n$ matrix, and q is a m-vector, for a certain integer m. To complete the proof, it suffices to show that

$$\left\{ x \in \mathbb{R}^n : Qx \leq q \right\} = \left\{ x \in \mathbb{R}^n : Qx \leq 0 \right\}.$$

The right-hand side in clearly included in the left-hand side, because P is a cone, hence $0 \in P$. As a result, $0 = Q0 \leq q$, so, if $Qx \leq 0$, then $Qx \leq q$.

To prove that the left-hand side is also included in the right-hand side, let us start by observing that, if $x \in P$, then by hypothesis we have $Qx \leq q$. Moreover, since P is a cone, we also have $\lambda x \in P$ for any $\lambda > 0$, and therefore $Q(\lambda x) \leq q$. Using the fact that Q is a linear mapping, and dividing both sides by λ yields $Qx \leq \lambda^{-1}q$ for every $\lambda > 0$. Finally, taking the infimum with respect to λ, gives $Qx \leq 0$, thereby proving the result. $\qquad\square$

(5.9) Example. The set $C = \{(x_1, x_2) \in \mathbb{R}^2 : -x_1 + x_2 \leq 0,\ -x_2 \leq 0\}$ is a polyhedral cone. A sketch of this polyhedral cone can be seen in Figure 5.2.

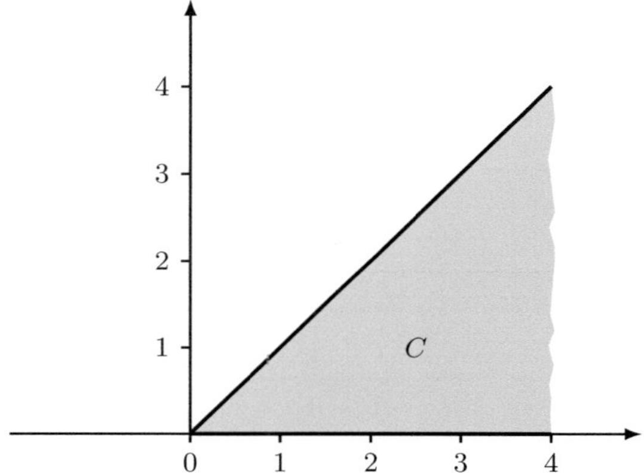

Figure 5.2. The figure shows the polyhedral cone, C, defined in Example (5.9).

(5.10) Definition. Let P be a polyhedron. A vector d in $\mathbb{R}^n$ is said to be a **direction of recession** for P if $x + \lambda d \in P$ for every $x \in P$ and $\lambda \geq 0$. In other words, d is a direction of recession for P if P contains all the rays with origin in any point of P in the direction d. The set of all directions of recession of P is denoted by $\mathrm{rec}(P)$ and it is called the **recession cone** of P.

(5.11) Example. Polyhedron P_1 in Example (5.5) does not have any direction of recession. On the other hand, the directions of recession of polyhedron P_2 are all vectors $d = (d_1, d_2)$ with $d_1 \geq 0$ and $d_2 \geq 0$. Thus, $\mathrm{rec}(P) = \{(x_1, x_2) \in \mathbb{R}^2 : x_1 \geq 0,\ x_2 \geq 0\}$.

The following theorem shows that $\mathrm{rec}(P)$ is a polyhedral cone.

(5.12) Theorem. *If* $P = \{x \in \mathbb{R}^n : Ax \leq b\}$, *then*

$$\mathrm{rec}(P) = \Big\{x \in \mathbb{R}^n : Ax \leq 0\Big\}.$$

Proof. Let d be a direction of recession for P. Thus we have $x + \lambda d \in P$ for every $x \in P$ and $\lambda \geq 0$. So, $A(x + \lambda d) \leq b$, hence $Ax + \lambda Ad \leq b$. if any component of the vector Ad were positive, the inequality $Ax + \lambda Ad \leq b$ would not hold for λ sufficiently large. As a result, $Ad \leq 0$. On the other hand, if d is a vector such that $Ad \leq 0$, then for every $x \in P$ and $\lambda \geq 0$ we have $A(x + \lambda d) = Ax + \lambda Ad \leq b + 0 = b$, hence $x + \lambda d \in P$. So, d is a direction of recession for P. $\qquad\square$

(5.13) Example. Consider the polyhedron P_2 in Example (5.5). We have

$$A = \begin{bmatrix} -1 & 0 \\ 0 & -1 \\ -1 & -1 \end{bmatrix}, \qquad b = \begin{bmatrix} -1 \\ -1 \\ -3 \end{bmatrix}.$$

Thus, by virtue of Theorem (5.12), we have

$$\mathrm{rec}(P_2) = \left\{ x \in \mathbb{R}^2 : Ax \leq 0 \right\} = \left\{ (x_1, x_2) \in \mathbb{R}^2 : x_1 \geq 0,\ x_2 \geq 0,\ x_1 + x_2 \geq 0 \right\}$$

so $\mathrm{rec}(P_2) = \{ (x_1, x_2) \in \mathbb{R}^2 : x_1 \geq 0,\ x_2 \geq 0 \}$ as we had already noticed in Example (5.11).

The next result shall require a notation from linear algebra. If A and B are two subsets of $\mathbb{R}^n$, let us recall that $A + B$ denotes the set of all sums $a + b$ with $a \in A$ and $b \in B$, that is to say,

$$A + B = \left\{ a + b : a \in A,\ b \in B \right\}.$$

This is sometimes called the **Minkowski addition**, after the famous mathematician HERMANN MINKOWSKI (1864–1909), described variously as German, Polish, Lithuanian-German, or Russian. With this notation at hand, we may state the following.

(5.14) Proposition. *Given a polyhedron P, we have $P + \mathrm{rec}(P) = P$.*

Proof. Let $x \in P$ and $d \in \mathrm{rec}(P)$. Then $x + d \in P$ by definition of direction of recession, so obviously $P + \mathrm{rec}(P) \subseteq P$. In order to prove the converse inclusion, it suffices to observe that since $0 \in \mathrm{rec}(P)$, then $x = x + 0 \in P + \mathrm{rec}(P)$. $\qquad\square$

Let us now introduce another important definition for the geometry of polyhedra.

(5.15) Definition. Let P be a polyhedron in $\mathbb{R}^n$. A **direction of lineality** for P is a vector d in $\mathbb{R}^n$ such that $d \in \mathrm{rec}(P)$ and $-d \in \mathrm{rec}(P)$. In other words, a vector d is a direction of lineality for P if P contains all straight lines in the direction d passing through every point of P. The set of all direction of lineality of P is denoted by $\mathrm{lin}(P)$, and it shall be called the **lineality space** associated to P.

(5.16) Theorem. *Let $A \in \mathfrak{M}(m, n, \mathbb{R})$ a matrix, $b \in \mathbb{R}^m$ a vector. Then, the lineality space of the polyhedron $P = \{x \in \mathbb{R}^n : Ax \leq b\}$ is*

$$\operatorname{lin}(P) = \left\{x \in \mathbb{R}^n : Ax = 0\right\}.$$

Proof. From Theorem (5.12) it follows that $\operatorname{rec}(P) = \{x \in \mathbb{R}^n : Ax \leq 0\}$. Thus,

$$\begin{aligned}
\operatorname{lin}(P) &= \operatorname{rec}(P) \cap (-\operatorname{rec}(P)) \\
&= \{x \in \mathbb{R}^n : Ax \leq 0,\ Ax \geq 0\} \\
&= \{x \in \mathbb{R}^n : Ax = 0\}.
\end{aligned}$$

This completes the proof. $\qquad\qquad\qquad\qquad\qquad\qquad\qquad\qquad\square$

(5.17) Example. In polyhedron

$$P = \left\{(x_1, x_2) \in \mathbb{R}^2 : 1 \leq x_2 \leq 3\right\},$$

every vector of the form $d = (d_1, 0)$, with $d_1 \in \mathbb{R}$, is a direction of lineality. (See Figure 5.3.) In other words, $\operatorname{lin}(P) = \{(x_1, x_2) \in \mathbb{R}^2 : x_2 = 0\}$.

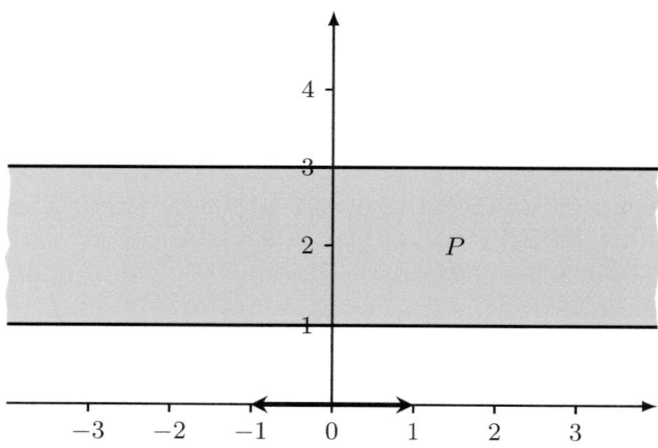

Figure 5.3. The figure shows the polyhedron P from Example (5.17). The arrows on the x-axis represent the two opposite directions of recession of the polyhedron P, and show that the x-axis is the lineality space, $\operatorname{lin}(P)$, of the polyhedron P.

(5.18) Remark. It is not difficult to check that all polyhedra of the form

$$P = \left\{x \in \mathbb{R}^n : Ax \leq b,\ x \geq 0\right\}$$

have $\operatorname{lin}(P) = \{0\}$. Indeed, if d is a non-zero direction of lineality, $x + td$ must lie in P, hence it must have all coordinates positive. However, for a suitable choice of t, the vector $x + td$ may have negative coordinates.

6. The polar of a polyhedral cone

In the rest of this section we want to show the relationship between polyhedra and their polar cones. We shall also show that the consequence of this relationship leads to an important result known as "Farkas's Lemma".

(6.1) Definition. Let C be a cone. We shall say that C is **finitely generated** if there exists a finite family $x^1, \ldots, x^m$ of points in C such that C coincides with the cone generated by $\{x^1, \ldots, x^m\}$. In other words, C is finitely generated if every point in C can be written as a conic combination of a finite number of its points $x^1, \ldots, x^m$.

(6.2) Remark. Let $C = \mathrm{cone}(q^1, \ldots, q^m)$ be a finitely generated cone, and observe that every element q in C can be written in the form

$$q = \sum_{j=1}^{m} \lambda_j q^j, \quad \text{where } \lambda_j \geq 0 \text{ for every index } j = 1, \ldots, m.$$

Now, if we denote by Q the $m \times n$ matrix whose jth row is the vector q^j, and by $\lambda = (\lambda_1, \ldots, \lambda_m)$ the vector of the coefficients of q, it is not difficult to recognise that $q = \sum_{j=1}^{m} \lambda_j q^j$ can be written as $q = Q^T \lambda$. Adopting this concise notation, the statement in Proposition (5.8) can be rephrased by saying that, if C is a finitely generated cone, there is a matrix $Q \in \mathfrak{M}(m, n, \mathbb{R})$ such that the set C can be written in the form

$$C = \left\{ y \in \mathbb{R}^n : y = Q^T \lambda, \ \lambda \in \mathbb{R}^m, \ \lambda \geq 0 \right\}.$$

(6.3) Proposition. *Let $q^1, \ldots, q^m$ be vectors in $\mathbb{R}^n$ and if C is the cone generated by these vectors, that is to say $C = \mathrm{cone}(q^1, \ldots, q^m)$, then*

$$C^* = \left\{ x \in \mathbb{R}^n : Qx \leq 0 \right\},$$

where Q is the $m \times n$ matrix whose rows are $q^1, \ldots, q^m$. In particular, the polar cone C^* of a finitely generated cone C is a polyhedral cone.

Proof. Let $C = \mathrm{cone}(q^1, \ldots, q^m)$ be a finitely generated cone, and let us denote by Q the $m \times n$ matrix whose rows are $q^1, \ldots, q^m$. We wish to prove that $C^* = \left\{ x \in \mathbb{R}^n : Qx \leq 0 \right\}$. To this end, let us start by observing that C^* is quite clearly included in the polyhedral cone $\left\{ x \in \mathbb{R}^n : Qx \leq 0 \right\}$. In fact, if $\langle q, x \rangle \leq 0$ for any $q \in C$, then, in particular, this happens when q is any of the vectors $q^1, \ldots, q^m$ that generate C. Thus, it suffices to prove the converse inclusion. To do this, let us take a vector x in $\mathbb{R}^n$ and assume that $Qx \leq 0$, and take a vector q in C. By means of Remark (6.2), q can be written in matrix form as $q = Q^T \lambda$, where $\lambda \geq 0$. Thus,

$$\langle q, x \rangle = \langle Q^T \lambda, x \rangle = \langle \lambda, Qx \rangle \leq 0,$$

because $\lambda \geq 0$ and $Qx \leq 0$. This fully proves the result. $\square$

(6.4) Proposition. *If C is a polyhedral cone, then C^* is finitely generated.*

Proof. Let us assume that $C = \{x \in \mathbb{R}^n : Qx \le 0\}$ is a polyhedral cone, where $Q \in \mathfrak{M}(m, n, \mathbb{R})$ is a $m \times n$ matrix. Denote by $q^1, \ldots, q^m$ the n-vectors representing the m rows of Q, and consider the set $D = \operatorname{cone}(q^1, \ldots, q^m)$. This is a finitely generated cone and $D^* = C$. Since D is a closed convex set, by Theorem (4.8), we have $D^{**} = C^* = D$, thus C^* is a finitely generated cone. $\qquad\square$

To sum up, Propositions (6.3) and (6.4) show that if a cone is finitely generated, then its polar cone is polyhedral, and, vice versa, if a cone is polyhedral, then its dual cone is finitely generated. This fact yields an important result that can be very useful in optimisation, and that is part of a greater collection of similar statements, known as the 'theorems of the alternative', that we are going to discuss in a later section.

(6.5) Theorem (Farkas's Lemma). *Let $A \in \mathfrak{M}(m, n, \mathbb{R})$ be a $m \times n$ matrix, and let $c \in \mathbb{R}^n$ be a n-vector. Then **either** there exists a vector $x \in \mathbb{R}^n$ such that $Ax \le 0$ and $\langle c, x \rangle > 0$, **or** there exists a vector $y \in \mathbb{R}^m$ such that $A^T y = c$ and $y \ge 0$.*

Proof. Let $C = \{A^T y : y \ge 0\}$ be the cone generated by the columns of the matrix A. Evidently, the statement

$$\text{there exists a vector } y \in \mathbb{R}^m \text{ such that } A^T y = c \text{ and } y \ge 0$$

is equivalent to $c \in C$. Furthermore, C is a closed convex cone and, by means of Proposition (6.3), its polar cone is $C^* = \{x \in \mathbb{R}^n : Ax \le 0\}$. Hence, condition $c \in C$ means that for all $x \in C^*$, we have $\langle c, x \rangle \le 0$. In other words, the statement

$$\text{there exists a vector } x \in \mathbb{R}^n \text{ such that } Ax \le 0 \text{ and } \langle c, x \rangle > 0$$

is equivalent to $c \notin C$. In summary, the statements of the theorem are equivalent to the mutually exclusive statements $c \notin C$ and $c \in C$, respectively; and the theorem is thereby fully proved. $\qquad\square$

7. Minkowski-Weyl Theorem

We have seen in the previous section that if a cone is polyhedral, its polar cone is finitely generated, and that, vice versa, if a cone is is finitely generated, its polar cone is polyhedral. It seems therefore that the two notions are dual one of the other. In reality, it turns out that they are actually equivalent conditions. The proof of this is rather subtle, and is the aim of this section.

(7.1) Theorem (Minkowski-Weyl). *A cone is polyhedral if and only if it is finitely generated.*

Proof. It suffices to prove that if a cone is finitely generated, then it is polyhedral. In fact, if this latter condition holds, and if C is a polyhedral cone, then by means of Proposition (6.4), the polar cone C^* is finitely generated, hence polyhedral. Applying again Proposition (6.4), it follows that C^{**} is finitely generated. On the other hand, by means of Theorem (4.8), $C = C^{**}$, whence if C is a polyhedral cone, it is also finitely generated.

It remains to prove that if C is a finitely generated cone, it is also polyhedral. To this end, assume that C is finitely generated, i.e. that there exist vectors $a^1, \ldots, a^m$ such that $C = \mathrm{cone}(a^1, \ldots, a^m)$. In other words, assume that

$$C = \left\{ x \in \mathbb{R}^n : x = A^T y, \, y \in \mathbb{R}^m, \, y \geq 0 \right\},$$

where $A \in \mathfrak{M}(m, n, \mathbb{R})$ is the $m \times n$ matrix whose rows are the vectors $a^1, \ldots, a^m$.

Observe that the vector space V generated by $a^1, \ldots, a^m$ has finite dimension and let r be its dimension. Clearly, $r \leq m$. This means that there exist r linearly independent vectors $e^1, \ldots, e^r$ that generate the vector space V. Denote by $E \in \mathfrak{M}(r, n, \mathbb{R})$ the $r \times n$ matrix whose rows are these vectors $e^1, \ldots, e^r$. Furthermore, we may always assume that the vectors $e^1, \ldots, e^r$ are an orthonormal basis of V (i.e. these vectors are mutually orthogonal and with norm 1). This means that the matrix E is semi-orthogonal, which means that EE^T is the identity matrix in $\mathbb{R}^r$, and $E^T E$ is the orthogonal projection of $\mathbb{R}^n$ into the space V generated by the rows of E, i.e. the vectors $e_1, \ldots, e_r$. Since both the rows of A and the rows of E are sets of generators of the vector space V, there exist two matrices $L \in \mathfrak{M}(m, r, \mathbb{R})$ and $M \in \mathfrak{M}(r, m, \mathbb{R})$ such that

$$A = LE, \qquad E = MA.$$

In particular, LM is the identity matrix in $\mathbb{R}^m$ and ML is the identity matrix in $\mathbb{R}^r$.

Now let us observe that $x \in C$ if and only if $x = A^T y$, where $y \in \mathbb{R}^m$ and $y \geq 0$. Hence, $x = A^T y = E^T L^T y = E^T w$, where $w = L^T y$ and $y \geq 0$. Now, since E is a semi-orthogonal matrix and since the cone C is included in the vector space V, generated by the rows of the matrix E, the restriction of the mapping E^T to the vector space V is bijective, and its inverse is E. Thus, multiplying both sides of the relation $x = E^T w$ by E, we then get $Ex = EE^T w = w$. In other words, the cone C can be written, in an equivalent fashion, as

$$C = \left\{ x \in \mathbb{R}^n : Ex = L^T y, \, y \in \mathbb{R}^m, \, y \geq 0 \right\}.$$

Let us set $Z = M^T E$. Then $Z \in \mathfrak{M}(m, n, \mathbb{R})$ and observe that $x \in C$ if and only if

$$Zx = M^T Ex = M^T L^T y = y \geq 0.$$

In other words, setting $Q = -Z$ we obtain that

$$C = \Big\{ x \in \mathbb{R}^n : Qx \leq 0 \Big\},$$

which means that the finitely generated cone C is also polyhedral. $\qquad \square$

8. Representation of polyhedra

The following theorem, together with its corollaries and consequences, is arguably the most important theorem surrounding polyhedra. In fact, this theorem provides a link between the algebraic structure of a polyhedron, in terms of linear inequalities, and its geometric definition, in terms of its vertices and its directions of recession. The proof of this theorem is not particularly hard, but the result, of its own, definitely deserves some attention.

(8.1) Theorem (Representation of polyhedra). *Given a polyhedron P in $\mathbb{R}^n$, there exist a finite subset $V = \{v^1, \ldots, v^m\}$ of P and a finite set of directions $D = \{d^1, \ldots, d^p\}$ such that*

$$P = \mathrm{conv}(V) + \mathrm{cone}(D).$$

Proof. Let P be a polyhedron. With no loss of generality, we can write P in the form $P = \{x \in \mathbb{R}^n : Ax \leq b\}$, and let us define the following polyhedron

$$C = \Big\{ (x, t) \in \mathbb{R}^{n+1} : t \geq 0,\ Ax - tb \leq 0 \Big\}.$$

By virtue of Minkowski-Weyl Theorem, we know that there exists a finite set $(u^1, t_1), \ldots, (u^s, t_s)$ of vectors in $\mathbb{R}^{n+1}$, such that C coincides with the cone generated by these points. In other words, we have

$$C = \mathrm{cone}\Big((u^1, t_1), \ldots, (u^s, t_s) \Big).$$

With no loss of generality, we can always assume that these vectors are normalised. In other words, for each $j = 1, \ldots, s$, either $t_j = 0$ or we can divide the full vector by t_j to get a new vector whose last component is 1. Thus, this set of generators for C can be written in an equivalent fashion as $(v^1, 1), \ldots, (v^m, 1), (d^1, 0), \ldots, (d^p, 0)$, where p is the number of such vectors whose last component in zero, and $m = s - p$.

Let us notice that $x \in P$ if and only if $(x, 1) \in C$, and this latter condition is satisfied if and only if there exist a m-vector of coefficients λ and a p-vector of coefficients μ such that

$$(x, 1) = \sum_{j=1}^{m} \lambda_j (v^j, 1) + \sum_{k=1}^{p} \mu_k (d^k, 0),$$

which is equivalent to

$$x = \sum_{j=1}^{m} \lambda_j v^j + \sum_{k=1}^{p} \mu_k d^k, \qquad \sum_{j=1}^{m} \lambda_m = 1.$$

This means that $x \in P$ if and only if $x \in \operatorname{conv}(v^1, \ldots, v^m) + \operatorname{cone}(d^1, \ldots, d^p)$, and $v^1, \ldots, v^m \in P$, because $(v^1, 1), \ldots, (v^m, 1) \in C$. The theorem is thereby fully proved. $\qquad\square$

(8.2) Remark. It is easy to remark that the representation provided by this theorem is not unique.

(8.3) Proposition. *Given a polyhedron P, there exists a finite subset V of P such that*

$$P = \operatorname{conv}(V) + \operatorname{rec}(P)$$

Proof. By virtue of Theorem (8.1), we already know that the polyhedron P can be written in the form $P = \operatorname{conv}(V) + \operatorname{cone}(D)$, where $V = \{v^1, \ldots, v^m\}$ is a finite subset of P, and $D = \{d^1, \ldots, d^p\}$ is finite. Thus, it will suffices to show that $\operatorname{cone}(D) = \operatorname{rec}(P)$. To this end, let us start by proving the inclusion $\operatorname{cone}(D) \subseteq \operatorname{rec}(P)$. In fact, since $\operatorname{rec}(P)$ is a convex cone, it suffices to show that $D \subseteq \operatorname{rec}(P)$. Now, if $x \in P$, there exist $v \in \operatorname{conv}(V)$ and $w \in \operatorname{cone}(D)$ such that $x = v + w$. Thus, for every $u \in D$, we have $w + \lambda u \in \operatorname{cone}(D)$ and therefore $x + \lambda u = v + w + \lambda u = v + (w + \lambda u) \in \operatorname{conv}(V) + \operatorname{cone}(D) = P$. This proves that u is a direction of recession for P, i.e. that $u \in \operatorname{rec}(P)$.

Let us now prove that $\operatorname{rec}(P) \subseteq \operatorname{cone}(D)$. To this end, let $d \in \operatorname{rec}(P)$ be a direction of recession for P. Then, by virtue of Lemma (5.8), we can always write $\operatorname{cone}(D)$ in the form

$$\operatorname{cone}(D) = \Big\{ x \in \mathbb{R}^n : Qx \leq 0 \Big\},$$

where Q is a suitable matrix. For any given point $x \in P$, and any scalar γ, by definition of direction of recession, we have $x + \gamma d \in P = \operatorname{conv}(V) + \operatorname{cone}(D)$, hence $x + \gamma d = \sum_{i=1}^{m} \lambda_i v^i + w$, where $\lambda_i \geq 0$, $\sum_{i=1}^{m} \lambda_i = 1$, and $w \in \operatorname{cone}(D)$. Since $v^1, \ldots, v^m$ are a finite set of vectors, we can always find a vector c such that $Qv^i \leq c$ for every $i = 1, \ldots, m$. In fact, for each component of c can be taken as the maximum value of the corresponding components of each of the vectors $v^1, \ldots, v^m$. Thus,

$$Q(x + \gamma d) = \sum_{i=1}^{m} \lambda_i Qv^i + Qw \leq \sum_{i=1}^{m} \lambda_i Qv^i \leq c \sum_{i=1}^{m} \lambda_i = c,$$

where the first step follows from the fact that $Qw \leq 0$ whenever $w \in \mathrm{cone}(D)$, and the last follows from the fact that $\sum_{i=1}^{m} \lambda_i = 1$. In other words, for all $\gamma \geq 0$, the vector $Q(x + \gamma d)$ is bounded from above by a constant vector. In order to complete the proof, it suffices to observe that, rearranging this inequality to the form $Qd \leq \gamma^{-1}(c - Qx)$, and taking the infimum with respect to γ, we finally obtain $Qd \leq 0$, which means that $d \in \mathrm{cone}(D)$. $\square$

(8.4) Example.

1. The polyhedron

$$P_1 = \left\{ (x_1, x_2) \in \mathbb{R}^2 : 1 \leq x_1 \leq 4, \, 1 \leq x_2 \leq 3 \right\}$$

   can be written in the form $P_1 = \mathrm{conv}\big((1,1),(1,3),(4,3),(4,1)\big)$.

2. The polyhedral cone

$$P_2 = \left\{ (x_1, x_2) \in \mathbb{R}^2 : x_2 \leq x_1, \, x_2 \geq 0 \right\}$$

   can be written in the form $P_2 = \mathrm{cone}\big((1,0),(1,1)\big)$.

3. The unbounded polyhedron

$$P_3 = \left\{ (x_1, x_2) \in \mathbb{R}^2 : x_1 \geq 1, \, x_2 \geq 1, \, x_1 + x_2 \geq 3 \right\}$$

   can be written in the form $P_3 = \mathrm{conv}\big((1,2),(2,1)\big) + \mathrm{cone}\big((1,0),(0,1)\big)$.

4. The unbounded polyhedron

$$P_4 = \left\{ (x_1, x_2) \in \mathbb{R}^2 : 1 \leq x_2 \leq 3 \right\}$$

   can be written in the form $P_4 = \mathrm{conv}\big((0,1),(0,3)\big) + \mathrm{cone}\big((1,0),(-1,0)\big)$. It is worthwhile to notice that, unlike the other polyhedra, in this case the points $(0,1)$ and $(0,3)$ are not vertices.

Example (8.4) shows that, in the representation $P = \mathrm{conv}(V) + \mathrm{cone}(D)$, where V is a finite subset of P and D is a finite set of directions, the elements of the set V may or may not be the vertices of the polyhedron. The following result shows the conditions under which this happens.

(8.5) Proposition. *If P is a polyhedron with $\mathrm{lin}(P) = \{0\}$, then*

$$P = \mathrm{conv}\big(\mathrm{vert}(P)\big) + \mathrm{rec}(P).$$

Proof. By means of Theorem (8.3), we already know that the polyhedron P can be written in the form $P = \mathrm{conv}(V) + \mathrm{rec}(P)$, where $V = \{v^1, \dots, v^n\}$ is a finite subset of P. Assume that the set V is minimal, i.e. that for every $U \subset V$, we have $\mathrm{conv}(U) + \mathrm{rec}(P) \subset P$, and let us prove the $V \subseteq \mathrm{vert}(P)$,

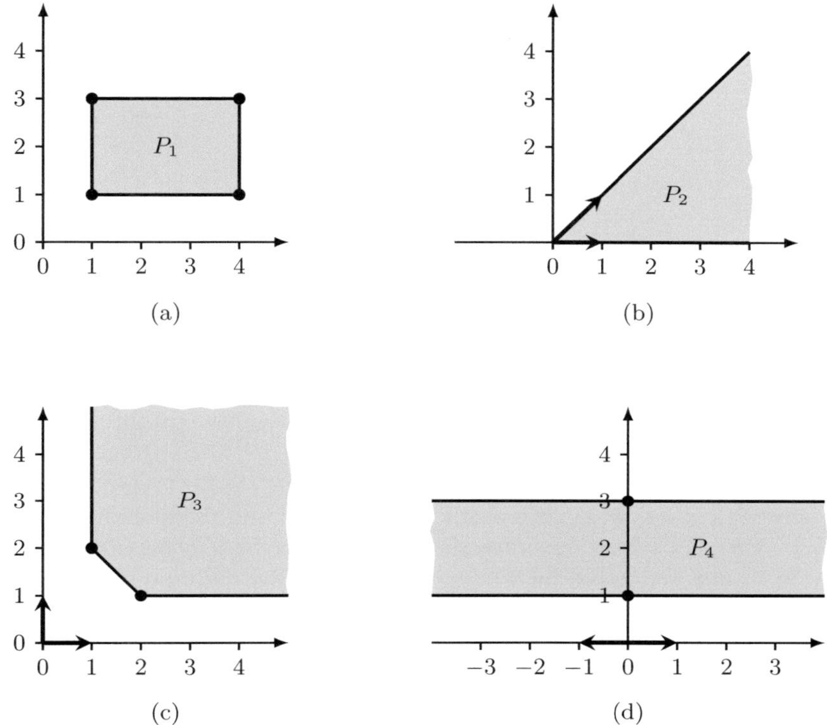

Figure 8.1. The figure shows the polyhedra described in Example (8.4). Each of these
sets can be written in the form $\mathrm{conv}(V)+\mathrm{cone}(D)$ where V is a finite subset of
of the polyhedron, and E is a finite set. The points ● represent the elements
of V, and the thick arrows, ➝, represent the elements of E.

i.e. that all vectors $v^1,\ldots,v^m$ are vertices of P. By contradiction, let us
suppose that there is an element of V that is not a vertex. With no loss of
generality, we may assume that this point is v^1. This means that there exist
two points u^1 and u^2 in P, with $u^1 \neq u^2$, and a scalar $\gamma \in (0,1)$, such that
$v^1 = \gamma u^1 + (1-\gamma)u^2$. Since u^1 and u^2 both belong to P, we can always write
them in the form $u^1 = w^1 + d^1$ and $u^2 = w^2 + d^2$, where $w^1, w^2 \in \mathrm{conv}(V)$
and $d^1, d^2 \in \mathrm{rec}(P)$. In particular, since w^1 and w^2 belong to the convex hull
of V, we can write them in the form

$$w^j = \sum_{i=1}^{m} \theta_{j,i} v^i, \qquad \theta_{j,i} \geq 0, \qquad \sum_{i=1}^{m} \theta_{j,i} = 1 \qquad (j = 1, 2),$$

Setting $\lambda_i = \gamma\theta_{1,i} + (1-\gamma)\theta_{2,i}$ yields

$$v^1 = \sum_{i=1}^{m} \lambda_i v^i + \left(\gamma d^1 + (1-\gamma)d^2\right),$$

where $\lambda_i \geq 0$ for each $i = 1, \ldots, m$,

$$\sum_{i=1}^{n} \lambda_i = \sum_{i=1}^{n} \left(\gamma\theta_{1,i} + (1 - \gamma)\theta_{2,i} \right) = \gamma \sum_{i=1}^{n} \theta_{1,i} + (1 - \gamma) \sum_{i=1}^{m} \theta_{2,i} = 1.$$

On the other hand, if it were $\lambda_1 < 1$, we would be able to write v^1 in the form

$$(8.6) \qquad v^1 = \sum_{i=2}^{m} \frac{\lambda_i}{1 - \lambda_1} v^i + \frac{1}{1 - \lambda_1} \left(\gamma d^1 + (1 - \gamma)d^2 \right)$$

thereby demonstrating that $v^1 \in \mathrm{conv}(V \setminus \{v^1\}) + \mathrm{rec}(P)$, but this would contradict the fact that V was assumed to be minimal. Hence, it must be $\lambda_1 = 1$, and therefore $\lambda_i = 0$ for all $i = 2, \ldots, m$. From (8.6), we obtain therefore $\gamma d^1 + (1 - \gamma)d^2 = 0$, which means that $d^2 = -(\gamma/(1 - \gamma))d^1$. In addition, since $0 < \gamma < 1$, by definition $\lambda_1 = \gamma\theta_{1,1} + (1 - \gamma)\theta_{2,1} = 1$, and therefore $\theta_{1,1} = \theta_{2,1} = 1$. As a result $w^1 = w^2 = v^1$ and therefore $u^1 = v^1 + d^1$ and $u^2 = v^1 + d^2$. Now, the direction of d^1 cannot be 0, otherwise d^2 would also be 0, and we would have $v^1 = u^1 = u^2$, which is impossible, because u^1 is assumed to be different from u^2. Thus, we have proved that both d^1 and $-d^1$ belong to the recession cone $\mathrm{rec}(P)$ which means that $\mathrm{lin}(P)$ has at least a direction, d^1. But this is a contradiction, because $\mathrm{lin}(P)$ was assumed that $\mathrm{lin}(P) = \{0\}$. As a result, every element of the set V must be a vertex of the polyhedron P, i.e. $V \subseteq \mathrm{vert}(P)$. Now this implies that $\mathrm{conv}(V) \subseteq \mathrm{conv}(\mathrm{vert}(P)) \subseteq P$ and therefore

$$P = \mathrm{conv}(V) + \mathrm{rec}(P) \subseteq \mathrm{conv}(\mathrm{vert}(P)) + \mathrm{rec}(P) \subseteq P + \mathrm{rec}(P) = P,$$

hence all these inclusions are actually equalities, and in particular

$$P = \mathrm{conv}(\mathrm{vert}(P)) + \mathrm{rec}(P)$$

as required. The proposition is thereby fully proved. $\qquad\square$

(8.7) Example. Consider the polyhedron

$$P = \left\{ (x_1, x_2) \in \mathbb{R}^2 : x_2 \leq 4, \; -x_1 - x_2 \leq 2, \; -x_1 + x_2 \leq 2, \; x_2 \geq 0 \right\}.$$

This set does not contain any straight-line, therefore $\mathrm{lin}(P) = \{0\}$. The vertices of P are $(2, 0)$, $(0, 2)$ and $(2, 4)$, whereas the recession cone is generated by the vector $(1, 0)$. Thus, the polyhedron P can be written in the form

$$P = \mathrm{conv}\big((2, 0), (0, 2), (2, 4)\big) + \mathrm{cone}\big((1, 0)\big).$$

Let us complete the section with a few consequences of Proposition (8.5).

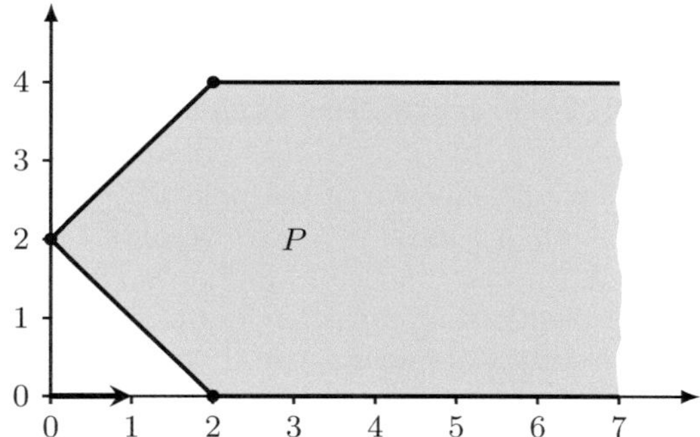

Figure 8.2. The figure shows the polyhedron P defined in Example (8.7). The points ●
mark the vertices of the polyhedron P and the arrow shows the only direction
of recession of the polyhedron. This polyhedron can be written as the sum
of the convex hull of its vertices, and its recession cone, which corresponds
to the half-line with direction $(1, 0)$.

(8.8) Corollary. *For a non-empty polyhedron P to have at least a vertex
it is necessary and sufficient that there is no directions of lineality.*

Proof. Let us start by proving that the condition is *necessary*. In other
words, assume that the polyhedron has at least one vertex, v, and let us prove
that $\lin(P) = \{0\}$. By contradiction, if there were a direction of lineality
$d \neq 0$, the two points $v + d$ and $v - d$ would both belong to P and therefore
$v = \frac{1}{2}(v + d) + \frac{1}{2}(v - d)$, which contradicts the hypothesis that v is a vertex
of P.

Let us now prove that the condition is also *sufficient*. To this end, assume that
$\lin(P) = \{0\}$. Then, by means of Proposition (8.5), $P = \operatorname{conv}\big(\operatorname{vert}(P)\big) +
\operatorname{rec}(P)$. Thus, either $P = \operatorname{rec}(P)$, in which case 0 is a vertex, or $\operatorname{vert}(P)$
contains at least a vector $v \neq 0$.   □

(8.9) Corollary. *A bounded polyhedron is the convex hull of its vertices.*

Proof. By Proposition (8.5), if P is a polyhedron,

$$P = \operatorname{conv}\big(\operatorname{vert}(P)\big) + \operatorname{rec}(P).$$

However, if P is bounded, then $\operatorname{rec}(P) = \{0\}$, thus $P = \operatorname{conv}\big(\operatorname{vert}(P)\big)$.   □

9. Motzkin's Theorem and theorems of the alternative

In this section we shall prove one of the most important 'theorems of the alternative' due to T. Motzkin. The theorems of the alternative are a collection
of statements, all similar, and all related to each other, which in a very rough

way all say that either a system of simultaneous equalities and/or inequalities has solutions, or another does, but not both. In other words, these theorems show pairs of mutually exclusive systems of simultaneous equalities and inequalities. We have already met one of these theorems (Farkas's Lemma – see Theorem (6.5)) as a consequence of the polarity between polyhedra. In this section, instead, we shall now show the theorem of the alternative due to Motzkin and another as a consequence of Motzkin's Theorem. The theorems of the alternative find plenty of applications in optimisation and therefore it is convenient to know their statement, even if they are all pretty much the same.

(9.1) Theorem (Motzkin). *Let $A \in \mathfrak{M}(m, n, \mathbb{R})$, $B \in \mathfrak{M}(p, n, \mathbb{R})$, and $C \in \mathfrak{M}(q, n, \mathbb{R})$ be three matrices, and suppose that $A \neq 0$. Then, **either** there is a vector $x \in \mathbb{R}^n$ such that $Ax \ll 0$, $Bx \leq 0$ and $Cx = 0$, **or** there are vectors $y_1 \in \mathbb{R}^m$, $y_2 \in \mathbb{R}^p$ and $y_3 \in \mathbb{R}^q$, with $y_1 > 0$, and $y_2 \geq 0$, such that*

$$A^T y_1 + B^T y_2 + C^T y_3 = 0.$$

Proof. Let us start by proving the second implication, i.e. that, if there are three vectors $y_1 \in \mathbb{R}^m$, $y_2 \in \mathbb{R}^p$, $y_3 \in \mathbb{R}^q$, with $y_1 > 0$, and $y_2 \geq 0$, such that $A^T y_1 + B^T y_2 + C^T y_3 = 0$, then there can be no vector $x \in \mathbb{R}^n$, such that $Ax \ll 0$, $Bx \leq 0$ and $Cx = 0$. This can be done by contradiction. In fact, suppose that such a vector x exists. Then, by the fact that the vectors y_1 and y_2 have non-negative coefficients, and $y_1 \neq 0$, we would have $\langle y_1, Ax \rangle < 0$, $\langle y_2, Bx \rangle \leq 0$ and $\langle y_3, Cx \rangle = 0$, whence

$$\langle A^T y_1 + B^T y_2 + C^T y_3, x \rangle = \langle A^T y_1, x \rangle + \langle B^T y_2, x \rangle + \langle C^T y_3, x \rangle$$
$$= \langle y_1, Ax \rangle + \langle y_2, Bx \rangle + \langle y_3, Cx \rangle < 0.$$

On the other hand, $\langle A^T y_1 + B^T y_2 + C^T y_3, x \rangle = \langle 0, x \rangle = 0$, and this, clearly, is indeed a contradiction.

Let us now prove that if there is a vector $x \in \mathbb{R}^n$ such that $Ax \ll 0$, $Bx \leq 0$ and $Cx = 0$, then we can never find three vectors $y_1 \in \mathbb{R}^m$, $y_2 \in \mathbb{R}^p$, $y_3 \in \mathbb{R}^q$, with $y_1 > 0$, and $y_2 \geq 0$, such that $A^T y_1 + B^T y_2 + C^T y_3 = 0$. To this end, consider the set

$$P = \left\{ (y_1, y_2, y_3) \in \mathbb{R}^m \times \mathbb{R}^p \times \mathbb{R}^q : y_1 \geq 0,\ y_2 \geq 0,\ \textstyle\sum_{k=1}^m y_{1k} = 1 \right\}.$$

This is obviously a polyhedron, hence a closed convex set. Thus, the set

$$D = \left\{ A^T y_1 + B^T y_2 + C^T y_3 : (y_1, y_2, y_3) \in P \right\}$$

is also a closed convex set, because it is the image of P under a linear map. With this new notation, the thesis is equivalent to proving that $0 \in D$. This can be done by contradiction. To this end, suppose that $0 \notin D$. Then, by means of Theorem (3.4), there exists a vector x such that

$$(9.2) \qquad\qquad \langle A^T y_1 + B^T y_2 + C^T y_3, x \rangle < 0$$

for all $(y_1, y_2, y_3) \in P$. Also, observe that the previous inequality remains true when we multiply the vector (y_1, y_2, y_3) by a positive real number. So, the inequality (9.2) can be extended to all $y_1 > 0$.

To complete the proof, it suffices to show that, by choosing wisely the three vectors y_1, y_2 and y_3 in the set P, we can show that the vector x is solution of $Ax \ll 0$, $Bx \leq 0$ and $Cx = 0$. To see this, denote by a_k $(k = 1, \ldots, m)$ the vectors corresponding to the rows of A, by b_i $(i = 1, \ldots, p)$ the vectors corresponding to the rows of B, and by c_j $(j = 1, \ldots, q)$ the vectors corresponding to the rows of C. Then, observe that, if y_1 is the kth element of the canonical basis in $\mathbb{R}^m$, $y_2 = 0$, and $y_3 = 0$, the inequality (9.2) simply becomes $\langle a_k, x \rangle < 0$ for every $k = 1, \ldots, m$, whence $Ax \ll 0$.

Now, take a unit vector $u > 0$ in $\mathbb{R}^m$ and a positive real number ε, and observe that, if $y_1 = \varepsilon u$, y_2 is the ith element of the canonical basis of $\mathbb{R}^p$, and $y_3 = 0$. With this choice, the inequality (9.2) becomes $\varepsilon \langle u, Ax \rangle + \langle b_i, x \rangle < 0$, for every $i = 1, \ldots, p$. Moreover, since this inequality holds for every real number $\varepsilon > 0$, taking the infimum yields $\langle b_i, x \rangle \leq 0$ for every $i = 1, \ldots, p$, whence $Bx \leq 0$. Finally, choose $y_1 = \varepsilon u$, $y_2 = 0$ and y_3 the jth element of the canonical basis of $\mathbb{R}^q$ first and its opposite afterwards. This leads to the inequalities $\langle c_j, x \rangle \leq 0$ and $\langle c_j, x \rangle \geq 0$ for every $j = 1, \ldots, q$, whence $Cx = 0$. Thus, x is simultaneously a solution of $Ax < 0$, $B \leq 0$ and $Cx = 0$. This fully proves the theorem. $\square$

A simple consequence of Motzkin's Theorem is another theorem of the alternative.

(9.3) Corollary (Gordan's Theorem). *Let $A \in \mathfrak{M}(m, n, \mathbb{R})$ be a matrix. Then, **either** there exists a vector $x \in \mathbb{R}^n$ such that $Ax \ll 0$, **or** there exists a vector $y \in \mathbb{R}^m$, with $y > 0$, such that $A^T x = 0$.*

Proof. This theorem is automatically proved noting that its statement coincides with the statement of Motzkin's Theorem setting $B = 0$ and $C = 0$. $\square$

Chapter II
Convex Functions

Convex (and concave) functions have many special and important properties. For instance, any local minimum of a convex function over a convex set is also a global minimum. This chapter is devoted to introduce the notion of convex (and concave) function and develop some of their most important properties. These properties will be utilised in developing suitable optimality conditions and algorithms to solve optimisation problems that involve convex (and concave) functions.

1. Definitions and basic properties

(1.1) Definition. Let $f : D \to \mathbb{R}$ be a real function, defined on a non-empty convex set D in $\mathbb{R}^n$. The function f is said to be **convex** on D if

$$f\big(\lambda x^1 + (1 - \lambda)x^2\big) \leq \lambda f(x^1) + (1 - \lambda)f(x^2)$$

for each $x^1, x^2 \in D$ and each $\lambda \in [0, 1]$. The function f is called **strictly** convex on D if the above inequality holds as a strict inequality for each distinct x^1 and x^2 in D and for each $\lambda \in (0, 1)$. Furthermore, the function f is said to be **concave** (resp. **strictly** concave) on D if $-f$ is convex (resp. strictly convex) on D.

Before delving into the properties of convex functions, it is worth to briefly discuss their geometric interpretation. Later on, we shall also see another geometric interpretation, which is more closely connected to the notion of convex sets. Meanwhile, however, let x^1 and x^2 be two distinct points in the domain of f, and consider the point $\lambda x^1 + (1 - \lambda)x^2$, where $\lambda \in (0, 1)$. This is a point in the convex set D lying on the segment with extremes x^1 and x^2, and the number $f(\lambda x^1 + (1 - \lambda)x^2)$ represents the value of the function f at this point. On the other hand, the number $\lambda f(x^1) + (1 - \lambda)f(x^2)$ is the weighted average of the values $f(x^1)$ and $f(x^2)$ that the function takes at the extremes. Thus, for a convex function, f, the value of f at every point on

© The Author(s), under exclusive license to Springer Nature Switzerland AG 2026

A. Carpignani and M. Pappalardo, *Theory and Methods of Optimisation*,

UNITEXT 175, https://doi.org/10.1007/978-3-032-01514-3_2

the segment line with extremes x^1 and x^2 is less than or equal to the height of the chord joining the points $(x^1, f(x^1))$ and $(x^2, f(x^2))$. For a concave function, instead, the chord joining the points $(x^1, f(x^1))$ and $(x^2, f(x^2))$ is below the function itself. This is shown in Figure 1.1.

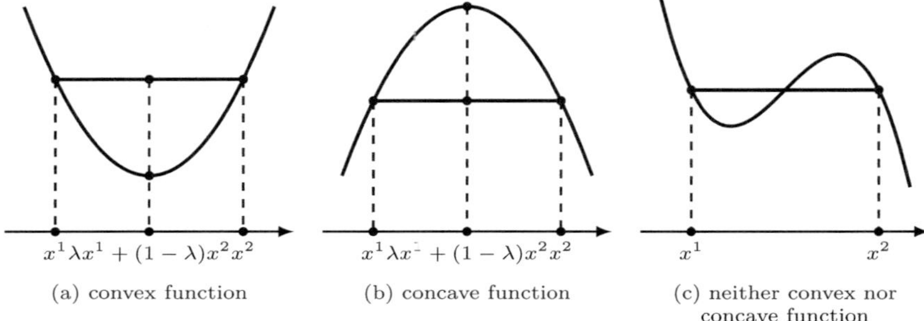

(a) convex function   (b) concave function   (c) neither convex nor concave function

Figure 1.1. The figure shows a convex function, a concave function and a function that is neither concave nor convex. In a convex function (see (a)), the points of the graph between x^1 and x^2 lie below the chord joining the two extremes, whereas in a concave function (see (b)), the points of the graph between x^1 and x^2 are all above the chord joining the two extremes.

(1.2) Example. The function $f : \mathbb{R} \to \mathbb{R}$ defined by $f(x) = |x|$ is convex. More generally, the function $f : \mathbb{R}^n \to \mathbb{R}$ defined by $f(x) = \|x\|$ is also convex. Indeed, this is immediately guaranteed by the triangle inequality.

(1.3) Example. The function $f : \mathbb{R} \to \mathbb{R}$ defined by $f(x) = x^2$ is convex. To show this, let x_1 and x_2 be two real numbers and let $\lambda \in [0, 1]$, and observe that the inequality $\lambda(1 - \lambda)(x_1 - x_2)^2 \geq 0$ holds. Expanding the binomial and rearranging, this leads immediately to the definition of convex function for $f(x) = x^2$.

Before we may give more complex examples of convex functions, it is convenient to introduce some results that enable us to create new convex functions from convex functions already known. The proofs of the three following results are simple checks that the definition of convexity applies.

(1.4) Lemma. Let $g : \mathbb{R}^m \to \mathbb{R}$ be a convex function. If $A \in \mathfrak{M}(m, n, \mathbb{R})$ is a $m \times n$ matrix, and $b \in \mathbb{R}^m$ is a m-vector, then the function $f : \mathbb{R}^n \to \mathbb{R}$ defined by $f(x) = g(Ax + b)$ is convex.

Proof. Let x^1 and x^2 be two points in $\mathbb{R}^n$ and $\lambda \in [0, 1]$. Then

$$\begin{aligned}
f\big(\lambda x^1 + (1 - \lambda)x^2\big) &= g\big(A\big(\lambda x_1 + (1 - \lambda)x_2\big) + b\big) \\
&= g\big(\lambda A x^1 + (1 - \lambda)A x^2 + \lambda b + (1 - \lambda)b\big) \\
&= g\big(\lambda(A x^1 + b) + (1 - \lambda)(A x^2 + b)\big) \\
&\leq \lambda g(A x^1 + b) + (1 - \lambda)g(A x^2 + b) \\
&= \lambda f(x^1) + (1 - \lambda)f(x^2).
\end{aligned}$$
$\qquad\square$

(1.5) Lemma. *Let $g : \mathbb{R} \to \mathbb{R}$ be a non-decreasing convex function and let $f : D \to \mathbb{R}$ be a convex function defined on a convex set D of $\mathbb{R}^n$. Then, the composite function $g \circ f : D \to \mathbb{R}$ is also convex.*

Proof. Let x^1 and x^2 be two points in D and let $\lambda \in (0,1)$. By definition of convex function for f, we have $f(\lambda x^1 + (1-\lambda)x^2) \leq \lambda f(x^1) + (1-\lambda)f(x^2)$. Similarly, applying the definition of convex function to g, and using the fact that g is non-decreasing and convex, we get

$$g\big(f(\lambda x^1 + (1-\lambda)x^2)\big) \leq g\big(\lambda f(x^1) + (1-\lambda)f(x^2)\big)$$
$$\leq \lambda g\big(f(x^1)\big) + (1-\lambda)g\big(f(x^2)\big)$$

which proves that $g \circ f$ is convex on D. $\qquad\square$

(1.6) Lemma. *Let $f_1, \ldots, f_m$ be convex functions on the same convex domain D. Then, the function $f = \sum_{i=1}^{m} \alpha_i f_i$, where $\alpha_i \geq 0$ for every $i = 1, \ldots, m$, is convex on D.*

Proof. Let x^1 and x^2 be two points in D and let $\lambda \in (0,1)$. Then,

$$f\big(\lambda x^1 + (1-\lambda)x^2\big) = \sum_{i=1}^{m} \alpha_i f_i\big(\lambda x^1 + (1-\lambda)x^2\big)$$
$$\leq \sum_{i=1}^{m} \alpha_i \Big[\lambda f_i(x^1) + (1-\lambda)f_i(x^2)\Big]$$
$$\leq \lambda \sum_{i=1}^{m} \alpha_i f_i(x^1) + (1-\lambda) \sum_{i=1}^{m} \alpha_i f_i(x^2)$$
$$= \lambda f(x^1) + (1-\lambda)f(x^2). \qquad\square$$

(1.7) Example. The real function f defined on $\mathbb{R}^n$ by $f(x) = \|x\|^2$ is convex, by Lemma (1.5). This is in fact an important function in non-linear optimisation, especially for the applications to machine learning.

(1.8) Example. The real function defined on $\mathbb{R}$ by $f(x) = ax^2 + bx + c$, with $a > 0$, is also convex. In fact, f can be written in completed-the-square form as $f(x) = a(x-p)^2 + q$, where p and q are two suitable real numbers, and Lemma (1.4) applied to $x \mapsto x^2$ and Lemma (1.6) ensure that this is indeed a convex function.

The next result shows another way to construct convex functions.

(1.9) Proposition. *Let $\mathcal{H}$ be a family of convex functions defined on a convex set $D \subseteq \mathbb{R}^n$. Let us set*

$$f(x) = \sup\Big\{g(x) : g \in \mathcal{H}\Big\} \qquad \text{for every } x \in D.$$

Then f is also a convex function on D.

Proof. For any pair of points x^1, and x^2 in D, and any real value $\lambda \in [0,1]$, observe that, since, by hypothesis, each function g in $\mathcal{H}$ is convex, we have

$$g(\lambda x^1 + (1-\lambda)x^2) \leq \lambda g(x^1) + (1-\lambda)g(x^2).$$

Taking the supremum over $\mathcal{H}$ on both sides of the previous inequality, and remembered that the supremum of the sum of two terms is always lower than or equal to the sum of the corresponding suprema, yields:

$$
\begin{aligned}
f\big(\lambda x^1 + (1-\lambda)x^2\big) &= \sup_{g\in\mathcal{H}} g\big(\lambda x^1 + (1-\lambda)x^2\big) \\
&\leq \sup_{g\in\mathcal{H}}\Big\{\lambda g(x^1) + (1-\lambda)g(x^2)\Big\} \\
&\leq \lambda \sup_{g\in\mathcal{H}} g(x^1) + (1-\lambda)\sup_{g\in\mathcal{H}} g(x^2) \\
&= \lambda f(x^1) + (1-\lambda)f(x^2).
\end{aligned}
$$

where the last step follows from the fact that, by definition, we have

$$
f(x^1) = \sup\Big\{g(x^1) : g \in \mathcal{H}\Big\}, \qquad f(x^2) = \sup\Big\{g(x^2) : g \in \mathcal{H}\Big\}.
$$

This shows that the function f, pointwise supremum of the family $\mathcal{H}$ of convex functions, is convex. $\qquad\square$

The previous result entails a similar one for concave function. Precisely,

(1.10) Proposition. *Let $\mathcal{H}$ be a family of concave functions defined on a convex set $D \subseteq \mathbb{R}^n$. Let us set*

$$
f(x) = \inf\Big\{g(x) : g \in \mathcal{H}\Big\} \qquad \text{for every } x \in D.
$$

Then f is also a concave function on D.

Proof. It suffices to observe that, f, g are concave functions if and only if $-f, -g$ are convex functions, for every $g \in \mathcal{H}$. The conclusion follows from Proposition (1.9). $\qquad\square$

The following result starts showing the links between convex functions and convex sets. To see this, let us start with a definition.

(1.11) Definition. Let D be a subset in $\mathbb{R}^n$ and let $f : D \to \mathbb{R}$ be a real function defined on D. For every real number α, the set

$$
D_\alpha = \Big\{x \in D : f(x) \leq \alpha\Big\}
$$

is called a **sub-level set**.

(1.12) Proposition. *In addition to the hypotheses of Definition (1.11), assume that the set D is convex and the function f is convex on D. Then, for every real number α, the sub-level set D_α is convex.*

Proof. Let x^1 and x^2 be two points in D_α. Then, by definition, $f(x^1) \leq \alpha$ and $f(x^2) \leq \alpha$. Now let $\lambda \in (0,1)$ and set $x = \lambda x^1 + (1-\lambda)x^2$. By convexity of D, we have $x \in D$. Furthermore, by convexity of f, we also have

$$f(x) \leq \lambda f(x^1) + (1-\lambda)f(x^2) \leq \lambda\alpha + (1-\lambda)\alpha = \alpha,$$

so $x \in D_\alpha$, and therefore D_α is convex. $\qquad\square$

Incidentally, let us observe that the converse of Proposition (1.12) is false; this can easily be checked using the function $f(x) = x^3$.

2. Epigraph and hypograph

A function $f : D \to \mathbb{R}$ can be fully described by the set

$$\Big\{ (x, f(x)) : x \in D \Big\}$$

which is generally called the graph of the function. We can also construct two other sets that are related to the graph of the function: the epigraph, which consists of the points above the graph, and the hypograph, which consists of the points below the graph. Precisely, let us give the following definition.

(2.1) Definition. Let D be a non-empty set in $\mathbb{R}^n$, and $f : D \to \mathbb{R}$. The **epigraph** of f, denoted by $\mathrm{epi}(f)$, is a subset of $\mathbb{R}^{n+1}$ defined as:

$$\mathrm{epi}(f) = \Big\{ (x, y) : x \in D,\ y \in \mathbb{R},\ f(x) \leq y \Big\}.$$

The **hypograph** of f, denoted by $\mathrm{hyp}(f)$, is a subset of $\mathbb{R}^{n+1}$ defined as:

$$\mathrm{hyp}(f) = \Big\{ (x, y) : x \in D,\ y \in \mathbb{R},\ y \leq f(x) \Big\}.$$

Figure 2.1 shows epigraph and hypograph of a convex function, (a), a concave function, (b), and function that is neither convex nor concave, (c). Notice that the epigraph of the convex function is a convex set, whereas the hypograph of a concave function is a convex set. For a function that is neither convex nor concave, however, neither the epigraph nor the hypograph is convex. This is a general result, which gives us the relationship between convex functions and convex sets that was previously mentioned. In fact, it turns out that for a real function f, defined on a non-empty convex subset D of $\mathbb{R}^n$, to be convex it is necessary and sufficient that its epigraph be convex. This is indeed the statement of the next theorem.

(2.2) Theorem. *Let D be a non-empty convex set in $\mathbb{R}^n$, and $f : D \to \mathbb{R}$. Then f is convex if and only if $\mathrm{epi}(f)$ is a convex set.*

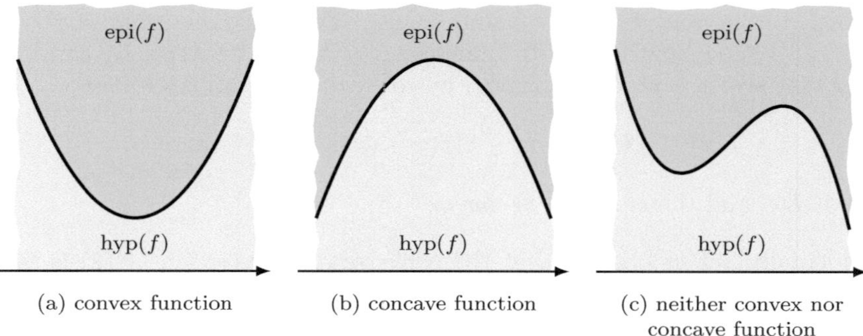

Figure 2.1. The epigraph and hypograph of a convex function, a concave function and a function that is neither concave, nor convex. Notice how the epigraph of a convex function is a convex set, whereas the hypograph of a concave function is a convex set.

Proof. Let us start by assuming that f is convex, and let (x^1, y_1) and (x^2, y_2) be two points in the epigraph of f. In other words, assume that $x^1 \in D$, $x^2 \in D$, and that $f(x^1) \leq y_1$ and $f(x^2) \leq y_2$. Let $\lambda \in (0, 1)$. Then

$$f\big(\lambda x^1 + (1 - \lambda)x^2\big) \leq \lambda f(x^1) + (1 - \lambda)f(x^2)$$
$$\leq \lambda y_1 + (1 - \lambda)y_2.$$

This proves that the point $\big(\lambda x^1 + (1 - \lambda)x^2, \lambda y_1 + (1 - \lambda)y_2\big)$ lies in the epigraph of the function f, hence epi(f) is convex. Conversely, let us assume now that the epigraph of f is convex, and let x^1 and x^2 be two point in the domain D of f. Then, the points $(x^1, f(x^1))$ and $(x^2, f(x^2))$ belong to epi(f), and by convexity of epi(f), it follows that the point

$$\big(\lambda x^1 + (1 - \lambda)x^2, \lambda f(x^1) + (1 - \lambda)f(x^2)\big)$$

also belongs to epi(f), for every $\lambda \in (0, 1)$. In other words, this means that

$$f\big(\lambda x^1 + (1 - \lambda)x^2\big) \leq \lambda f(x^1) + (1 - \lambda)f(x^2),$$

that is, f is convex. This completes the proof. $\square$

Let us notice immediately that this result could be stated, in an equivalent manner, for the hypograph. In fact, the transformation $f \mapsto -f$ also maps the hypograph of f to the epigraph of $-f$, and vice versa. Therefore, from Theorem (2.2), it turns out that a function $f : D \to \mathbb{R}$, where D is convex set in $\mathbb{R}^n$, is concave if and only if its hypograph is convex.

Thus, the result of Theorem (2.2) enables us to establish at a glance whether a function is convex by simply watching its epigraph to see whether it is a convex set. Using this result, we can immediately establish that, with reference to Figure 1.1, the function (a) is convex, the function (b) is concave, because its opposite function is convex, and the function in (c) is neither convex nor concave, because neither its epigraph nor its hypograph are convex.

3. Upper and lower semi-continuity

Before being able to describe the topological properties of convex functions, we need to recall a definition that generalises the notion of continuity for a real function of several variables. To this end, let us start to introduce some notation. First off, if $f : D \to \mathbb{R}$ is a real function, defined on an open set D of $\mathbb{R}^n$, and $x^* \in D$, recall that by $N(x^*)$ we denote the set of all neighbourhoods of x^*.

(3.1) Definition. In the framework described above, we call the **limit inferior** of the function f at the point x^* the (possibly infinite) number

$$\liminf_{x \to x^*} f(x) = \sup_{V \in N(x^*)} \inf_{y \in V \setminus \{x^*\}} f(y).$$

Similarly, we call the **limit superior** of the function f at the point x^* the (possibly infinite) number

$$\limsup_{x \to x^*} f(x) = \inf_{V \in N(x^*)} \sup_{y \in V \setminus \{x^*\}} f(y).$$

Observe immediately that, contrary to the limit of a function at a point, the limit inferior and limit superior always exist, and the following relationship holds:

$$\liminf_{x \to x^*} f(x) \leq \limsup_{x \to x^*} f(x).$$

We are now able to give the following definition.

(3.2) Definition. Let D be an open set in $\mathbb{R}^n$, $f : D \to \mathbb{R}$ be a real function defined on D, and x^* a point in D. We shall say that f is **lower semi-continuous** at x^* if

$$f(x^*) \leq \liminf_{x \to x^*} f(x).$$

Similarly, we shall say that f is **upper semi-continuous** at x^* if

$$\limsup_{x \to x^*} f(x) \leq f(x^*).$$

In addition, f is said to be lower (resp. upper) semi-continuous, if it is lower (resp. upper) semi-continuous at each point x^* in its domain.

It is simple to show that the transformation $f \mapsto -f$ maps a lower semi-continuous function to an upper semi-continuous function, and vice versa. In addition, as a result of the definition of semi-continuity, and the fact that the limit inferior is always lower than or equal to the limit superior, it is immediate to prove the following result.

(3.3) Proposition. *In the hypotheses of Definition (3.2), the function f is continuous at x^* if and only if the function f is both lower semi-continuous and upper semi-continuous. Furthermore, if one of (hence both) these conditions holds, then*

$$\liminf_{x \to x^*} f(x) = \limsup_{x \to x^*} f(x) = \lim_{x \to x^*} f(x) = f(x^*).$$

The next result, instead, gives an important characterisation of lower semi-continuity, in terms of notions that are already familiar.

(3.4) Proposition. *Let D be an open set in $\mathbb{R}^n$, and $f : D \to \mathbb{R}$ be a real function defined on D. Then, the following conditions are equivalent:*

(a) *f is lower semi-continuous.*

(b) *The sub-level set $D_\alpha = \{x \in D : f(x) \leq \alpha\}$ is closed for every real number α.*

(c) *$\mathrm{epi}(f)$ is closed in $D \times \mathbb{R}$.*

Proof. To start off, let us prove the implication (a) $\Rightarrow$ (b). To this end, assume that f is lower semi-continuous, and let us prove that $D_\alpha^c = \mathbb{R}^n \setminus D_\alpha$ is open. Take $x^* \notin D_\alpha$. By definition, this means that $f(x^*) > \alpha$. Now let us observe that, for any positive real number ε smaller than $f(x^*) - \alpha$, from the fact that f is lower semi-continuous at x^*, it follows that there exists a neighbourhood $V \in N(x^*)$ such that $f(x) \geq f(x^*) - \varepsilon > \alpha$, for every $x \in V \setminus \{x^*\}$. This is clearly true for $x = x^*$ as well, and this proves that $V \subseteq D_\alpha^c$. This proves that D_α^c is open.

Let us now prove the implication (b) $\Rightarrow$ (c). Assume that every sub-level set D_α is open, and let us prove that $\mathrm{epi}(f)^c$ is open. Take $(x, \alpha) \notin \mathrm{epi}(f)$. By definition, this means that $f(x) > \alpha$. Observe that, for any positive real number ε smaller than $f(x^*) - \alpha$, since $f(x^*)$ lies in $D_{\alpha+\varepsilon}^c$, which is assumed by hypothesis to be open, there exists a neighbourhood $V \in N(x^*)$ such that $f(x) > \alpha + \varepsilon$, for all $x \in V$. Hence, for every $(x, \beta) \in V \times (\alpha - \varepsilon, \alpha + \varepsilon)$, we must also have $f(x) > \beta$, which means that $V \times (\alpha - \varepsilon, \alpha + \varepsilon) \subseteq \mathrm{epi}(f)^c$. This proves that $\mathrm{epi}(f)^c$ is open.

Let us finally prove the implication (c) $\Rightarrow$ (a). Assume that $\mathrm{epi}(f)^c$ is open, take a point x^* in D, and let us prove that f is lower semi-continuous at x^*. Let α be a real number such that $f(x^*) > \alpha$. Thus, $(x, \alpha) \in \mathrm{epi}(f)^c$. Since this is assumed to be an open set by hypothesis, there exist a neighbourhood $V \in N(x^*)$ and a positive real number ε such that $V \times (\alpha - \varepsilon, \alpha + \varepsilon) \subseteq \mathrm{epi}(f)^c$. This means that, if $x \in V$ and $\alpha - \varepsilon < \beta < \alpha + \varepsilon$, then $f(x) > \beta$. Taking the infimum of the left-hand side and the supremum of the right-hand side, this gives $\inf_{x \in V} f(x) \geq \alpha + \varepsilon > \alpha$. Thus,

$$\liminf_{x \to x^*} f(x) = \sup_{V \in N(x^*)} \inf_{x \in V} f(x) > \alpha.$$

Since α is arbitrary, this proves that

$$\liminf_{x \to x^*} f(x) \geq f(x^*),$$

i.e. that f is lower semi-continuous at x^*. $\square$

4. Continuity of convex functions

Convex functions have special topological properties. In particular, every convex function is continuous in the interior of its domain. This is the content of the next theorem. In order to prove the theorem, however, it will be useful to prove the following simple topological lemma.

(4.1) Lemma. *Let D be a non-empty open convex set in $\mathbb{R}^n$, and let u and v be two distinct points in D. Then there exists a third point w in D such that u can be written as a convex combination of v and w. In other words, $u = \lambda v + (1 - \lambda)w$, with $\lambda \in (0,1)$.*

Proof. Let u and v be two distinct points in D and denote by $\varrho = \|u-v\|$ the distance between u and v. Since D is an open set, there exists a positive real number δ such that the open ball $B(u,\delta)$ is included in D. Take a number ε with $0 < \varepsilon < \delta$ and consider the point $w = u + (\varepsilon/\varrho)(u - v)$. By definition, we have $\|u - w\| = (\varepsilon/\varrho)\|u - v\| = \varepsilon$ and this ensures that $w \in B(u,\delta) \subseteq D$. Rearranging the definition of w to make u the subject, we have

$$u = \left(\frac{\varepsilon}{\varrho + \varepsilon}\right) v + \left(\frac{\varrho}{\varrho + \varepsilon}\right) w$$

and the conclusion follows from the fact that the two coefficients of the right-hand side of the previous equality add up to 1. $\square$

Using the above lemma, we can finally prove the following important result.

(4.2) Theorem. *Let D be a non-empty convex set, and let $f : D \to \mathbb{R}$ be a convex function. Then, f is continuous on the interior of D.*

Proof. Let f be a convex function defined on a convex set D, and assume that x^* is a point in the interior of D. This means that there exists a neighbourhood $V \in N(x^*)$ such that $V \subseteq D$. With no loss of generality, we can assume that V is convex as well. Take $x \in V$ and observe that, by virtue of Lemma (4.1), always are two points, x^1 and x^2 in V such that

$$x = \lambda x^1 + (1 - \lambda)x^*, \qquad x^* = \mu x^2 + (1 - \mu)x$$

with $0 < \lambda < 1$ and $0 < \mu < 1$. Since f is convex, these two equalities give

$$f(x) \leq \lambda f(x^1) + (1 - \lambda)f(x^*), \qquad f(x^*) \leq \mu f(x^2) + (1 - \mu)f(x).$$

Taking the limit superior of the first, and the limit inferior of the second, as $x \to x^*$, and then taking the limit $\lambda \to 0^+$ in the first expression, and the limit $\mu \to 0^+$ in the second, gives

$$\limsup_{x \to x^*} f(x) \le f(x^*), \qquad f(x^*) \le \liminf_{x \to x^*} f(x)$$

which proves that the function f is both upper and lower semi-continuous at the point x^*, hence continuous as well at the same point. $\qquad\square$

5. Sub-gradients of convex functions

We are now ready to introduce an important concept that is going to be very helpful in problems of optimisation involving convex functions. In fact, if convex functions are topologically well-behaved (indeed, they are continuous in the interior of their domains), they may not be smooth. For instance, think about the real function $f(x) = |x|$. The main feature of a smooth function is that we can always find a tangent hyper-plane to the graph. For a convex function, this may not be possible. However, for a convex function, the epigraph is a convex set, and this ensures, by virtue of Theorem I(3.6), that what we can always find a supporting hyper-plane at every point of its boundary, which means at each point of the graph. Rephrasing this idea in the language of convex functions leads to the following definition.

(5.1) Definition. Let D be a non-empty convex set in $\mathbb{R}^n$, and $f : D \to \mathbb{R}$ be a real convex function on D. A vector $\xi \in \mathbb{R}^n$ is called a **sub-gradient** of f at a point $x^* \in D$ if

$$f(x) \ge f(x^*) + \langle \xi, x - x^* \rangle \qquad \text{for all } x \in D.$$

The collection of all sub-gradients of the function f at the point x^* is called the **sub-differential** of f at x^* and it shall be denoted by $\partial f(x^*)$.

(5.2) Example. The Euclidean norm, $f(x) = \|x\|$, has sub-gradient at every $x \in \mathbb{R}^n$. In particular, it has sub-gradient at $x = 0$, and the set $\partial f(0)$ consists of all vectors ξ such that $\|z\| \ge \langle \xi, z \rangle$ for all z; in other words, it is the Euclidean unit ball. Conversely, when $x \ne 0$, the set $\partial f(x)$ consists of the single vector $\|x\|^{-1}x$.

Obviously, the sub-differential of a convex function f at any point x^* is a closed convex set. In fact, if ξ^1 and ξ^2 are any two sub-gradients of f at x^*, and if $\lambda \in (0,1)$, then $\xi = \lambda\xi^1 + (1-\lambda)\xi^2$ is also a sub-gradient. Indeed, we have

$$\begin{aligned}
\langle \xi, x - x^* \rangle &= \langle \lambda\xi^1 + (1-\lambda)\xi^2, x - x^* \rangle \\
&= \lambda\langle \xi^1, x - x^* \rangle + (1-\lambda)\langle \xi^2, x - x^* \rangle \\
&\le \lambda\big(f(x) - f(x^*)\big) + (1-\lambda)\big(f(x) - f(x^*)\big) \\
&= f(x) - f(x^*).
\end{aligned}$$

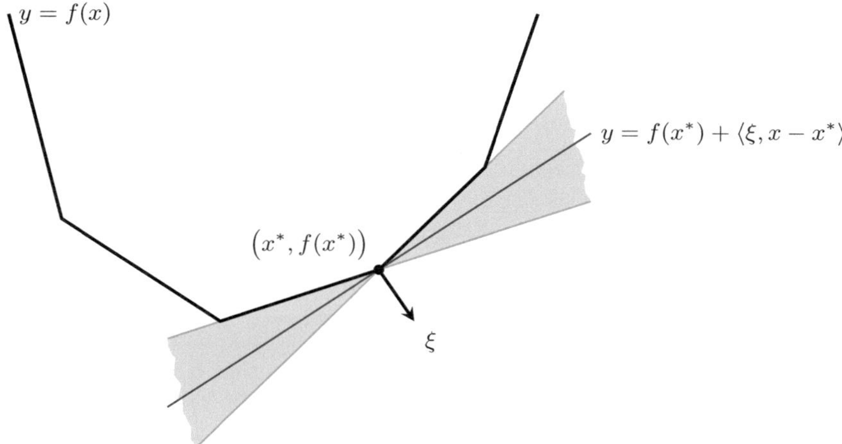

Figure 5.1. The diagram shows a convex function f. A point with coordinates $(x^*, f(x^*))$ is highlighted, and the double cone shaded represents all possible tangent lines at the graph of the function f at the point $(x^*, f(x^*))$. In particular, one of these tangent lines, with equation $y = f(x^*) + \langle \xi, x - x^* \rangle$, is highlighted. Note that, the graph of the function f lies all on the upper semi-plane generated by this line, whence the vector ξ that generates the line is a sub-gradient of f at x^*.

To show that $\partial f(x^*)$ is also closed, consider a sequence $\{\xi^\nu\}$ of elements of $\partial f(x)$. Then, for every index ν, one has

$$\langle \xi^\nu, x - x^* \rangle \leq f(x) - f(x^*) \qquad \text{for all } x \in D \text{ and every index } \nu.$$

If we further assume that $\{\xi^\nu\}$ is convergent towards a vector ξ, then, by taking the limit in the previous inequality, we get $\langle \xi, x - x^* \rangle \leq f(x) - f(x^*)$, which shows that ξ is a sub-gradient of f at x^* as well.

Further, the function $g(x) = f(x^*) + \langle \xi, x - x^* \rangle$ corresponds to a supporting hyper-plane of the epigraph, epi(f), of the function f at the point $(x^*, f(x^*))$. Correspondingly, the vector ξ represents the slope of such a supporting hyper-plane, as shown in Figure 5.1. This is described formally in the next theorem where it is also proved that every convex and concave function has at least one sub-gradient at any point in the interior of its domain. The proof relies on the fact that a convex set always has a supporting hyper-plane at every point of the boundary.

(5.3) Theorem. *Let D be a non-empty convex set in $\mathbb{R}^n$ and let $f : D \to \mathbb{R}$ be a convex function defined on D. Then, for every x^* in the interior of D, there exists a vector ξ such that the hyper-plane*

$$H = \left\{ (x, y) \in \mathbb{R}^n \times \mathbb{R} : y = f(x^*) + \langle \xi, x - x^* \rangle \right\}$$

supports epi(f) at the point $\big(x^*, f(x^*)\big)$. In particular,

$$f(x) \geq f(x^*) + \langle \xi, x - x^* \rangle \qquad \text{for every } x \in D.$$

Hence, ξ is a sub-gradient of f at x^*.

Proof. By virtue of Theorem (2.2), the set epi(f) is convex. In addition, the point $\big(x^*, f(x^*)\big)$ lies on the boundary of epi(f). Thus, Theorem I(3.6) ensures the existence of a non-zero vector (ζ, η) in $\mathbb{R}^n \times \mathbb{R}$ such that

$$(5.4) \qquad \langle \zeta, x - x^* \rangle + \eta\big(y - f(x^*)\big) \leq 0 \qquad \text{for all } (x, y) \in \text{epi}(f).$$

Observe that η cannot be positive. Otherwise, the inequality would fail by simply taking y sufficiently large. We shall now prove that $\eta \neq 0$. This can be done by contradiction. Indeed, assume $\eta = 0$. Then the previous inequality becomes $\langle \zeta, x - x^* \rangle \leq 0$ for all $x \in D$. However, by hypothesis, x^* lies in the interior of D. Thus, there exists a number $\lambda > 0$ such that $x^* + \lambda\zeta \in D$. Now, applying (5.4) with $x = x^* + \lambda\zeta$, we get $\lambda\langle \zeta, \zeta \rangle \leq 0$, and this implies that $\zeta = 0$. Hence $(\zeta, \eta) = (0, 0)$, contradicting the fact that the vector (ζ, η) is a non-zero vector. Thus, $\eta < 0$. Now, by setting $\xi = |\eta|^{-1}\zeta$ and dividing the inequality (5.4) by $|\eta|$, we finally get

$$(5.5) \qquad \langle \xi, x - x^* \rangle - y + f(x^*) \leq 0 \qquad \text{for all } (x, y) \in \text{epi}(f).$$

This shows that

$$H = \left\{ (x, y) \in \mathbb{R}^n \times \mathbb{R} : y = f(x^*) + \langle \xi, x - x^* \rangle \right\}$$

is a supporting hyper-plane of epi(f) at the point $\big(x^*, f(x^*)\big)$. Furthermore, by letting $y = f(x)$ in the inequality (5.5) and rearranging to make $f(x)$ the subject of the formula, we also get $f(x) \geq f(x^*) + \langle \xi, x - x^* \rangle$ for all $x \in D$, and this completes the proof. $\qquad\qquad\square$

(5.6) Corollary. *Let D be a non-empty convex set in $\mathbb{R}^n$, and let $f : D \to \mathbb{R}$ be a strictly convex function on D. Then, for every $x \in \text{int } D$, there exists a vector ξ such that*

$$f(x) > f(x^*) + \langle \xi, x - x^* \rangle \qquad \text{for every } x \in D \text{ with } x \neq x^*.$$

Proof. By virtue of Theorem (5.3), there exists a vector ξ such that

$$(5.7) \qquad f(x) \geq f(x^*) + \langle \xi, x - x^* \rangle \qquad \text{for every } x \in D.$$

By contradiction, let us suppose that there exists a point $\tilde{x} \neq x^*$ such that $f(\tilde{x}) = f(x^*) + \langle \xi, \tilde{x} - x^* \rangle$. Since f is a strictly convex real function, for every real number λ, such that $\lambda \in (0, 1)$, we get

$$(5.8) \qquad \begin{aligned} f\big(\lambda\tilde{x} + (1 - \lambda)x^*\big) &< \lambda f(\tilde{x}) + (1 - \lambda)f(x^*) \\ &= f(x^*) + \lambda\langle \xi, \tilde{x} - x^* \rangle. \end{aligned}$$

On the other hand, setting $x = \lambda \tilde{x} + (1 - \lambda)x^*$ in (5.7), gives

$$f\big(\lambda \tilde{x} + (1 - \lambda)x^*\big) \geq f(x^*) + \lambda \langle \xi, \tilde{x} - x^* \rangle,$$

which in turn contradicts (5.8). The corollary is thereby proved. $\square$

To sum up, Theorem (5.3) establishes that a convex function always has a sub-gradient. The converse of this theorem is generally false: there may be functions defined on a non-empty convex set D with a sub-gradient at each point but that are not overall convex. This does not happen, however, when the domain is open. Precisely, the following result holds.

(5.9) Theorem. *Let D be a non-empty convex set in $\mathbb{R}^n$ and let $f : D \to \mathbb{R}$ be a real function defined on D. Assume that, for each point x^* in D, there exists a sub-gradient ξ, that is a vector such that*

$$f(x) \geq f(x^*) + \langle \xi, x - x^* \rangle \qquad \text{for each } x \in D.$$

Then f is convex.

Proof. Let x^1 and x^2 be two points in D and let $\lambda \in (0, 1)$. Since D is assumed to be convex, $x = \lambda x^1 + (1 - \lambda)x^2 \in D$. By assumption, there exists a sub-gradient ξ of f at this point x. This means that the following two inequalities hold:

$$f(x^1) \geq f\big(\lambda x^1 + (1 - \lambda)x^2\big) + (1 - \lambda)\langle \xi, x^1 - x^2 \rangle,$$
$$f(x^2) \geq f\big(\lambda x^1 + (1 - \lambda)x^2\big) + \lambda \langle \xi, x^2 - x^1 \rangle,$$

Multiplying these two inequalities by λ and $(1 - \lambda)$, respectively, and adding up the results, we get

$$\lambda f(x^1) + (1 - \lambda)f(x^2) \geq f\big(\lambda x^1 + (1 - \lambda)x^2\big),$$

which proves that f is a convex function. $\square$

6. The Moreau-Rockafellar Theorem

Quite surprisingly, sub-gradients enjoy properties that are very similar to the familiar ones of differential calculus, but their proof tends to be more challenging. The following result is particularly remarkable and is due to the French mathematician JEAN JACQUES MOREAU (1923–2014) and the American mathematician R. TYRRELL ROCKAFELLAR, one of the leading scholars in optimisation theory and author of the major reference textbook in convex analysis.

(6.1) Theorem (Moreau–Rockafellar). *Let D be a non-empty convex subset of $\mathbb{R}^n$, and let f_1 and f_2 be convex real functions defined on D. Define $f = f_1 + f_2$. Then, the function f is convex, and for every $x^* \in D$, the following relation holds:*

$$\partial f(x^*) = \partial f_1(x^*) + \partial f_2(x^*),$$

where the sum on the right-hand side denotes the Minkowski sum.

Proof. Since both f_1 and f_2 are convex on D, it is easy to check that their sum f is also convex. Thus, it suffices to prove the identity of the subdifferentials.

Let us start by proving the inclusion $\partial f_1(x^*) + \partial f_2(x^*) \subseteq \partial f(x^*)$. To this end, let $\xi = \xi_1 + \xi_2$ with $\xi_1 \in \partial f_1(x^*)$ and $\xi_2 \in \partial f_2(x^*)$. Then, for all $x \in D$, we have

$$\begin{aligned}
f(x) &= f_1(x) + f_2(x) \\
&= f_1(x^*) + \langle \xi_1, x - x^* \rangle + f_2(x^*) + \langle \xi_2, x - x^* \rangle \\
&= f(x^*) + \langle \xi_1 + \xi_2, x - x^* \rangle \\
&= f(x^*) + \langle \xi, x - x^* \rangle
\end{aligned}$$

which shows that then $\xi \in \partial f(x^*)$.

It remains to prove the converse inclusion. To this end, let ξ be an element of $\partial f(x^*)$, and consider the following two sets in $\mathbb{R}^n \times \mathbb{R}$:

$$K_1 = \left\{ (x - x^*, y) : x \in D, \, y > f_1(x) - f_1(x^*) - \langle \xi, x - x^* \rangle \right\},$$
$$K_2 = \left\{ (x - x^*, y) : x \in D, \, y \leq f_2(x^*) - f_2(x) \right\}.$$

It is not difficult to recognise that both sets K_1 and K_2 are convex and non-empty. Further, $K_1 \cap K_2$ is empty, because otherwise there would exist a point $(x - x^*, y)$ belonged to both, such that

$$f_1(x) - f_1(x^*) + f_2(x) - f_2(x^*) < \langle \xi, x - x^* \rangle,$$

contradicting the assumption $\xi \in \partial f(x^*)$. Thus, by means of Theorem I(3.7), there exists a non-zero vector (ζ, η) and a scalar $\alpha \in \mathbb{R}$ such that

$$(6.2) \quad \begin{aligned}
\langle \zeta, x - x^* \rangle + \eta y &\leq \alpha \quad \text{for } x \in D, \, y > f_1(x) - f_1(x^*) - \langle \xi, x - x^* \rangle, \\
\langle \zeta, x - x^* \rangle + \eta y &\geq \alpha \quad \text{for } x \in D, \, y \leq f_2(x^*) - f_2(x).
\end{aligned}$$

Setting $x = x^*$ and $y = 0$ in the second inequality gives $\alpha \leq 0$. Setting $x = x^*$ and $y = \varepsilon > 0$ in the first inequality, it follows that $\eta \varepsilon \leq \alpha$. Since this is true for every $\varepsilon > 0$ and for $\alpha \leq 0$, this ensures that $\alpha = 0$ and $\eta \leq 0$. Let us prove that $\eta \neq 0$. By contradiction, if it were $\eta = 0$, using again the first inequality of (6.2), we would get $\langle \zeta, x - x^* \rangle \leq 0$, for each $x \in D$. Now, taking $\lambda > 0$ sufficiently small to ensure $x = x^* + \lambda \zeta \in D$, and substituting this value of x in the inequality, we would have $\lambda\langle\zeta,\zeta\rangle \le 0$, hence $\zeta = 0$. Since the vector (ζ,η) is assumed to be non-zero, this proves that $\eta < 0$.

To complete the proof, let us set $\xi_2 = -\eta^{-1}\zeta$ and $\xi_1 = \xi - \xi_2$, let us show that, then, $\xi_2 \in \partial f_2(x^*)$ and $\xi_1 \in \partial f_1(x^*)$. Indeed, from the second inequality in (6.2), for every point x in D, we have

$$\langle\zeta, x - x^*\rangle + \eta y \ge 0 \quad \text{with } x \in D \text{ and } y \le f_2(x^*) - f_2(x).$$

Substituting into the second inequality in (6.2) yields

$$\langle\zeta, x - x^*\rangle - \eta\Big(f_2(x) - f_2(x^*)\Big) \ge 0,$$

which rearranges to

$$f_2(x) \ge f_2(x^*) + \langle -\eta^{-1}\zeta, x - x^*\rangle = f_2(x^*) + \langle\xi_2, x - x^*\rangle,$$

so $\xi_2 \in \partial f_2(x^*)$. To show that $\xi_1 \in \partial f_1(x^*)$, let us now take the first inequality in (6.2),

$$\langle\zeta, x - x^*\rangle + \eta y \le 0 \quad \text{with } x \in D \text{ and } y > f_1(x) - f_1(x^*) - \langle\xi, x - x^*\rangle,$$

and set $y = f_1(x) - f_1(x^*) - \langle\xi, x - x^*\rangle + \varepsilon$, where ε is a strictly positive real number. Thus,

$$\langle\zeta, x - x^*\rangle + \eta\Big[f_1(x) - f_1(x^*) - \langle\xi, x - x^*\rangle + \varepsilon\Big] \le 0.$$

Rearranging this expression, and using the fact that, by definition, $\xi_1 = \xi - \xi_2$,

$$f_1(x) \ge f_1(x^*) + \langle\xi, x - x^*\rangle - \eta^{-1}\langle\zeta, x - x^*\rangle - \varepsilon$$
$$= f_1(x^*) + \langle\xi_1, x - x^*\rangle - \varepsilon.$$

Since this holds for all $\varepsilon > 0$, taking the infimum with respect to ε gives

$$f_1(x) \ge f_1(x^*) + \langle\xi_1, x - x^*\rangle,$$

so $\xi_1 \in \partial f_1(x^*)$. Therefore, $\xi = \xi_1 + \xi_2$ with $\xi_i \in \partial f_i(x^*)$ for $i = 1, 2$, which completes the proof. $\qquad\square$

7. Differentiable convex functions

We have mentioned how convex functions are particularly well behaved with continuity, and we have also shown that even though we cannot guarantee that a convex function is always smooth, we have established that at least the graph of a convex function always has support hyper-planes at all points in its domain's interior. Now we shall focus on smooth convex function, showing how the extra hypothesis to be differentiable induces a wealth of useful properties. Beforehand, however, let us recall, for convenience and to fix notations, the definition of differentiable function.

(7.1) Definition. Let D be an open set in $\mathbb{R}^n$, $f : D \to \mathbb{R}$ a real function defined on D and x^* a point in D. We shall say that f is **differentiable at** x^* if there exists a (necessarily unique) vector which is denoted by $\nabla f(x^*)$ and is called the **gradient** of f at x^*, such that

$$f(x) = f(x^*) + \langle \nabla f(x^*), x - x^* \rangle + o(x - x^*) \qquad \text{for all } x \in D,$$

where $o(x-x^*)$ denotes an infinitesimal term, with order higher than $\|x-x^*\|$, as $x \to x^*$. In addition, if a function $f : D \to \mathbb{R}$ is differentiable at every point $x \in D$, we shall simply say that f is a **differentiable function**.

(7.2) Remark. If $f : D \to \mathbb{R}$ is a differentiable function at $x^* \in D$, the gradient $\nabla f(x^*)$ can be written as the vector of the partial derivatives at x^* of f, that is to say

$$\nabla f(x^*) = \left(\frac{\partial f}{\partial x_1}(x^*), \dots, \frac{\partial f}{\partial x_n}(x^*) \right).$$

The algebra of all differentiable functions on D, whose partial derivatives are all continuous on D shall be denoted by $\mathcal{C}^1(D)$ and the real functions belonging to this class shall be called functions **of class $\mathcal{C}^1$** on D.

Differentiable functions have a smooth graph. The following proposition shows that if a convex function is differentiable, then it has only one sub-gradient, and it coincides with the gradient of the function. This ought to be a rather intuitive result, because a smooth function has a unique tangent hyper-plane at each point, and the tangent hyper-plane acts also as a supporting hyper-plane when the function is convex. The next proposition only makes this intuition formal.

(7.3) Proposition. *Let D be an open non-empty convex set, $f : D \to \mathbb{R}$ a convex real function defined on D, and x^* a point in D. If the function f is differentiable at x^*, then $\nabla f(x^*)$ is the only sub-gradient of f at x^*. Hence, the sub-differential $\partial f(x^*)$ of f at x^* is reduced to the singleton set $\{\nabla f(x^*)\}$.*

Proof. Since the function f is convex, by virtue of Theorem (5.3), there exists a sub-gradient ξ of f at x^*. This means that, for all $x \in D$, we have $f(x) \geq f(x^*) + \langle \xi, x - x^* \rangle$. On the other hand, since the function f is differentiable at x^*, for all $x \in D$, we also have

$$f(x) = f(x^*) + \langle \nabla f(x^*), x - x^* \rangle + o(x - x^*).$$

It suffices to prove that $\xi = \nabla f(x^*)$. To this end, take a direction $v \neq 0$ and a sufficiently small real number λ that the point $x^* + \lambda v$ lies in D. Fix a positive real number ε. Setting $x = x^* + \lambda v$ and taking λ sufficiently small such that the term $o(\lambda\|v\|)$ is smaller than $\lambda\varepsilon$, gives

$$f(x^* + \lambda v) \geq f(x^*) + \lambda\langle \xi, v \rangle$$
$$f(x^* + \lambda v) = f(x^*) + \lambda\langle \nabla f(x^*), v \rangle + o(\lambda v)$$
$$< f(x^*) + \lambda\langle \nabla f(x^*), v \rangle + \lambda\varepsilon$$

Subtracting these two expressions and tidying up the algebra, we eventually get $\langle \xi - \nabla f(x^*), v \rangle < \varepsilon$. Taking the infimum with respect to ε, this inequality becomes $\langle \xi - \nabla f(x^*), v \rangle \leq 0$. Now, let us choose $v = \xi - \nabla f(x^*)$, and observe that, with this particular choice of v, the previous inequality takes the form $\|\xi - \nabla f(x^*)\|^2 \leq 0$ which yields $\xi = \nabla f(x^*)$. This fully proves the proposition. $\qquad\square$

As a consequence of this proposition, we can now state a criterion to establish when a differentiable function is convex using its gradient. This is summarised in the theorem below, not because it is a very complicated matter to actually deserve the label 'theorem' (it is actually a mere rephrase of previous results in light of the new information that, for a differentiable function there is only one sub-gradient), but because of the importance that this rephrase has in the applications.

(7.4) Theorem. *Let D be a non-empty open convex set in $\mathbb{R}^n$, $f : D \to \mathbb{R}$ be a differentiable real function defined on D. Then, f is convex if and only if, for any $x^* \in D$, we have*

$$f(x) \geq f(x^*) + \langle \nabla f(x^*), x - x^* \rangle \qquad \text{for all } x \in D.$$

Similarly, f is strictly convex if and only if, for any $x^* \in D$, we have

$$f(x) > f(x^*) + \langle \nabla f(x^*), x - x^* \rangle \qquad \text{for all } x \in D, \text{ with } x \neq x^*.$$

Proof. These two implications are indeed the statements of Theorem (5.3) and Theorem (5.9) in light of the fact that, by means of Proposition (7.3), the only sub-gradient of a differentiable function f at x^* is $\nabla f(x^*)$, and therefore the inequalities are rewritten setting $\xi = \nabla f(x^*)$. $\qquad\square$

8. Twice differentiable convex functions

An even richer class of convex functions are those that are not only differentiable, but twice differentiable. Again, for convenience with the terminology and the notations, let us first recall the meaning of a twice differentiable function. It is assumed, however, that the reader already knows elements of multivariate calculus.

(8.1) Definition. Let D be a non-empty open subset of $\mathbb{R}^n$, $f : D \to \mathbb{R}$ a real function defined on D, and x^* a point in D. We shall say that f is **twice differentiable at** x^* if there exist a vector, which is denoted by $\nabla f(x^*)$ and shall be called the **gradient** of f, and a $n \times n$ matrix, which is denoted by $Hf(x^*)$ and shall be called the **Hessian matrix** of f at x^*, such that

$$f(x) = f(x^*) + \langle \nabla f(x^*), x - x^* \rangle + \tfrac{1}{2}\langle x - x^*, Hf(x^*)(x - x^*) \rangle + o\big(\|x - x^*\|^2\big)$$

for all $x \in D$. In addition, if $f : D \to \mathbb{R}$ is twice differentiable at every point $x \in D$, we shall simply say that f is a **twice differentiable function**.

(8.2) Remark. It is clear that if a function $f : D \to \mathbb{R}$ is twice differentiable, then it is also differentiable, and in this case the definition of gradient remain unaltered. In particular, the gradient is again uniquely determined. Similarly, the Hessian matrix is uniquely determined as well. In fact, it is the matrix of all second order partial derivatives of f at x^*. In formulae:

$$Hf(x^*) = \begin{bmatrix} \dfrac{\partial^2 f}{\partial x_1^2}(x^*) & \dfrac{\partial^2 f}{\partial x_2 \partial x_1}(x^*) & \cdots & \dfrac{\partial^2 f}{\partial x_n \partial x_1}(x^*) \\[2ex] \dfrac{\partial^2 f}{\partial x_1 \partial x_2}(x^*) & \dfrac{\partial^2 f}{\partial x_2^2}(x^*) & \cdots & \dfrac{\partial^2 f}{\partial x_n \partial x_2}(x^*) \\[2ex] \vdots & \vdots & \ddots & \vdots \\[2ex] \dfrac{\partial^2 f}{\partial x_1 \partial x_n}(x^*) & \dfrac{\partial^2 f}{\partial x_2 \partial x_n}(x^*) & \cdots & \dfrac{\partial^2 f}{\partial x_n^2}(x^*) \end{bmatrix}$$

The algebra of all twice differentiable functions on D, whose second partial derivatives are all continuous on D shall be denoted by $\mathcal{C}^2(D)$ and the real functions belonging to this class shall be called functions **of class** $\mathcal{C}^2$ on D.

More generally, the algebra of all real functions, defined on a non-empty open set D, which have partial derivatives up to k and including the order k, and whose kth partial derivatives are all continuous on D shall be denoted by $\mathcal{C}^k(D)$ and every function belonging to this class shall be said to be functions **of class** $\mathcal{C}^k$ on D.

It turns out that all the information about the convexity of a twice differentiable function is stored in the Hessian matrix, and more precisely within the algebraic structure of this matrix when is thought of as a 'quadratic form'. For this reason, let us recall the following definitions.

(8.3) Definition. Let $A \in \mathfrak{M}(n,n,\mathbb{R})$ be a $n \times n$ matrix. We shall say that A is **positive semi-definite** (PSD) if

$$\langle x, Ax \rangle \geq 0 \qquad \text{for all } x \in \mathbb{R}^n.$$

Symmetrically, we shall say that A is **negative semi-definite** (NSD) if

$$\langle x, Ax \rangle \leq 0 \qquad \text{for all } x \in \mathbb{R}^n.$$

We shall say that A is **indefinite** (ID) if it is neither positive nor negative semi-definite.

A slightly stronger definition is the following.

(8.4) Definition. Let $A \in \mathfrak{M}(n,n,\mathbb{R})$ be a $n \times n$ matrix. We shall say that A is **positive definite** (PD) if

$$\langle x, Ax \rangle > 0 \qquad \text{for all } x \in \mathbb{R}^n \text{ with } x \neq 0.$$

Symmetrically, we shall say that A is **negative definite** (ND) if

$$\langle x, Ax \rangle < 0 \qquad \text{for all } x \in \mathbb{R}^n \text{ with } x \neq 0.$$

It is clear that if a matrix is positive or negative definite, respectively, then it is also positive or negative semi-definite, respectively. Furthermore, with these definitions, we can now show that a twice differentiable function f defined on a non-empty open convex set is convex if and only if its Hessian matrix is positive semi-definite at every point in D, and that, symmetrically, f is concave if and only if its Hessian matrix is negative semi-definite at every point in D. This is the content of the next theorem.

(8.5) Theorem. *Let D be a non-empty open convex subset of $\mathbb{R}^n$, and let $f : D \to \mathbb{R}$ be a function of class $\mathcal{C}^2$ defined on D. Then, f is convex if and only if the Hessian matrix of f at every point in D is positive semi-definite, that is, for every $x^* \in D$, we have*

$$\langle x, Hf(x^*)x \rangle \geq 0 \qquad \text{for all } x \in \mathbb{R}^n.$$

Proof. Let us start by proving that the condition is *necessary*. To this end, assume that f is a convex twice-differentiable real function defined on D, fix a point $x^* \in D$, and let us show that $\langle x, Hf(x^*)x \rangle \geq 0$ for every $x \in D$. To this end, fix x and observe that, since D is open, we can always find a positive real number λ small enough that $\tilde{x} = x^* + \lambda x \in D$. Furthermore, if ε is a positive real number, taking an even smaller λ, if necessary, we can also assume that the term $o(\|\tilde{x} - x^*\|^2) = o(\lambda^2 \|x\|^2)$ in the definition of twice differentiable function, is smaller than $\lambda^2 \varepsilon$. Since f is, in particular, a differentiable function, by virtue of Theorem (7.4), and by definition of twice differentiable function, we get

$$f(x^* + \lambda x) \geq f(x^*) + \lambda \langle \nabla f(x^*), x \rangle$$
$$f(x^* + \lambda x) = f(x^*) + \lambda \langle \nabla f(x^*), x \rangle + \tfrac{1}{2}\lambda^2 \langle x, Hf(x^*)x \rangle + o(\lambda^2 \|x\|^2)$$
$$< f(x^*) + \lambda \langle \nabla f(x^*), x \rangle + \tfrac{1}{2}\lambda^2 \langle x, Hf(x^*)x \rangle + \lambda^2 \varepsilon.$$

Subtracting these two expressions and tidying up the algebra, we then get

$$\tfrac{1}{2}\langle x, Hf(x^*)x \rangle + \varepsilon > 0.$$

Taking the infimum with respect to ε, this gives $\tfrac{1}{2}\lambda^2 \langle x, Hf(x^*)x \rangle \geq 0$, this shows that the Hessian matrix is positive semi-definite at every point $x \in D$.

Let us now show that the condition is also *sufficient*. To this end, assume that the Hessian matrix $Hf(x^*)$ is positive semi-definite for all x^*. By means of the mean value theorem, there exists a vector $\hat{x} = \lambda x^* + (1 - \lambda)x$, with $\lambda \in (0, 1)$, such that

$$f(x) = f(x^*) + \langle \nabla f(x^*), x - x^* \rangle + \tfrac{1}{2}\langle x - x^*, Hf(\hat{x})(x - x^*) \rangle.$$

Since D is convex, $\hat{x} \in D$, and since $Hf(\hat{x})$ is positive semi-definite,

$$\langle x - x^*, Hf(\hat{x})(x - x^*) \rangle \geq 0 \qquad \text{for all } x \in D.$$

Using this information, the previous equality leads to

$$f(x) \geq f(x^*) + \langle \nabla f(x^*), x - x^* \rangle \qquad \text{for all } x \in D.$$

This suffices to prove that f is convex by virtue of Theorem (7.4). □

This theorem is extremely useful to check whether or not a function is convex by only checking whether the Hessian matrix is positive semi-definite. In particular, when

$$f(x) = \tfrac{1}{2}\langle x, Qx \rangle + \langle c, x \rangle + d$$

is a quadratic function, where Q is a symmetric matrix, then the Hessian matrix $Hf(x) = Q$ does not depend on the point x. Hence, checking convexity for a quadratic function reduces to checking positive semi-definiteness of one single matrix.

Finally, a similar result may be obtained for strictly convex functions. Indeed, with the same argument used in the case of differentiable functions, we can prove that, on a non-empty open convex set D, for a twice differentiable function $f : D \to \mathbb{R}$ to be strictly convex is sufficient that the Hessian matrix at each point $x^* \in D$ is positive definite, that is

$$\langle x, Hf(x^*)x \rangle > 0 \qquad \text{for all } x \in \mathbb{R}^n \text{ with } x \neq x^*.$$

However, this provides only with a sufficient condition, because there can be a strictly convex function with Hessian matrix positive semi-definite but not positive definite everywhere in D.

9. Strongly convex functions

An important extension of the concept of convex functions is that of strongly convex functions.

(9.1) Definition. Given a non-empty convex set D in $\mathbb{R}^n$, and a real function $f : D \to \mathbb{R}$. If there exists a strictly positive real number τ such that:

$$f\big(\lambda x^1 + (1 - \lambda)x^2\big) \leq \lambda f(x^1) + (1 - \lambda)f(x^2) - \tfrac{1}{2}\tau\lambda(1 - \lambda)\|x^1 - x^2\|^2$$

for each $x^1, x^2 \in D$ and each $\lambda \in [0, 1]$. the function f is said to be **strongly convex** with parameter τ on D.

The additional quadratic term $\tfrac{1}{2}\tau\lambda(1-\lambda)\|x^1 - x^2\|^2$ ensures that the function is not only convex but also has a unique minimum. It is easy to prove that the strong convexity implies the strict convexity. Strongly convex functions have better convergence properties in optimisation algorithms, as they provide a stronger guarantee that the function's curvature is bounded away from zero.

There is a fundamental reason that justify the introduction of this class of functions. Strongly convex functions have curvature bounded below by a positive constant. This means that optimisation algorithms, such as gradient descent, converge more quickly to the minimum. The additional quadratic term in the definition of strong convexity helps avoid situations where the algorithm might slow down or oscillate near the minimum and it contributes to the numerical stability of optimisation algorithms. In practice, this means that algorithms are less sensitive to rounding errors and perturbations in the data, making solutions more reliable. In particular, many real-world problems, such as regression and classification in machine learning, benefit from the properties of strongly convex functions. For example, the cost function used in logistic regression is strongly convex, ensuring that the optimisation algorithm quickly finds the unique optimal solution.

We now give a characterization of strongly convex functions:

(9.2) Theorem. *Let $f : D \to \mathbb{R}$ be a real function, defined on a non-empty convex set D in $\mathbb{R}^n$. Then, f is strongly convex with parameter τ if and only if the function $g : D \to \mathbb{R}$ defined on D as $g(x) = f(x) - \frac{1}{2}\tau\|x\|^2$ is convex.*

Proof. The convexity of g is

$$g\big(\alpha y + (1-\alpha)x\big) \leq \alpha g(y) + (1-\alpha)g(x)$$

which can be written as

$$f\big(\alpha y + (1-\alpha)x\big) - \tfrac{1}{2}\tau\|\alpha y + (1-\alpha)x\|^2 \leq$$
$$\leq \alpha\big[f(y) - \tfrac{1}{2}\tau\|y\|^2\big] + (1-\alpha)\big[f(x) - \tfrac{1}{2}\tau\|x\|^2\big].$$

By indicating with LHS and RHS, respectively, the left-hand side and the right-hand side of the previous inequality, we have

$$\mathrm{LHS} = f\big(\alpha y + (1-\alpha)x\big) - \tfrac{1}{2}\tau\big[\alpha^2\|y\|^2 + (1-\alpha)^2\|x\|^2 + 2\alpha(1-\alpha)\langle x,y\rangle\big]$$

$$\mathrm{RHS} = \alpha f(y) + (1-\alpha)f(x) - \tfrac{1}{2}\tau\big[\alpha\|y\|^2 + (1-\alpha)\|x\|^2\big]$$
$$= \alpha f(y) + (1-\alpha)f(x) - \tfrac{1}{2}\tau\big[\alpha^2\|y\|^2 + (1-\alpha)^2\|x\|^2 +$$
$$+ \alpha(1-\alpha)(\|y\|^2 + \|x\|^2)\big].$$

Subtracting by both sides the quantity

$$-\tfrac{1}{2}\tau\big[\alpha^2\|y\|^2 + (1-\alpha)^2\|x\|^2 + 2\alpha(1-\alpha)\langle x,y\rangle\big]$$

we have that LHS $\leq$ RHS if and only if

$$f\big(\alpha y + (1-\alpha)x\big) \leq \alpha f(y) + (1-\alpha)f(x)$$
$$- \tfrac{1}{2}\tau\alpha(1-\alpha)\big[\|y\|^2 + \|x\|^2 - 2\langle x,y\rangle\big]$$
$$= \alpha f(y) + (1-\alpha)f(x) - \tfrac{1}{2}\tau\alpha(1-\alpha)\|y - x\|^2,$$

that is the definition of strong convexity of f. $\qquad\square$

(9.3) Remark. The statement of the previous theorem can be rephrased by saying that a function $f : D \to \mathbb{R}$, defined on a non-empty convex set D in $\mathbb{R}^n$, is strongly convex with parameter τ if and only if there exists a convex function ψ and a strictly positive real number τ such that $f(x) = \psi(x) + \frac{1}{2}\tau\|x\|^2$.

(9.4) Example. It is not difficult to recognise that the real function f defined on $\mathbb{R}$ by $f(x) = x^4$ is convex but not strongly convex, because its second derivative, $f''(x) = 12x^2$, is not strictly positive.

Part II

Theory of Optimisation

Chapter III
Introduction to Optimisation

In this chapter we can finally introduce the most basic concepts in optimisation, the notion of maximum and minimum, and we shall define what an optimisation problem is. The rest of this chapter will then look at the more general and elementary necessary and/or sufficient conditions for optimality for convex functions and we shall introduce a special class of functions, the 'coercive' functions, that are particularly useful in optimisation problems, because they ensure, sometimes with minimal extra hypotheses, the existence of optimal solutions. First of all, however, we shall delve into the definition of the various types of maxima and minima, and define what linear and non-linear programming problems are.

1. Optimisation problems

The content of mathematical programming consists of the theory and methods of solving problems of finding extremes of functions on sets defined by linear and non-linear constraints (equalities and inequalities). To describe this formally, let us start by introducing the definition of extremes. Essentially, the word 'extreme' means 'either a maximum or a minimum'.

We assume that the reader is already familiar with the notion of maximum and minimum for a real function of one or several variables, but it shall be useful to recall the formal definitions nonetheless, with the main intent to ensure uniformity in the language.

(1.1) Definition. Let D be a non-empty subset of $\mathbb{R}^n$, and let $f : D \to \mathbb{R}$ be a real function defined on D. We shall say that a point $x^* \in D$ is a **global minimum** for f, if

$$f(x^*) \leq f(x) \qquad \text{for all } x \in D.$$

Similarly, we will say that x^* is a **global maximum** for f, if

$$f(x^*) \geq f(x) \qquad \text{for all } x \in D.$$

© The Author(s), under exclusive license to Springer Nature Switzerland AG 2026
A. Carpignani and M. Pappalardo, *Theory and Methods of Optimisation*,
UNITEXT 175, https://doi.org/10.1007/978-3-032-01514-3_3

If f has a global minimum, the value of f at that point is called its **minimum value**, and it shall be denote by $\min\{f(x) : x \in D\}$ or by $\min_{x \in D} f(x)$. Similarly, if the function f has a global maximum, the value of f at that point is called its **maximum value**, and it shall be denoted by $\max\{f(x) : x \in D\}$ or by $\max_{x \in D} f(x)$.

(1.2) Remark. The two notions of maximum and minimum are deeply related, because a point x^* is obviously a global minimum for f if and only if it is a global maximum for $-f$, and the minimum value of f is the opposite of the maximum value of $-f$, that is

$$\max\Big\{f(x) : x \in D\Big\} = -\min\Big\{-f(x) : x \in D\Big\}.$$

As a result, if we wish to apply a principle of parsimony, we may only give definitions regarding minima, obtaining a symmetrical result for maxima by applying the transformation $f \mapsto -f$. It is important, however, to warn that in doing so, if f is a convex function, $-f$ will be a concave function, and vice versa. Hence, any statement involving the minimum of a convex function may be substituted with the maximum of a concave function, and vice versa. For this reason, when we wish to study problems of maxima and minima of a convex function, we will have argue about maxima and minima distinctly.

Utilising the definition above, we are now able to state the main problem in optimisation.

(1.3) Definition. Assume we are given a subset D of $\mathbb{R}^n$ and a real function $f : D \to \mathbb{R}$ defined on D. The problem of finding the minimum or maximum value of f on D, that is to say the problem

$$\min\Big\{f(x) : x \in D\Big\} \qquad \text{or} \qquad \max\Big\{f(x) : x \in D\Big\},$$

is called a **mathematical programming problem**, or a **mathematical program**, or, in a more modern terminology, an **optimisation problem**, and it is generally denoted as follows:

$$\begin{array}{ll} \text{minimise} & f(x) \\ \text{subject to} & x \in D \end{array} \qquad \text{or} \qquad \begin{array}{ll} \text{maximise} & f(x) \\ \text{subject to} & x \in D \end{array}$$

Given an optimisation problem, the function f shall be called the **objective function**, the condition $x \in D$ shall be called the **constraint**. Furthermore, the set D shall be called the **feasible region** or the **feasible domain**, and every point $x \in D$ shall be called a **feasible solution**. If there exists a feasible solution x^* such that $f(x^*)$ is, depending on the type of optimisation problem, the minimum or the maximum of the objective function f on D, then x^* shall be called an **optimal solution** and the real value $f(x^*)$ shall be called the **optimal value** of the problem.

An optimisation problem is a very general problem, where there are no restrictions on the set D of constraints or the objective function $f : D \to \mathbb{R}$. As a result, a very general problem may be incredibly hard to solve, or it may have no solutions at all. For this reason, it it will be important to introduce suitable restrictions on the constraints and the objective functions. This creates different types of problems, which can be solved with different techniques. The four main areas that we may encounter are the following.

1. To maximise (or minimise) a linear function f, where the feasible set D is a polyhedron. This particular type of problems are referred to as **linear programming** (LP). In other words, a linear programming problem can be written in the form

$$\begin{aligned} \text{maximise} \quad & \langle c, x \rangle \\ \text{subject to} \quad & Ax \leq b \end{aligned}$$

   where $A \in \mathfrak{M}(m, n, \mathbb{R})$, $b \in \mathbb{R}^m$ and $c \in \mathbb{R}^n$.

2. To minimise (or maximise) an objective function f that is a quadratic function, and the feasible set is a polyhedron. This particular type of problems is called **quadratic programming** (QP). In other words, a quadratic programming problem can be written in the form

$$\begin{aligned} \text{minimise} \quad & \tfrac{1}{2}\langle x, Qx \rangle + \langle c, x \rangle \\ \text{subject to} \quad & Ax \leq b \end{aligned}$$

   where $Q \in \mathfrak{M}(n, n, \mathbb{R})$ is symmetric, $A \in \mathfrak{M}(m, n, \mathbb{R})$, $b \in \mathbb{R}^m$, and $c \in \mathbb{R}^n$.

3. To minimise (or maximise) an objective function f on a feasible set described by equalities and/or inequalities, where either the objective function or some of the constraints (or both) is non-linear. This particular type of problems is called **non-linear programming** (NLP). In other words, a non-linear programming problem can be written in the form

$$\begin{aligned} \text{minimise} \quad & f(x) \\ \text{subject to} \quad & g_i(x) \leq 0 \text{ for all } i = 1, \ldots, m \\ & h_j(x) = 0 \text{ for all } j = 1, \ldots, p \end{aligned}$$

   where f, g_i $(i = 1, \ldots, m)$, h_j $(j = 1, \ldots, p)$ are real functions, defined on a suitable common domain, and at least one of them is non-linear.

4. To minimise an objective function f that is convex, on a convex feasible set D (or maximise an objective function f that is concave, on a convex feasible set D). This type of problems are referred to as **convex programming** (CP). Precisely, a convex programming problem can be written in the form

$$\begin{aligned} \text{minimise} \quad & f(x) \\ \text{subject to} \quad & x \in D \end{aligned}$$

where D is a non-empty convex set in $\mathbb{R}^n$, and $f : D \to \mathbb{R}$ is a convex function defined on D; or in the form

$$\begin{aligned} \text{maximise} \quad & f(x) \\ \text{subject to} \quad & x \in D \end{aligned}$$

where D is a non-empty convex set in $\mathbb{R}^n$, and $f : D \to \mathbb{R}$ is a concave function defined on D.

(1.4) Remark. Even though a quadratic programming problem can always be seen as a special type of non-linear programming problem, the interest of quadratic programming is twofold: on one side, quadratic programming may have access to simpler methods that do not work in the general case of non-linear programming, leading to simpler and faster algorithms, but on the other side, quadratic programming problems often occur in applications, making this type of problems relatively important as a self-standing theory.

For example, a special and especially important type of quadratic programming consists in the so called 'least square' problems, that is to say problems in which we seek to minimise a function of the form $f(x) = \frac{1}{2}\|Rx + q\|^2$, subject to the constraint $Ax \leq b$ (where R and A are suitable matrices and r and b suitable vectors). This type of problems is extremely common for example in Machine Learning, where the function $f(x)$ usually represents the cost function, i.e. the error made by an algorithm to predict (or classify) a result. Expanding the expression of $f(x)$ it is easy to recognise that this is indeed a quadratic programming problem, with $Q = \frac{1}{2}R^T R$, and $c = R^T q$. Let us observe that in the expansion of the function $f(x)$ where would be an additional constant $d = \frac{1}{2}\|q\|^2$, but since the minimum does not change when to the objective function is added a constant, this term may be omitted. This shall be clarified further in section 2.

Global optimal solutions are somewhat very strict notions, and it may be convenient to loosen up the definitions. For example, let us give the following definition.

(1.5) Definition. For the optimisation problem

$$(1.6) \qquad \begin{aligned} \text{minimise} \quad & f(x) \\ \text{subject to} \quad & x \in D \end{aligned}$$

a feasible solution x^* is called a **local** optimal solution if there is a neighbourhood $V \in N(x^*)$ such that $f(x^*) \leq f(x)$ for every $x \in V \cap D$. In other words, a feasible solution x^* is a local optimal solution for the optimisation problem (1.6) if it is an optimal solution of the optimisation problem with objective function f and feasible set $V \cap D$. If, in addition, $f(x^*) < f(x)$ for every $x \in D \cap V \setminus \{x^*\}$, then x^* shall be called a **strict** local optimal solution. Finally, a local optimal solution x^* is said to be **locally unique** if it is the unique local optimal solution in a neighbourhood of x^*, i.e. if there there exists a neighbourhood $V \in N(x^*)$ such that x^* is the only local optimal solution in $V \cap D$.

(1.7) Remark. It is not difficult to show that a locally unique optimal solution is also a strict local optimal solution but the converse is generally false, as the following example shows: on the domain $D = [0, 1]$ take the function f defined by $f(0) = 0$ and $f(x) = \sin^2(1/x) + x^2$ for $x \neq 0$. Then $x^* = 0$ is a strict local minimum, but not a locally unique minimum. Hence, the notion of locally unique optimal solution is stronger than that of strict local optimal solution. However, for a global optimal solution, these two notions are equivalent.

2. Equivalent transformations

Optimisation problems arise naturally in many applied fields, such as engineering, economics, or statistics, and, depending on the field and the problem at hand, they turn up to be naturally in one specific format rather than another. For example, in economics, optimisation problems often focus on maximising profit, whereas in statistics they they often focus on minimising a cost or an error. But the differences between problems is not only limited to the focus on maximisation or minimisation, but the shape of the feasible set often looks different, depending on context where the problem had come to light. For a systematic study of a problem, however, it is more useful to focus on a specific 'standard' format and re-conduct any problem to that one. For this reason, it is particularly important to develop some routine techniques that enable to pass from a given form to another in a simple and consistent way. A word of caution becomes necessary before proceeding any further. The 'standard' format of an optimisation problem is largely subjective and different authors tend to choose different formats as their 'standard', generally depending on their focus, but sometimes even more irrationally on their taste. The purpose of this section is to give the tools to pass from one to the other in a systematic way.

Let us start by defining what we mean by 'equivalent' problems.

(2.1) Definition. Let D_A a non-empty subset in $\mathbb{R}^n$ and $f_A : D_A \to \mathbb{R}$ a real function defined on D_A. Let D_B be a non-empty subset in $\mathbb{R}^m$, and $f_B : D_B \to \mathbb{R}$ a real function defined on D_B. Consider the following two optimisation problems:

$(\mathcal{A})$: Minimise (or maximise) the function $f_A(x)$ subject to $x \in D_A$,

$(\mathcal{B})$: Minimise (or maximise) the function $f_B(z)$ subject to $z \in D_B$.

We shall say that $(\mathcal{A})$ is **equivalent** to $(\mathcal{B})$, if there exist two continuous function $\phi : D_A \to D_B$ and $\psi : D_B \to D_A$ such that, for each optimal solution x^* of problem $(\mathcal{A})$, $\phi(x^*)$ is an optimal solution of $(\mathcal{B})$, and for each optimal solution z^* of problem $(\mathcal{B})$, $\psi(z^*)$ is an optimal solution of $(\mathcal{A})$. In addition, given an optimisation problem, say $(\mathcal{A})$, if $(\mathcal{B})$ is another problem, equivalent to $(\mathcal{A})$, we shall say that $(\mathcal{B})$ is obtained from problem $(\mathcal{A})$ applying an **equivalent transformation**.

(2.2) Remark. The relation introduced in Definition (2.1) is easily an equivalence relation on the class of all optimisation problems, and we can say that problems $(\mathcal{A})$ and $(\mathcal{B})$ are **equivalent**. To understand how the notion of equivalence for optimisation problems can help, consider an optimisation problem $(\mathcal{A})$ to minimise (or maximise) the function $f(x)$ subject to the constraint $x \in D$, and assume that the problem is too complex to be solved directly. Assume however, that we can find an optimisation problem, $(\mathcal{B})$, equivalent to problem $(\mathcal{A})$, but that can be solved directly to find an optimal solution z^*, or such that we can find a sequence $\{z^k\}$ of feasible points that tend to an optimal solution z^* as n approaches $+\infty$. Then, if we happen to know the function ψ, defined on the feasible set of $(\mathcal{B})$ with values in the feasible set of $(\mathcal{A})$, then we can always 'retrieve' an optimal solution $x^* = \psi(z^*)$ of $(\mathcal{A})$, or find a sequence $\{x^k\}$ of points in the feasible set of $(\mathcal{A})$, defined as $x^k = \psi(z^k)$, which tends to an optimal solution of $(\mathcal{A})$.

(2.3) Lemma. *Let D be a non-empty set in $\mathbb{R}^n$, $f : D \to \mathbb{R}$ a real function defined on D, and let g be a real function defined on the range of f. If g is increasing on the range of f, then the problems*

$$\begin{array}{cc} \text{minimise} & f(x) \\ \text{subject to} & x \in D \end{array} \quad \text{and} \quad \begin{array}{cc} \text{minimise} & g \circ f(x) \\ \text{subject to} & x \in D \end{array}$$

are equivalent.

Similarly, if the function g is decreasing on the range of f, then the problems

$$\begin{array}{cc} \text{minimise} & f(x) \\ \text{subject to} & x \in D \end{array} \quad \text{and} \quad \begin{array}{cc} \text{maximise} & g \circ f(x) \\ \text{subject to} & x \in D \end{array}$$

are equivalent.

Proof. Let us assume that the optimisation problem is to minimise $f(x)$ subject to $x \in D$ and assume that g is increasing on the range of f. Consider the two problems

$$(\mathcal{A}): \left\{ \begin{array}{cc} \text{minimise} & f(x) \\ \text{subject to} & x \in D \end{array} \right. \quad \text{and} \quad (\mathcal{B}): \left\{ \begin{array}{cc} \text{minimise} & g \circ f(x) \\ \text{subject to} & x \in D \end{array} \right.$$

If x^* is an optimal solution of $(\mathcal{A})$, then $f(x^*) \le f(x)$ for all $x \in D$. Since g is increasing, this means that $g(f(x^*)) \le g(f(x))$ for all $x \in D$, hence x^* is an optimal solution of $(\mathcal{B})$. Conversely, if x^* is an optimal solution of $(\mathcal{B})$, then $g(f(x^*)) \le g(f(x))$ for all $x \in D$ and, again, the fact that g is increasing yields $f(x^*) \le f(x)$ for all $x \in D$, i.e. x^* is an optimal solution of $(\mathcal{A})$.

The proof of the second part is similar, with the only difference that, when g is a decreasing function, the relation $y_1 \le y_2 \Leftrightarrow g(y_1) \ge g(y_2)$ leads to the minimum of the first problem becoming the maximum of the second problem, and vice versa. $\qquad\square$

(2.4) Remark. Before moving on and showing the several applications of this lemma, let us also observe that, in the same hypotheses and with the same notation as in Lemma (2.3), by subsequently applying the first and second part of the lemma, we can also prove that the problems

$$
\begin{array}{ll}
\text{maximise} & f(x) \\
\text{subject to} & x \in D
\end{array}
\qquad \text{and} \qquad
\begin{array}{ll}
\text{maximise} & g \circ f(x) \\
\text{subject to} & x \in D
\end{array}
$$

are equivalent, and similarly, that the problems

$$
\begin{array}{ll}
\text{maximise} & f(x) \\
\text{subject to} & x \in D
\end{array}
\qquad \text{and} \qquad
\begin{array}{ll}
\text{minimise} & g \circ f(x) \\
\text{subject to} & x \in D
\end{array}
$$

are also equivalent.

Switching maximum and minimum

A simple corollary of Lemma (2.3) ensures that one can always transform a minimum problem into a maximum problem and vice versa, by considering a naturally equivalent problem. Precisely, the following results holds.

(2.5) Proposition. *Let D be a non-empty set in $\mathbb{R}^n$, and $f : D \to \mathbb{R}$ a real function defined on D. Then, the following two optimisation problems are equivalent:*

$$
\begin{array}{ll}
\text{minimise} & f(x) \\
\text{subject to} & x \in D
\end{array}
\qquad \text{and} \qquad
\begin{array}{ll}
\text{maximise} & -f(x) \\
\text{subject to} & x \in D
\end{array}
$$

Proof. It is sufficient to apply Lemma (2.3) with the decreasing function $g(y) = -y$. This proves that the two optimisation problems are indeed equivalent. $\square$

Regularisation of the objective function

In many applied problems, especially in statistics and machine learning, the objective function is a "norm function". This function suffers from a major flaw: it is not differentiable at some points. This means that any method to find an optimal solution based on differentiable functions cannot be applied. However, by taking an equivalent problem, this can can be readily overcome, as shown in the following result.

(2.6) Proposition. *The optimisation problem to minimise (resp. maximise) the function $\|x\|$ subject to $x \in D$ is equivalent to the optimisation problem to minimise (resp. maximise) the function $\frac{1}{2}\|x\|^2$ subject to $x \in D$.*

Proof. Since the modulus is a function with positive values, the function $g(y) = \frac{1}{2}y^2$ is therein increasing and the result follows from Lemma (2.3). $\square$

Transforming linear constraints

A slightly more complicated problem concerns changing the shape of an optimisation problem. Let us focus first on linear constraints, i.e. when the feasible set is a polyhedron, which is described by a finite number of linear constraints. A linear constraint is either an equality of the form $Ax = b$, or an inequality of the form $Ax \leq b$ or $Ax \geq b$. Let us start by observing that the two inequalities are completely equivalent, because it suffices to multiply throughout by -1 to pass from one to another. Also, an equality constraint can immediately be transformed into an inequality, observing that $Ax = b$ is the same as $Ax \leq b$ and $Ax \geq b$. The following proposition shows that, conversely, we can always pass from an inequality constraint to an equality one with an equivalent transformation.

(2.7) Proposition. *Let $A \in \mathfrak{M}(m, n, \mathbb{R})$ be a matrix, $b \in \mathbb{R}^m$, and f a real function, defined on the polyhedron $P = \{x \in \mathbb{R}^n : Ax \leq b\}$. Then the optimisation problem*

$$\begin{aligned} \text{maximise} \quad & f(x) \\ \text{subject to} \quad & Ax \leq b \end{aligned}$$

with variable $x \in \mathbb{R}^n$, is equivalent to the problem

$$\begin{aligned} \text{maximise} \quad & f(x) \\ \text{subject to} \quad & Ax + s = b, \\ & s \geq 0. \end{aligned}$$

with variables $x \in \mathbb{R}^n$ and $s \in \mathbb{R}^m$.

Proof. Set $g(x, s) = f(x)$, for every $x \in P$ and every $s \in \mathbb{R}^m$. Consider the problem to maximise the objective function $g(x, s) = f(x)$, subject to the two constraints $Ax + s = b$ and $s \geq 0$. It suffices to show that we can find two function ϕ and ϕ that map the optimal solutions of one problem into the optimal solutions of the other. To this end, let us start by observing that, if (x^*, s^*) is an optimal solution of this second problem, then $g(x^*, s^*) = f(x^*)$ is an optimal value of f, and from the fact that $Ax^* + s^* = b$ and $s^* \geq 0$, it follows that $Ax^* \leq b$. This proves that x^* is an optimal solution of the first problem. Conversely, if x^* is an optimal solution of the first problem, set $s^* = b - Ax^*$. Then, clearly $s^* \geq 0$, because x^* satisfies $Ax^* \leq b$. In addition $g(x^*, s^*) = f(x^*)$ is evidently the optimal value of g. Thus, (x^*, s^*) is optimal solution of the second problem. Taking the functions $\phi(x) = (x, b - Ax)$ and $\psi(x, s) = x$, it is proved that the problems are equivalent. $\square$

Similarly, the following proposition holds with an identical proof.

(2.8) Proposition. *Let $A \in \mathfrak{M}(m, n, \mathbb{R})$ be a matrix, $b \in \mathbb{R}^m$, and f a real function, defined on a non-empty set in $\mathbb{R}^n$ containing the polyhedron $P = \{x \in \mathbb{R}^n : Ax \geq b\}$. Then the optimisation problem*

$$\begin{aligned} \text{maximise} \quad & f(x) \\ \text{subject to} \quad & Ax \geq b \end{aligned}$$

with variable $x \in \mathbb{R}^n$, is equivalent to the problem

$$\begin{array}{ll} \text{maximise} & f(x) \\ \text{subject to} & Ax - s = b \\ & s \geq 0 \end{array}$$

with variables $x \in \mathbb{R}^n$ and $s \in \mathbb{R}^m$.

(2.9) Definition. The variables $s \in \mathbb{R}^m$ that appear in statement of the second problem in both Proposition (2.7) and Proposition (2.8) are referred to as **slack** variables.

Another useful transformation consists in making all variables positive. To achieve this, it suffices to split every variable x into two positive variables x^+ and x^-, called the **positive** and **negative** part, respectively, so that

$$x = x^+ - x^-.$$

More generally, the following results hold.

(2.10) Proposition. *Let $A \in \mathfrak{M}(m, n, \mathbb{R})$ be a matrix, $b \in \mathbb{R}^m$ a vector, and f a real function, defined on a non-empty set in $\mathbb{R}^n$ containing the hyper-plane $\{x \in \mathbb{R}^n : Ax = b\}$. Suppose that the variable x is split into two groups $x = (x_1, x_2)$ with $x_1 \in \mathbb{R}^p$ and $x_2 \in \mathbb{R}^{n-p}$, and that the first is constrained to be non-negative, whereas the second is unconstrained. In addition, split correspondingly, $A = \begin{bmatrix} A_1 & A_2 \end{bmatrix}$, where $A_1 \in \mathfrak{M}(m, p, \mathbb{R})$ and $A_2 \in \mathfrak{M}(m, n - p, \mathbb{R})$. Then the optimisation problem*

$$\begin{array}{ll} \text{maximise} & f(x_1, x_2) \\ \text{subject to} & A_1 x_1 + A_2 x_2 = b \\ & x_1 \geq 0 \end{array}$$

with variable $(x_1, x_2) \in \mathbb{R}^n$, is equivalent to the problem

$$\begin{array}{ll} \text{maximise} & f(x_1, u_1 - u_2) \\ \text{subject to} & A_1 x_1 + A_2 u_1 - A_2 u_2 = b \\ & x_1, u_1, u_2 \geq 0 \end{array}$$

with variables $x_1 \in \mathbb{R}^p$, $u_1, u_2 \in \mathbb{R}^{n-p}$.

Proof. Set $g(x_1, u_1, u_2) = f(x_1, u_1 - u_2)$, with $x_1 \in \mathbb{R}^p$ and $u_1, u_2 \in \mathbb{R}^{n-p}$, and consider the optimisation problem to maximise $g(x_1, u_1, u_2)$ subject to $A_1 x_1 + A_2 u_1 - A_2 u_2 = b$, $x_1, u_1, u_2 \geq 0$. If (x_1^*, u_1^*, u_2^*) is an optimal solution of this problem, then $(x_1^*, u_1^* - u_2^*)$ is a solution of the initial problem. Conversely, if (x_1^*, x_2^*) is a solution of the initial problem, then $(x_1^*, (x_2^*)^+, (x_2^*)^-)$ is a solution of the second problem. By setting $\phi(x_1, x_2) = (x_1, x_2^+, x_2^-)$ and $\psi(x_1, u_1, u_2) = (x_1, u_1 - u_2)$, it is proved that the two problems are indeed equivalent, and the proof is completed. $\qquad \square$

3. Minimising convex functions

We know that convex functions are remarkably well behaved. In particular, their good nature shows up in optimisation problems, when we are seeking the global minimum of a convex function on a convex set. Indeed, the following theorem holds.

(3.1) Theorem. *Let D be a non-empty convex set in $\mathbb{R}^n$ and let $f : D \to \mathbb{R}$ be a convex function on D. Suppose that $x^* \in D$ is a local minimum of f. Then, x^* is a global minimum. Furthermore, if either x^* is a strict local minimum, or f is strictly convex, then x^* is the unique global minimum.*

Proof. Let x^* be a local minimum for f. By definition, this means that there exists a neighbourhood $V \in N(x^*)$, with $V \subseteq D$, such that $f(x^*) \leq f(x)$ for all $x \in V$. By contradiction, suppose that x^* is not a global minimum. In other words, suppose that there exists another point, $w \in D$, such that $f(w) < f(x^*)$. By convexity of f, then, for each real number $\lambda \in [0, 1]$, we have

$$
\begin{aligned}
f\big(\lambda w + (1 - \lambda)x^*\big) &\leq \lambda f(w) + (1 - \lambda)f(x^*) \\
&< \lambda f(x^*) + (1 - \lambda)f(x^*) \\
&= f(x^*).
\end{aligned}
$$

This proves that $f(x) < f(x^*)$ for every point $x = \lambda x^* + (1 - \lambda)w$ in the segment with extremes w and x^*. On the other hand, taking λ small enough gives $x = \lambda x^* + (1 - \lambda)w \in V$, and this is a contradiction, because, for every $x \in V$, we know that $f(x^*) \leq f(x)$ by hypothesis. This proves that x^* must be a global minimum.

It remains to prove the second part of the theorem. To this end, suppose first that x^* is a strict local minimum. Since this is, in particular, a local minimum, we have just proved that x^* is thereby a global minimum. This means that $f(x^*) < f(x)$ for every $x \in D$ with $x \neq x^*$. By contradiction, suppose that x^* is not unique. In other words, suppose that there is a different point $w \in D$, such that $f(w) = f(x^*)$. Then, $x_\lambda = \lambda x^* + (1 - \lambda)w \in D$ for every $\lambda \in (0, 1)$, because D is assumed to be convex, and from the convexity of f, we obtain $f(x_\lambda) \leq \lambda f(x^*) + (1 - \lambda)f(w) = \lambda f(x^*) + (1 - \lambda)f(x^*) = f(x^*)$. At this point, observe that, for any $V \in N(x^*)$, we can always choose the number λ to be small enough such that $x_\lambda \in V$, and this contradicts the fact that x^* is, by hypothesis, a strict local minimum.

Finally, suppose that the function f is strictly convex and that x^* is a local minimum. Since a strictly convex function is, in particular, convex, we already know that x^* is indeed a global minimum. By contradiction, then, suppose that this global minimum is not unique, i.e. that there is a different point $w \in D$, such that $f(x^*) = f(w)$. Taking once again a point $x = \lambda x^* + (1 - \lambda)w$, with $\lambda \in (0, 1)$, and observe that $x \in D$ because D is convex. Thus, using the fact that f is a strictly convex function, we get $f(x) < \lambda f(x^*) + (1 - \lambda)f(w) = \lambda f(x^*) + (1 - \lambda)f(x^*) = f(x^*)$, which is impossible, because x^* is the global minimum of f in D. Hence, x^* is the unique global minimum, and the theorem is fully proved. $\qquad\square$

We can now develop a necessary and sufficient condition for the existence of a global optimal solution (i.e. a global minimum) for a convex function. The following theorem shows in particular how the notion of sub-gradient is strongly related to optimality, in a very intuitive manner. In fact, at a minimum, we would expect a supporting hyper-plane to be horizontal, as a supporting hyper-plane can be thought as a wall that a convex function cannot cross, but only touch. Such a horizontal wall would indeed signify that all the points that do not touch the wall must be at a higher level, i.e. that the point that touch the wall is indeed at the minimum level. This intuitive idea is made formal in the next theorem.

(3.2) Theorem. *Let E be an open convex set in $\mathbb{R}^n$, $f : D \to \mathbb{R}$ a convex function on E, and $D \subseteq E$ a non-empty convex set. A point $x^* \in D$ is a global minimum of f if and only if f has a sub-gradient ξ at x^* such that*

$$\langle \xi, x - x^* \rangle \geq 0 \qquad \text{for all } x \in D.$$

Proof. Let us start by proving that the condition is *sufficient*. To this end, suppose that there is a sub-gradient ξ of f at x^* such that $\langle \xi, x - x^* \rangle \geq 0$ for all $x \in D$. Since the function f is convex, and ξ is a sub-gradient, we have by definition

$$f(x) \geq f(x^*) + \langle \xi, x - x^* \rangle \geq f(x^*) \qquad \text{for all } x \in D,$$

hence x^* is a global minimum for f.

It remains to prove that the condition is also *necessary*. To this end, let us suppose that the point x^* is a global minimum of f and consider the following two sets in $\mathbb{R}^n \times \mathbb{R}$:

$$K_1 = \Big\{ (x - x^*, y) : x \in E,\ f(x) - f(x^*) < y \Big\},$$
$$K_2 = \Big\{ (x - x^*, y) : x \in D,\ y \leq 0 \Big\}.$$

It is not difficult to recognise that both the sets K_1 and K_2 are convex. In fact, K_2 is convex being D convex. To prove that K_1 is also convex, take two points $(x^1 - x^*, y_1)$ and $(x^2 - x^*, y_2)$ in K_1 and observe that, by convexity of f,

$$f\big(\lambda x^1 + (1 - \lambda)x^2\big) - f(x^*) \leq \lambda f(x^1) + (1 - \lambda)f(x^2) - f(x^*)$$
$$< \lambda y_1 + (1 - \lambda)y_2.$$

Thus, $\lambda(x^1 - x^*, y_1) + (1 - \lambda)(x^2 - x^*, y_2) \in K_1$. Further, $K_1 \cap K_2$ is empty, because otherwise there would exist a point $(x - x^*, y)$ such that

$$f(x) - f(x^*) < y \leq 0,$$

contradicting the hypothesis that x^* is a minimum. Thus, by virtue of Theorem I(3.7), there exists a non-zero vector (ζ, η) and a scalar α such that

$$(3.3) \quad \begin{aligned} \langle \zeta, x - x^* \rangle + \eta y &\leq \alpha \qquad \text{for all } x \in E \text{ and } f(x) - f(x^*) < y, \\ \langle \zeta, x - x^* \rangle + \eta y &\geq \alpha \qquad \text{for all } x \in D \text{ and } y \leq 0. \end{aligned}$$

Setting $x = x^*$ and $y = 0$ in the second inequality gives $\alpha \leq 0$. Setting $x = x^*$ and $y = \varepsilon > 0$ in the first inequality, it follows that $\eta \varepsilon \leq \alpha$. Since this is true for every $\varepsilon > 0$ and for $\alpha \leq 0$, this ensures both that $\alpha = 0$ and $\eta \leq 0$. Let us prove that $\eta \neq 0$. By contradiction, if it were $\eta = 0$, using again the first inequality of (3.3), we would get $\langle \zeta, x - x^* \rangle \leq 0$, for each $x \in E$. Now, taking $\lambda > 0$ sufficiently small to ensure $x = x^* + \lambda \zeta \in E$, and substituting this value of x in the inequality, we would have $\lambda \langle \zeta, \zeta \rangle \leq 0$, hence $\zeta = 0$. Since the vector (ζ, η) is assumed to be non-zero, this proves that $\eta < 0$.

Now, dividing the two inequalities in (3.3) by $|\eta|$, setting $\xi = -\eta^{-1}\zeta$, and tidying up the expressions, we finally get to the following inequalities:

$$\begin{aligned} \langle \xi, x - x^* \rangle &\leq y \qquad \text{if } x \in E, \ f(x) - f(x^*) < y, \\ \langle \xi, x - x^* \rangle &\geq y \qquad \text{if } x \in D, \ y \leq 0, \end{aligned}$$

Setting $y = 0$ in the second inequality, we get $\langle \xi, x - x^* \rangle \geq 0$ for every $x \in D$, whereas from the first inequality, taking the infimum with respect to y, it follows that

$$f(x) \geq f(x^*) + \langle \xi, x - x^* \rangle \qquad \text{for all } x \in E,$$

So, ξ is a sub-gradient of f at x^* with the property that $\langle \xi, x - x^* \rangle \geq 0$ for all $x \in D$, and this completes the proof. $\qquad \square$

An immediate consequence of Theorem (3.2) is that, when a convex function, defined on an open convex set, has a minimum, one of the sub-gradients of the function at the minimum is necessarily 0. Precisely, the following corollary holds.

(3.4) Corollary. *In the same hypotheses as Theorem (3.2), assume that x^* lies in the interior of D. Then x^* is a minimum if and only zero is a sub-gradient of f at x^*.*

Proof. By means of Theorem (3.2), x^* is a minimum if and only if there exists a sub-gradient ξ of f at x^* such that $\langle \xi, x - x^* \rangle \geq 0$. However, since x^* lies in the interior of D, there exists a sufficiently small positive real number λ such that $x = x^* - \lambda \xi \in D$. With this choice of x, the condition above becomes $-\lambda \langle \xi, \xi \rangle \geq 0$, which implies $\xi = 0$. $\qquad \square$

Furthermore, in the special case that the function is differentiable, Theorem (3.2) can be rephrased in an even simpler way as follows.

(3.5) Corollary. *In the same hypotheses as Theorem (3.2), assume that f is differentiable. Then, x^* is a minimum if and only if $\langle \nabla f(x^*), x - x^* \rangle \geq 0$ for all $x \in D$. Furthermore, if x^* lies in the interior of D, then x^* is a local minimum if and only if $\nabla f(x^*) = 0$.*

Proof. From Proposition II(7.3) we know that, if f is differentiable at x^*, the only sub-gradient of f at x^* is $\nabla f(x^*)$. Applying Theorem (3.2) to this sub-gradient gives the result of the first part of the corollary. The second part follows from Corollary (3.5) and from the fact that the sub-gradient of a differentiable function is unique. $\qquad\square$

4. Maximising convex functions

Unlike the minimisation case, maximising a convex function is in generally a much harder a task to carry out. Indeed, in such case a function can have many local maxima. Fortunately, however, there is an important result that shows that a maximum of a convex function cannot be located anywhere, but only in the boundary. The rest of this section is devoted to showing some simple necessary condition for a convex function to have a local maximum, and we shall delve into the proof of this important result. To this end, let us begin with the following theorem. As we have already mentioned, maximising a convex function on a convex set is equivalent to minimising a concave function on the same set. It is for this reason that maximising a convex function is sometimes referred to as 'concave programming'.

(4.1) Theorem. *Let E be a non-empty open convex set in $\mathbb{R}^n$, $f : D \to \mathbb{R}$ a convex function on E, and $D \subseteq E$ a convex set and $x^* \in D$. If, x^* is a local maximum of f, then,*

$$\langle \xi, x - x^* \rangle \leq 0 \qquad \text{for all } x \in D$$

for any sub-gradient ξ of f at x^*.

Proof. Assume that x^* is a local maximum for f, ξ is a sub-gradient of f at x^*, and consider a neighbourhood $V \in N(x^*)$, with $V \subseteq D$, such that $f(x) \leq f(x^*)$ for every $x \in V$. We wish to prove that, for every $x \in D$, we have $\langle \xi, x - x^* \rangle \leq 0$. To this end, take $x \in D$ and observe that, taking λ sufficiently small, we may always assume that the point $x_\lambda = \lambda x + (1 - \lambda)x^*$ lies in V, so $f(x_\lambda) \leq f(x^*)$. Now, putting this together with the fact that ξ is a sub-gradient of f at x^*, gives $0 \geq f(x_\lambda) - f(x^*) \geq \langle \xi, x_\lambda - x^* \rangle = \lambda \langle \xi, x - x^* \rangle$, and this proves that $\langle \xi, x - x^* \rangle \leq 0$. The theorem is thereby fully proved. $\quad\square$

In the special case that the function is differentiable, the previous theorem can be rephrased in an even simpler way as follows.

(4.2) Corollary. *In the same hypotheses as Theorem (4.1), assume that f is differentiable. If x^* is a local maximum of f, then*

$$\langle \nabla f(x^*), x - x^* \rangle \leq 0 \qquad \text{for all } x \in D.$$

Proof. From Proposition II(7.3) we know that, if f is differentiable at x^*, the only sub-gradient of f at x^* is $\nabla f(x^*)$. Applying Theorem (4.1) to this sub-gradient gives the desired result. $\qquad\qquad\qquad\qquad\qquad\square$

Let us now complete the section by delving into the following important theorem, that shows that the global maximum of a convex function, if it exists, must lie on the boundary. This is clearly a bold result from an optimisation point of view. In fact, if an optimisation problem is to maximise a convex function f, subject to a constraint $x \in D$, where D is a convex set, by means of the next theorem, we can immediately restrict our attention to the boundary of D, that is those lying in the set $D \cap \partial D$, because the solution of this optimisation problem, if it exists, must lie in this set.

(4.3) Theorem. *Let D be a non-empty convex set, and $f : D \to \mathbb{R}$ a convex non-constant function. If f admits a global maximum, then it lies on the boundary of D.*

Proof. Let x^* be a global maximum of f on D. Since the function f is assumed not to be constant by hypothesis, there must be a point $w \in D$ with $f(w) < f(x^*)$. By contradiction, suppose that the point x^* lies in the interior of D. In other words, suppose that there exists a neighbourhood V of x^* such that $V \subseteq D$ which may be assumed to be convex. Thus, there exists a point v in V such that $x^* = \lambda v + (1 - \lambda)w$, for a suitable value λ in $[0, 1]$. Clearly $f(v) \leq f(x^*)$, for x^* is a maximum. Since, by hypothesis, the function f is convex, we have

$$\begin{aligned}
f(x^*) &= f\big(\lambda v + (1 - \lambda)w\big) \\
&\leq \lambda f(v) + (1 - \lambda)f(w) \\
&< \lambda f(x^*) + (1 - \lambda)f(x^*) = f(x^*)
\end{aligned}$$

and this is a clear contradiction. Hence, x^* cannot be in the interior of D and therefore it must belong to the boundary. This proves the theorem. $\quad\square$

(4.4) Example. Suppose that the problem is to maximise the objective function $f(x, y) = x^2 + 3y^2 - x$ subject to the constraint $x^2 + 2y^2 \leq 5$. Since f is convex, by virtue of Theorem (4.3), we can restrict the search of a maximum to the set $\{(x, y) \in \mathbb{R}^2 : x^2 + 2y^2 = 5\}$. Restricted to this set, the objective function becomes $f(x) = -\frac{1}{2}x^2 - x + \frac{15}{2}$ and its maximum value is achieved at $x = -1$ and, correspondingly, $y = \sqrt{2}$ or $y = -\sqrt{2}$. The points $(-1, \sqrt{2})$ and $(-1, -\sqrt{2})$ are both feasible and they both are the global maximum of our problem.

Observe that the previous theorem only guarantees that a global maximum lies on the boundary. Indeed, a local maximum can be in the interior of the domain, as the following example shows.

(4.5) Example. Suppose that the problem is to maximise the convex function $f : [-2, 2] \to \mathbb{R}$, defined by $f(x) = -x$, if $x \in [-2, -1]$, $f(x) = 1$, if $x \in [-1, 1]$, and $f(x) = x$, if $x \in [1, 2]$. It is not difficult to recognise that the point $x^* = 0$ is a local maximum and it is a point in the interior of the feasible set, $[-2, 2]$, of the function.

When the feasible set is a bounded polyhedron, not only is the global maximum at the boundary, but it is actually on one of the vertices of the polyhedron, as shown in the following theorem.

(4.6) Theorem. *Let $f : D \to \mathbb{R}$ be a convex function on a non-empty open convex set D, and P a bounded polyhedron included in D. Then, the optimisation problem to maximise the objective function f subject to the constraint $x \in P$ admits a global maximum which lies in at least a vertex of the polyhedron P.*

Proof. Let us start by observing that, since the function f is convex, by means of Theorem II(4.2), it is also continuous on D. Since P is a closed and bounded set in $\mathbb{R}^n$, hence compact, by means of Weierstrass theorem, the function f admits a global maximum on P. Let us denote a point of global maximum by x^*, and let us prove that $f(x^*)$ coincides with the value of f in one of the vertices. To this end, from Corollary I(8.9), we know that P is the convex hull of its vertices $v^1, \ldots, v^m$. In other words, we know that $P = \mathrm{conv}(v^1, \ldots, v^m)$. Let us denote by γ the maximum of $f(v^1), \ldots, f(v^m)$, the values that f takes at each vertex. Then, the thesis of the theorem can be rephrased by saying that $f(x^*) = \gamma$. Now, since x^* is a maximum point, we have $f(x^*) \geq \gamma$. So, it ultimately suffices to show that the converse inequality holds. To this end, observe that, since $x^* \in P = \mathrm{conv}(v^1, \ldots, v^m)$, it can be written as $x^* = \sum_{i=1}^{m} \lambda_i v^i$, with $\sum_{i=1}^{m} \lambda_i = 1$ and $\lambda_i \geq 0$ for all $i = 1, \ldots, m$. Now, using the fact that f is convex, and that γ is the maximum of the values $f(v^i)$, with $i = 1, \ldots, m$, we obtain

$$f(x^*) = f\left(\sum_{i=1}^{m} \lambda_i v^i\right) \leq \sum_{i=1}^{m} \lambda_i f(v^i) \leq \gamma \sum_{i=1}^{m} \lambda_i = \gamma,$$

and this proves the theorem. $\square$

5. Coercive functions

Let us now introduce a special type of functions, which are particularly useful in optimisation. To understand the reason behind their definition, let us start by observing that, when the feasible region is not compact we cannot be sure about the existence of minimum. However, the following theorem gives a simple sufficient condition for a global minimum to exist.

(5.1) Theorem. *Suppose that the set D is unbounded and that f is continuous. If there exists $s \in \mathbb{R}$ such that the sub-level set*

$$D_s = \left\{ x \in D : f(x) \leq s \right\}$$

is not empty and compact, then there exists the global minimum of f on D.

Proof. Let us begin by observing that, if $\inf_{x \in D} f(x) = s$, then every point of D_s is a point of global minimum. Conversely, if $\inf_{x \in D} f(x) < s$, then a minimising sequence $\{x^k\}$ is contained in sub-level set D_s for every $k > \bar{k}$, for a suitable $\bar{k}$. Since the set D_s is, by assumption, compact, the minimising sequence admits a sub-sequence which converges towards a global minimum point, and this completes the proof. $\square$

So, Theorem (5.1) shows that if we can guarantee that all sub-level sets D_s are compact, then we are certain that the function at hand has a global minimum. This means that a class of functions that have this property becomes particularly interesting from an optimisation point of view. We shall see that the class of 'coercive' functions have this special feature, and this explains why we are now giving the following definition.

(5.2) Definition. Let D be a non-empty closed set in $\mathbb{R}^n$ and $f : D \to \mathbb{R}$ be a real function defined on D. The function f is said to be **coercive** if for every sequence $\{x^k\}$ in D with $\|x^k\| \to +\infty$, we have $f(x^k) \to +\infty$ as well.

At first glance, may look like there is no relationship between a coercive function and the property of compactness of the sub-level sets. However, when a function is both continuous and coercive, it turns out that it actually possesses this very property, as shown in the next theorem.

(5.3) Theorem. *Let D be a non-empty closed unbounded set in $\mathbb{R}^n$, and f be a continuous real function defined on D. Then, f is coercive if and only if, for every number s, the sub-level set $D_s = \{x \in D : f(x) \leq s\}$ is compact.*

Proof. Let us start by proving that the condition is *necessary*, that is, let us suppose that the function f is coercive and let us prove that every sub-level set D_s is compact. Let us start by observing that since f is continuous, then the set $D_s = \{x \in D : f(x) \leq s\}$ is closed. We only need to prove that D_s is also bounded. This can be achieved by contradiction. Precisely, assume that there is a real number s such that the sub-level set D_s is unbounded. Thus, there must be a sequence $\{x^k\}$ of elements of D_s such that $\|x^k\| \to +\infty$ as $k \to +\infty$. Since the function f is coercive, this implies $f(x^k) \to +\infty$, which contradicts the fact that, by assumption, for every index k, we have $x^k \in D_s$, hence $f(x^k) \leq s$. Thus, the set D_s must be bounded, whence compact.

Let us now prove that the condition is also *sufficient*. To this end, suppose that for every real number s, the sub-level set D_s is compact, and let us prove that the function f must be coercive. By contradiction, assume that f

is not coercive, that is there exists a sequence $\{x^k\}$ of points in D such that $\|x^k\| \to +\infty$ as $k \to +\infty$, and $f(x^k)$ is bounded from above uniformly in k. In other words, let us assume that $\{x^k\}$ is a sequence of points in D such that $\|x^k\| \to +\infty$ and there is a real number s with $x^k \in D_s$ for every index k. Now, since D_s is assumed to be compact, we may extract from the sequence $\{x^k\}$ a sub-sequence that converges to a point in D_s, and this is a contradiction because, such a sub-sequence would be bounded from above (by s), whereas, by hypothesis, any sub-sequence extracted from $\{x^k\}$ must be unbounded. $\qquad\square$

(5.4) Corollary.   *In the same hypotheses as Theorem (5.3), if the function f is coercive, then f has a global minimum on D.*

Proof. Since the function f is coercive, by Theorem (5.3), every sub-level set is compact. Thus, applying Theorem (5.1) we find that f has a global minimum in D. $\qquad\square$

(5.5) Remark.   In some important classes of optimisation problems, the set D is not closed and the function f may contain terms that go to infinity on the boundary of D. In cases like these, we can still prove that the sub-level sets are compact by adding the condition that for every sequence $\{x^k\}$ in D with $\{x^k\} \to \bar{x}$, and $\bar{x} \notin D$, we have $f(x^k) \to +\infty$, and this yields the same result as Corollary (5.4).

(5.6) Example.   The function $f(x) = x^4 - x^2$ is coercive but not convex; while the function $f(x) = e^x$ is convex but not coercive.

(5.7) Remark.   To complete this section, let us recall that a quadratic function $f(x) = \frac{1}{2}\langle x, Qx \rangle + \langle c, x \rangle$ is coercive if and only it is strongly convex, which is equivalent to say that the smallest eigenvalue of the matrix Q is positive.

Chapter IV
Linear Programming

Linear programming addresses the problem of optimising a linear function, either by minimising or maximising it, subject to linear constraints. The feasible region in such problems is therefore a polyhedron. In spite of the strong mathematical assumption of linearity, both for the objective function and the constraints, linear programming problems are extremely common in applications and, from a theoretical point of view, they also present techniques that can be exported to the more complex case of non-linear programs.

1. Linear programming problems

A linear programming problem, or simply a 'linear program', is a problem of minimising or maximising a linear function subject to linear constraints of the inequality and/or equality type. Before delving into the theory of linear programming, let us start by showing a few simple examples of problems that can be solved using linear programming.

(1.1) Example (Industrial production). A firm manufactures two different types of fabric. To produce $100\,\mathrm{kg}$ of fabric A, $28\,\mathrm{kg}$ of wool and $7\,\mathrm{kg}$ of cotton are required, whereas to produce $100\,\mathrm{kg}$ of fabric B, $7\,\mathrm{kg}$ of wool and $14\,\mathrm{kg}$ of cotton are required. Suppose that it takes 3 hours of work to produce $100\,\mathrm{kg}$ of either fabric, and assume that the firm has $168\,\mathrm{kg}$ of wool and $84\,\mathrm{kg}$ of cotton available per week, and that there are 42 working hours available for production. If the profit on fabric A is £20 per kilogram, and that on fabric B is £10 per kilogram, how many of each fabrics should be produced in order to maximise the profit per week?

In order to solve this problem, let us denote by x_1 and x_2 the amount, in kilogram, of fabric A and fabric B, respectively, produced in a week. Thus, the total profit per week is represented by the function $f(x_1, x_2) = 20x_1 + 10x_2$. The problem to maximise this profit, respecting all the production constraints

© The Author(s), under exclusive license to Springer Nature Switzerland AG 2026
A. Carpignani and M. Pappalardo, *Theory and Methods of Optimisation*,
UNITEXT 175, https://doi.org/10.1007/978-3-032-01514-3_4

is the following:

$$\begin{aligned}
\text{maximise} \quad & 20x_1 + 10x_2 \\
\text{subject to} \quad & 28x_1 + 7x_2 \le 168 \\
& 7x_1 + 14x_2 \le 84 \\
& 3x_1 + 3x_2 \le 42 \\
& x_1 \ge 0,\ x_2 \ge 0
\end{aligned}$$

(1.2) Example (Diet problem). A university student is trying to plan his meals on a tight budget. He needs to choose from four available foods to meet his daily nutritional requirements: rice, chicken breast, broccoli and milk. Each food has a specific cost and provides calories and protein. The student's goal is to select the quantities of each food to minimise the total cost while ensuring that he meets his dietary needs. We know that rice costs £2 per serving, provides 250 calories, and 5 grams of protein; chicken breast costs £4 per serving, provides 150 calories, and 25 grams of protein; broccoli costs 50p per serving, provides 50 calories, and 10 grams of protein; milk costs £1.50 per serving, provides 100 calories, and 8 grams of protein. The daily nutritional requirements that the student should meet are: at least 600 calories per day, and at least 50 grams of protein per day.

In order to solve this problem let us denote by x_1, x_2, x_3 and x_4 the amount of rice, chicken breast, broccoli and milk, respectively, that the student should buy. Thus, the total student's cost for the food, in pence, is $f(x_1, x_2, x_3, x_4) = 200x_1 + 400x_2 + 50x_3 + 150x_4$. The problem to minimise this cost, respecting all the constraints given by the problem is the following:

$$\begin{aligned}
\text{minimise} \quad & 200x_1 + 400x_2 + 50x_3 + 150x_4 \\
\text{subject to} \quad & 250x_1 + 150x_2 + 50x_3 + 100x_4 \ge 600 \\
& 5x_1 + 25x_2 + 10x_3 + 8x_4 \ge 50 \\
& x_1, x_2, x_3, x_4 \ge 0
\end{aligned}$$

As we can see from the previous examples, linear programs arise naturally in mathematical modelling, and they may be either maximisation problems, or minimisation problems, depending on whether one is focused on maximising a profit or minimising a cost (or a loss). Nonetheless, in order to tackle a linear problem from a theoretical point of view, it might be convenient to reduce every linear program to a specific 'standard' form, and always state results for a problem in that specific form. As we have already mentioned in Chapter III, Section 2, up to an equivalent transformation, a linear program can be changed (e.g. by transforming a minimisation problem into a maximisation problem, or reshaping the constraints to a specific format), so this is not really a complicated thing to do. However, mathematicians have not agreed on a specific 'standard' form for a linear program, and different authors make different choices depending on their customs and traditions. The following definition provides the 'standard' form that is going to be utilised in the sequel.

(1.3) Definition. Let $A \in \mathfrak{M}(m, n, \mathbb{R})$ be a non-zero matrix, $b \in \mathbb{R}^m$, $c \in \mathbb{R}^m$ be vectors, and assume that c is not zero. The optimisation problem

$$\begin{aligned} \text{maximise} \quad & \langle c, x \rangle \\ \text{subject to} \quad & Ax \leq b \end{aligned}$$

is called a linear program in **standard form** or simply the **standard** linear program.

(1.4) Example. The problem

$$\begin{aligned} \text{minimise} \quad & 3x_1 - 5x_2 + 6x_3, \\ \text{subject to} \quad & x_1 - 2x_2 + 4x_3 = 7, \\ & x_2 \geq 0. \end{aligned}$$

can be easily transformed in standard form as follows:

$$\begin{aligned} \text{maximise} \quad & -3x_1 + 5x_2 - 6x_3, \\ \text{subject to} \quad & x_1 - 2x_2 + 4x_3 \leq 7, \\ & -x_1 + 2x_2 - 4x_3 \leq -7, \\ & -x_2 \leq 0. \end{aligned}$$

Linear programs are very common in mathematical modelling, and their real power is that the optimal solutions are not free to lie anywhere in the feasible region, but they can only sit on particular points. Before delving into the more general criteria for solving a linear program, it is worth to describe, in the rest of this section, a very simple geometric procedure. Even though this method is really only feasible for two dimensional problems, with a relatively simple feasible set, it provides a great deal of insight into the structure of a linear program, hence it deserves to be analysed. Incidentally, this will give us the possibility to introduce some terms that are used especially in mathematical modelling of economical and financial problems.

(1.5) Let $A \in \mathfrak{M}(m, n, \mathbb{R})$ be a non-zero matrix, $b \in \mathbb{R}^m$ and $c \in \mathbb{R}^n$, and assume that c is not zero. Consider the linear programming problem

$$\begin{aligned} \text{maximise} \quad & \langle c, x \rangle \\ \text{subject to} \quad & Ax \leq b. \end{aligned}$$

In order to solve this problem, note that the points with the same objective value z satisfy the equation $\langle c, x \rangle = z$. This is the equation of a hyper-plane, and more precisely of the hyper-plane perpendicular to the vector c. In the special and especially simple case of a two-dimensional problem, this hyper-plane is the straight-line perpendicular to the vector c. This justifies the following definition taken from the economics terminology.

(1.6) Definition. In the same hypotheses as in (1.5), if the problem is to maximise the function $\langle c, x \rangle$, the hyper-plane $\{x \in \mathbb{R}^n : \langle c, x \rangle = z\}$ is referred to as the **iso-profit** hyper-plane. On the other hand, if the problem is to minimise the function $\langle c, x \rangle$, the same hyper-plane is now referred to as the **iso-cost** hyper-plane.

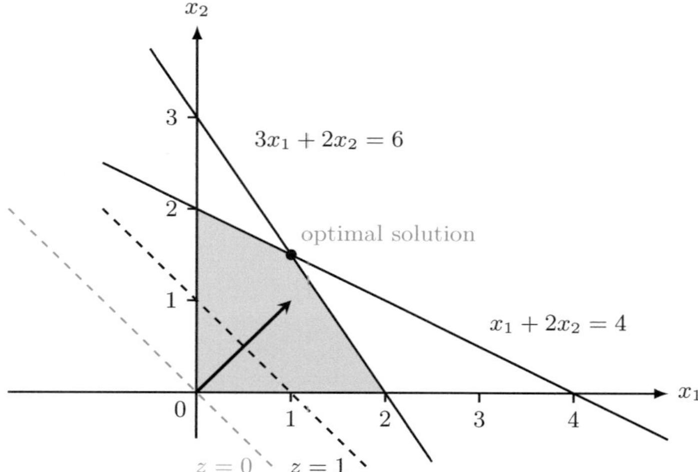

Figure 1.1. The graph shows the feasible set described in Example (1.7). The dashed line represents the iso-profit line with equation $x_1 + x_2 = 1$. Moving forward this line, we can recognise that the optimal solution lies at the vertex indicated with a solid bullet.

The idea of the solution is now to move the iso-profit hyper-plane in the direction that maximises its numeric coefficient as much as possible, with the only constraint that the iso-profit hyper-plane intersects the feasible set. When an optimal solution is reached, the iso-profit hyper-plane cannot be moved any further, because further increasing of the coefficient z would lead to only points outside of the feasible set. This can be proved by contradiction: in fact, if we could increase further the coefficient of the iso-profit hyper-plane and the resultant hyper-plane had a point in common with the feasible set, the value of the objective function on this new point would be greater than its value at the optimal solution.

Let us show how to use this method with an example.

(1.7) Example. Consider the problem to maximise the objective function $x_1 + x_2$ subject to the constraints $3x_1 + 2x_2 \le 6$, $x_1 + 2x_2 \le 4$ and with $x_1, x_2 \ge 0$. To this end, let us start by sketching the feasible set on a grid, and draw the iso-profit lines, starting from a (more or less) random one, and moving perpendicularly to the vector $c = (1, 1)$, until the line reaches the optimum value. This is shown in Figure 1.1. As we can see from the graph, that shows the case when $z = 1$, moving the equation $x_1 + x_2 = z$ forward, the optimal solution is unique and it lies on the intersection of lines $3x_1 + 2x_2 = 6$ and $x_1 + 2x_2 = 4$.

(1.8) Example. Consider the problem to maximise the objective function $x_1 + 2x_2$ subject to the constraints $3x_1 + 2x_2 \le 6$, $x_1 + 2x_2 \le 4$

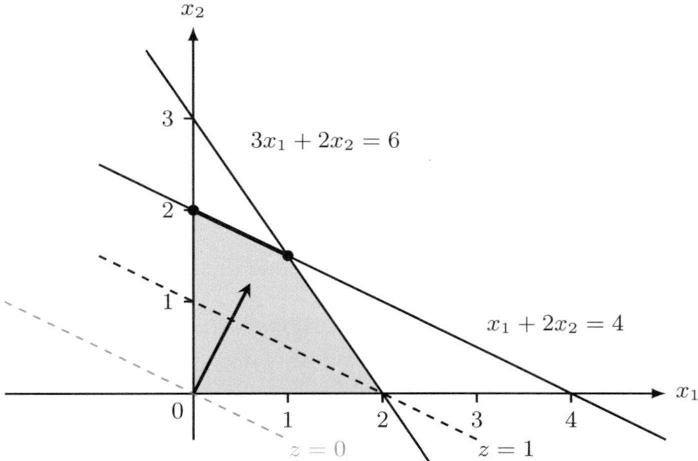

Figure 1.2. The graph shows the feasible set described in Example (1.8). The dashed line represents the iso-profit line $x_1 + 2x_2 = 1$. Moving forward this line, we can recognise that the optimal solutions are all points in the segment of the feasible set lying on the equation $x_1 + 2x_2 = 4$.

and $x_1, x_2 \geq 0$. Again, let us start by sketching the feasible set on a grid, and draw the iso-profit lines, starting from a (more or less) random one, and moving perpendicularly to the vector $c = (1, 2)$, until the line reaches the optimum value. This is shown in Figure 1.2. As we can see from the graph, in this case the iso-profit line is parallel to one of the constraints, and it achieves the maximum possible value when it coincides with that constraint, $x_1 + 2x_2 = 4$. Thus, all points in the segment of this line that belongs to the feasible set are optimal solutions for this problem, and the solutions are infinite.

2. Geometry of linear programming

Linear programs may count on a very rich geometric structure, and this structure will massively simplify the search for an optimal solution. Precisely, given a linear program to maximise the function $\langle c, x \rangle$ subject to $Ax \leq b$, let us observe that the feasible region,

$$P = \left\{ x \in \mathbb{R}^n : Ax \leq b \right\}$$

is nothing but a polyhedron. By exploiting the full range of properties of polyhedra developed in Chapter I, we can see that there are limitations on where the optimal solution of a linear program can sit. Let us start with the following simple, but really important, result.

(2.1) Theorem. *Let $A \in \mathfrak{M}(m, n, \mathbb{R})$ be a $m \times n$ matrix, $b \in \mathbb{R}^m$, $c \in \mathbb{R}^n$ be vectors, and assume that $c \neq 0$. Denote by P the polyhedron*

$$\left\{ x \in \mathbb{R}^n : Ax \leq b \right\},$$

and consider the optimisation problem $(\mathcal{P})$ to maximise $\langle c, x \rangle$ subject to the constraint $x \in P$. Then, the following statements hold:

(a) *An optimal solution of this problem is not in the interior of P.*

(b) *The local optimal solutions of this problem are global solutions as well.*

(c) *If the problem has two distinct optimal solutions, then it has infinite solutions.*

Proof. To prove statement (a) it suffices to observe that, if an optimal solution were internal, the gradient of the objective function $f(x) = \langle c, x \rangle$ at that point would be 0. Since $\nabla f(x) = c \neq 0$, it follows that the optimal solution cannot be internal.

To prove statement (b), assume that z is a local optimal solution and, by contradiction, assume that z is not a global solution. Then, there exists in P a point x such that $\langle c, x \rangle > \langle c, z \rangle$. Now, if $\lambda \in (0, 1)$, consider the point $\lambda x + (1 - \lambda)z \in P$ (because P is a polyhedron, hence convex), and observe that

$$\begin{aligned}
\langle c, \lambda x + (1 - \lambda)z \rangle &= \lambda \langle c, x \rangle + (1 - \lambda)\langle c, z \rangle \\
&> \lambda \langle c, z \rangle + (1 - \lambda)\langle c, z \rangle \\
&= \langle c, z \rangle.
\end{aligned}$$

This means that, at each point in the segment with extremes z and x the objective function takes a value greater than the value at z. In particular, if $V \in N(z)$ is a neighbourhood of z, such that $\langle c, x \rangle \leq \langle c, z \rangle$ for every $x \in V$, we can take λ small enough to ensure $\lambda x + (1 - \lambda)z \in V$, and this contradicts the fact that z is a local optimal solution.

Finally, to prove statement (c), let x^1 and x^2 be two distinct optimal solutions for $(\mathcal{P})$, let $\lambda \in (0, 1)$, and denote by $\bar{v}$ the optimal value of problem $(\mathcal{P})$. Then $\langle c, x^1 \rangle = \langle c, x^2 \rangle = \bar{v}$. Observe that

$$\begin{aligned}
\langle c, \lambda x^1 + (1 - \lambda)x^2 \rangle &= \lambda \langle c, x^1 \rangle + (1 - \lambda)\langle c, x^2 \rangle \\
&= \lambda \bar{v} + (1 - \lambda)\bar{v} \\
&= \bar{v},
\end{aligned}$$

which proves that every convex combination of these two optimal solutions is also an optimal solution, whence there are infinite optimal solutions. $\qquad \square$

The previous theorem already gives some important insight: the optimal solutions are not in the interior of the feasible region, which reduces the search of the optimal solutions to the boundary of the feasible region, and local optimal solutions are also global. However, the previous theorem only scratches the surface of the problem, using only the fact that the feasible region is convex and that the objective function is linear, i.e. convex as well. To exploit even further the properties of the feasible region, let us also observe that, by virtue of Minkowski-Weyl Theorem (see Theorem I(7.1)), we can always write the polyhedron P in the form

$$P = \operatorname{conv}(v^1, \dots, v^r) + \operatorname{cone}(d^1, \dots, d^p).$$

With the aid of this decomposition, we can now prove that an optimal solution of the problem certainly lies in the set $\{v^1, \dots, v^r\}$. In particular, we shall prove that there is only a finite number of feasible solutions among which we can search for the optimal.

(2.2) Theorem (Fundamental Theorem of Linear Programming).
Let $v^1, \dots, v^r$, and $d^1, \dots, d^p$ be two finite sequences of vectors in $\mathbb{R}^n$, let

$$P = \operatorname{conv}(v^1, \dots, v^r) + \operatorname{cone}(d^1, \dots, d^p),$$

and let $c \in \mathbb{R}^n$. Consider the optimisation problem to maximise the linear function $f(x) = \langle c, x \rangle$ subject to the constraint $x \in P$. If this problem has a finite optimal value, then there exists an index k in $1, \dots, r$ such that v^k is an optimal solution for this problem.

Proof. By definition, $x \in P$ if and only if

$$x = \sum_{i=1}^{r} \lambda_i v^i + \sum_{j=1}^{p} \mu_j d^j$$

for some $\lambda_1, \dots, \lambda_r \geq 0$ and $\mu_1, \dots, \mu_p \geq 0$, with $\sum_{i=1}^{r} \lambda_i = 1$. To start off, let us observe that, if there existed an index $j = 1, \dots, p$ such that $\langle c, d^j \rangle > 0$, then, by setting to zero all variables $\mu_1, \dots, \mu_p$, with the exception of μ_j, the objective function would take the form

$$\left\langle c, \sum_{i=1}^{r} \lambda_i v^i + \mu_j d^j \right\rangle,$$

and it would tend to infinite, as μ_j tends to infinite. As a result, the problem would have an infinite optimal value. On the other hand, since we have assumed that the problem has a finite optimal value, we may assume that $\langle c, d^j \rangle \leq 0$ for every index $j = 1, \dots, p$.

Let us denote by γ the maximum value of $\langle c, v^1 \rangle, \dots, \langle c, v^r \rangle$. In order to prove the theorem, we wish to show that γ is the optimal value of this problem.

Indeed, if $x = \sum_{i=1}^{r} \lambda_i v^i + \sum_{j=1}^{p} \mu_j d^j$, we have

$$\left\langle c, \sum_{i=1}^{r} \lambda_i v^i + \sum_{j=1}^{p} \mu_j d^j \right\rangle = \sum_{i=1}^{r} \lambda_i \langle c, v^i \rangle + \sum_{j=1}^{p} \mu_j \langle c, d^j \rangle$$
$$\leq \sum_{i=1}^{r} \lambda_i \langle c, v^i \rangle$$
$$\leq \sum_{i=1}^{r} \lambda_i \gamma$$
$$= \gamma$$

where the first inequality follows from the fact that, for every $j = 1, \ldots, p$, one has $\langle c, d^j \rangle \leq 0$, and the second from the fact that, by definition of γ, $\langle c, v^i \rangle \leq \gamma$ for every $i = 1, \ldots, r$. Thus, $\max\{\langle c, x \rangle : x \in P\} \leq \gamma$.

On the other hand, since γ is the value of the function $x \mapsto \langle c, x \rangle$ in a point of the polyhedron, one also has $\gamma \leq \max\{\langle c, x \rangle : x \in P\}$, and this completes the proof of the theorem. $\qquad\square$

Let us see a couple of consequences of this important theorem, and we shall comment them all afterwards.

(2.3) Corollary. *In the hypotheses of Theorem (2.2), if P is non-empty and has no directions of lineality, and if the optimal value of $\langle c, x \rangle$ is finite, then the problem to maximise $\langle c, x \rangle$ subject to $x \in P$ has an optimal solution and one of the vertices of P is an optimal solution.*

Proof. Corollary I(8.8) ensures that, in the hypotheses of the corollary, there exists at least one vertex for P. In particular, $P = \operatorname{conv}(\operatorname{vert}(P)) + \operatorname{rec}(P)$. Thus, applying Theorem (2.2) gives that there exists an optimal solution and it must be in one of the vertices. $\qquad\square$

(2.4) Corollary. *In the hypotheses of Theorem (2.2), assume that P is a bounded non-empty polyhedron. Then the problem to maximise $\langle c, x \rangle$ subject to $x \in P$ has an optimal solution and one of the vertices of P is an optimal solution.*

Proof. Since a bounded polyhedron is compact, by virtue of Weierstrass Theorem, the optimisation problem has a finite optimal solution. From Corollary I(8.9), since P is a bounded polyhedron, then $P = \operatorname{conv}(\operatorname{vert}(P))$. Then, by means of Theorem (2.2), an optimal solution lies in $\operatorname{vert}(P)$. $\qquad\square$

As always, after an important result, it is time to gather the consequence. In fact, the fundamental theorem of linear programming limits the search for the optimal solutions to a finite number of feasible solutions. From a theoretical point of view, this does not even need any optimality criterion, because it would suffice to evaluate the objective function on each of the candidates offered by the fundamental theorem of linear programming, and pick the one with the highest value. Unfortunately, however, the Weyl-Minkowski Theorem does not give any clue on how to find the points $v^1, \ldots, v^r$. This is when the geometry completes its job, and leave the scene to algebra. Before discussing the algebra of linear programming, however, we are going to need to use duality to create an effective optimality condition. This is the goal of the next two sections.

3. Introduction to duality

The aim of this section is to develop a criterion to state whether or not a feasible solution is an optimal solution. This can be achieved by associating to the initial problem another linear program, ancillary to the given one, and whose solutions are entangled with the solutions of the initial problem. To this auxiliary linear program we shall give the name of 'dual', borrowing this term from linear algebra. Similarly, we shall call the original problem the 'primal' linear program, to emphasise the pairing of these two programs as well as to avoid cumbersome terminology like 'the original problem' or 'the initial problem'. Incidentally, the term 'primal' was suggested as an appropriate Latin antonym by Dantzig's father, Tobias Dantzig, who was also a mathematician, in order to refer to 'the original problem of which this is the dual'. We shall see, however, that the dichotomy between 'primal' and 'dual' is not actually so absolute, because we shall prove shortly that the dual of the dual is the primal linear program.

Also, a note of caution should be made at this point: in the sequel we shall introduce two notions of 'dual' linear program. One regards the shape of the linear program, and the other, much deeper and arguably more important, regards the connection with the 'primal' linear program. It is paramount to distinguish these two notions, as one is purely aesthetic, while the other is much more deeply correlated with the original problem.

To this end, let us start by recalling that a linear programming problem can be written in different forms, using suitable equivalent transformations, to change the form of the objective function and the constraints. Precisely, we wish to focus on two special types of linear programs, which are highlighted in the following definition. Notice however that the first part of this definition coincides with Definition (1.3), with a simple update in the terminology that aims to highlight the future notion of 'duality'.

(3.1) Definition. Let $A \in \mathfrak{M}(m, n, \mathbb{R})$ be a non-zero matrix, $b \in \mathbb{R}^m$ and $c \in \mathbb{R}^n$, and assume that $c \neq 0$. A linear program in **primal standard form** is

$$\begin{aligned} \text{maximise} \quad & \langle c, x \rangle \\ \text{subject to} \quad & Ax \leq b. \end{aligned}$$

A linear program in **dual standard form** is

$$\begin{aligned} \text{minimise} \quad & \langle c, x \rangle \\ \text{subject to} \quad & Ax = b, \ x \geq 0. \end{aligned}$$

The idea (which is actually very common in mathematics) is to associate to a problem which is written in primal form a suitable problem. This problem will be naturally written in dual form, and will show characteristics that are interlinked to the primal problem, so that the information gathered from the dual problem can help establishing whether a feasible solution of the primal problem is optimal or not. Incidentally, the criterion that we are going to develop will do even more: it will at the same time find an optimal solution for the primal and the dual problem. The definition of the primal/dual pair is suggested by the theorems of alternative, and in particular by Farkas's Lemma (see Theorem I(6.5)). Formally, let us give the following definition.

(3.2) Definition. Let $A \in \mathfrak{M}(m, n, \mathbb{R})$ be a non-zero matrix, $b \in \mathbb{R}^m$ and $c \in \mathbb{R}^n$, and assume that $c \neq 0$. The dual problem of

$$(\mathcal{P}) : \begin{cases} \text{maximise} & \langle c, x \rangle \\ \text{subject to} & Ax \leq b \end{cases}$$

is

$$(\mathcal{D}) : \begin{cases} \text{minimise} & \langle b, y \rangle \\ \text{subject to} & A^T y = c, \ y \geq 0. \end{cases}$$

(3.3) Remark. It is important to notice immediately that the dual of a linear program in primal standard form is in dual standard form, but the dual of a linear program need not be in dual standard form. Similarly, a linear program need not be in primal standard form, but it can always be written in primal standard form up to a finite sequence of equivalent transformations, which entails that any linear program has a dual. In particular, a linear program can always be written in dual standard form by means of a finite sequence of equivalent transformation, but this does not mean that the problem thereby obtained is its dual. It is just a linear program in dual standard form. Contrarily, the dual of a linear program in primal standard form is by definition a linear program in dual standard form, but these are two different liner programs, with different objective functions and with feasible regions living in different spaces.

The content of the previous remark is extremely important, and it is imperative to have this idea clear before going any further. The following examples should contribute to clarify it.

(3.4) Example. Consider the following problem:

$$\begin{aligned} \text{maximise} \quad & 3x_1 + 5x_2, \\ \text{subject to} \quad & x_1 + 2x_2 \leq 7, \\ & 4x_1 + 3x_2 \leq 4. \end{aligned}$$

This problem is equivalent to

$$(3.5) \quad \begin{array}{ll} \text{minimise} & -3u_1 + 3u_2 - 5u_3 + 5u_4, \\ \text{subject to} & u_1 - u_2 + 2u_3 - 2u_4 + s_1 = 7, \\ & 4u_1 - 4u_2 + 3u_3 - 3u_4 + s_2 = 4, \\ & u_1,\, u_2,\, u_3,\, u_4,\, s_1,\, s_2 \geq 0. \end{array}$$

Observe that this is not the dual of the initial problem, but it is a problem in dual standard form which is equivalent to the given one. Instead, the dual of the initial problem is the following:

$$\begin{array}{ll} \text{minimise} & 7y_1 + 4y_2, \\ \text{subject to} & y_1 + 4y_2 = 3, \\ & 2y_1 + 3y_2 = 5, \\ & y_1,\, y_2 \geq 0. \end{array}$$

Interestingly, if by means of equivalent transformations, we transformed this problem in primal standard form, its dual would be (3.5).

As we have mentioned earlier on, the dual of the dual turns out to be the original primal problem. The following proposition shows this important and useful fact.

(3.6) Proposition. *With the same hypotheses and with the same notations as in Definition (3.2), the dual of $(\mathcal{D})$ is $(\mathcal{P})$.*

Proof. In order to use Definition (3.2) to $(\mathcal{D})$ we first need to put it in primal standard form. This can be done through equivalent transformations. To start off, by virtue of Proposition III(2.5), and taking the transpose of all expressions, problem $(\mathcal{D})$ is equivalent to

$$\begin{array}{ll} \text{maximise} & -\langle b, y \rangle \\ \text{subject to} & A^T y \leq c, \\ & -A^T y \leq -c, \\ & -I^T y \leq 0 \end{array}$$

Applying the definition of dual to this problem gives

$$\begin{array}{ll} \text{minimise} & \langle u_1, c \rangle - \langle u_2, c \rangle \\ \text{subject to} & A u_1 - A u_2 - I u_3 = -b \\ & u_1,\, u_2,\, u_3 \geq 0 \end{array}$$

Taking the transpose of every expression in the previous problem and setting $x = u_2 - u_1$ gives

$$\begin{array}{ll} \text{minimise} & -\langle c, x \rangle \\ \text{subject to} & Ax + u_3 = b,\, u_3 \geq 0 \end{array}$$

Now, it suffices to apply Proposition III(2.7) and again Proposition III(2.5) to show that this problem is in fact equivalent to $(\mathcal{P})$. This proves that the dual of the dual of a problem is indeed the given problem. $\qquad\square$

4. Duality and optimality conditions

As we mentioned, our final aim consists in utilising the notion of dual of a linear programming problem to establish whether or not a feasible solution is optimal. In the sequel, we shall derive an efficient criterion based on the notion of dual. Indeed, this criterion would say more that: given a linear program (the primal) and taken its dual linear program (the dual), it will establish whether a pair of feasible solution, the first for the primal, and the second for the dual, is an optimal pair, i.e. consists of an optimal solution for the primal and an optimal solution for the dual.

To this end, let us start by fixing the hypotheses and notations that are going to be used throughout this section.

(4.1) Hypotheses. Let $A \in \mathfrak{M}(m, n, \mathbb{R})$ be a non-zero matrix, let $b \in \mathbb{R}^m$ and $c \in \mathbb{R}^n$. Consider the following pair of linear programs, the first in primal standard form, and the second, the dual of the former, in dual standard form:

$$(\mathcal{P}) : \begin{cases} \text{maximise} & \langle c, x \rangle \\ \text{subject to} & Ax \leq b, \end{cases} \qquad (\mathcal{D}) : \begin{cases} \text{minimise} & \langle b, y \rangle \\ \text{subject to} & A^T y = c, \; y \geq 0. \end{cases}$$

Furthermore, denote by P the feasible region of $(\mathcal{P})$, and by D the feasible region of $(\mathcal{D})$. In other words, let us set:

$$P = \left\{ x \in \mathbb{R}^n : Ax \leq b \right\}, \qquad D = \left\{ y \in \mathbb{R}^m : A^T y = c, \; y \geq 0 \right\}.$$

Finally, denote by $v_\mathcal{P}$ and $v_\mathcal{D}$ the optimal value of $(\mathcal{P})$ and $(\mathcal{D})$, respectively.

(4.2) Remark. Let us observe that so long as the polyhedron P is not empty, one can always assume that $\mathrm{lin}(P) = \{0\}$ and therefore that the polyhedron does have at least one vertex. Indeed, if there existed a non-zero direction of lineality, we could always take an equivalent transformation to force the variable to be positive, therefore embedding the problem in a higher dimensional space, and this would ensure that the resulting polyhedron be included in the *positive orthant* (that is to say the subset of $\mathbb{R}^n$ described by the equation $x \geq 0$) thereby preventing any non-zero directions of lineality.

(4.3) Proposition. *Under the hypotheses (4.1),*

(a) *If P is not empty, and D is empty, then $v_\mathcal{P} = v_\mathcal{D} = +\infty$,*

(b) *If P is empty, and D is not empty, then $v_\mathcal{P} = v_\mathcal{D} = -\infty$.*

Proof. Let us start by proving statement (a). To this end, assume that the polyhedron P is not empty. This means that there exists a vector $\bar{x}$ such that $A\bar{x} \leq b$. If the polyhedron D is empty, the system $A^T y = c$ with $y \geq 0$ is impossible. Therefore, by virtue of Farkas's Lemma (see Theorem I(6.5)), there exists a vector $d \in \mathbb{R}^n$ such that $Ad \leq 0$ and $\langle c, d \rangle > 0$. In other words, Farkas's Lemma ensures that we can always find a direction of recession for P such that $\langle c, d \rangle > 0$. Thus, calculating the objective function along the ray through $\bar{x}$ in the direction d, we eventually get

$$\langle c, \bar{x} + \lambda d \rangle = \langle c, \bar{x} \rangle + \lambda \langle c, d \rangle,$$

and this approaches infinity as $\lambda \to +\infty$. On the other hand, the minimum of a function over an empty set is, by definition, $+\infty$, and this proves the chain of equalities $v_{\mathcal{P}} = v_{\mathcal{D}} = +\infty$.

It remains to prove statement (b). This however follows directly from statement (a), because, by means of Proposition (3.6), the primal problem can be thought as the dual of the dual. $\square$

The previous result covers the case when one of the feasible regions is empty, but the real meaningful results follow when both regions are non-empty. In this special, and especially important case, the following result holds.

(4.4) Proposition (Weak duality). *Under the hypotheses (4.1), if both P and D are non-empty, then $\langle c, x \rangle \leq \langle b, y \rangle$ for all $x \in P$ and all $y \in D$.*

Proof. Assume that P and D are non-empty; let $x \in P$ and $y \in D$. Then

$$\langle c, x \rangle = \langle A^T y, x \rangle = \langle y, Ax \rangle \leq \langle b, y \rangle$$

because, by hypothesis, $Ax \leq b$ and $y \geq 0$. $\square$

Weak duality says that the value of the objective function of the dual problem, calculated on any feasible point of the dual problem, gives an upper bound for the primal optimum value. Conversely, the value of the objective function of the primal problem, calculated on any feasible point of the primal problem, gives a lower bound for the dual optimal value. Hence, by weak duality, when the feasible regions of the primal and the dual problems are both non-empty, the optimal values of the primal and the dual problems are both finite, and the optimal value of the primal problem is smaller than or equal to the optimal value of the dual problem. In actual facts, we can prove something stronger than that. Precisely,

(4.5) Theorem (Strong duality). *Under the hypotheses (4.1), if both P and D are non-empty, then $v_{\mathcal{P}}$ and $v_{\mathcal{D}}$ are both finite and $v_{\mathcal{P}} = v_{\mathcal{D}}$.*

Proof. By means of Proposition (4.4) it follows that the primal problem has finite optimal value. Hence, by means of the Fundamental Theorem of Linear Programming (Theorem (2.2)), there exists an optimal solution $\bar{x}$. Let us denote by $I(\bar{x})$ the set of all 'active constraints' at $\bar{x}$, that is to say,

$$I(\bar{x}) = \left\{ i : A_i \bar{x} = b_i \right\},$$

and consider the matrix $A_{I(\bar{x})}$ obtained selecting from A only the rows A_i of A with index i in $I(\bar{x})$. Consider a vector $d \in \mathbb{R}^n$ such that $A_{I(\bar{x})}d \leq 0$. If, for each $i \notin I(\bar{x})$, we have $A_i d < 0$, then d is a direction of recession for the polyhedron P, hence $\bar{x} + \lambda d \in P$ for every $\lambda \geq 0$. On the other hand, if there is at least an index $i \notin I(\bar{x})$ such that $A_i d > 0$, from the fact that $A_i \bar{x} < b_i$, it follows that for a sufficiently small $\lambda > 0$,

$$A_i(\bar{x} + \lambda d) = A_i \bar{x} + \lambda A_i d < b_i + \lambda A_i d \leq b_i,$$

hence $\bar{x} + \lambda d \in P$. Either way, we can find feasible points of the form $\bar{x} + \lambda d$ with $\lambda > 0$. Since $\bar{x}$ is an optimal solution of the primal problem, we must have $\langle c, \bar{x} + \lambda d \rangle \leq \langle c, \bar{x} \rangle$, hence $\langle c, d \rangle \leq 0$. In other words, the system of equations $A_{I(\bar{x})}d \leq 0$ and $\langle c, d \rangle > 0$ is impossible. Thus, by means of Farkas's Lemma, the system of simultaneous equations $A_{I(\bar{x})}^T u = c$ with $u \geq 0$ has at least one solution $\bar{u}$. Let us define vector $\bar{y}$ in $\mathbb{R}^m$ as $\bar{y}_i = \bar{u}_i$ when $i \in I(\bar{x})$, and $\bar{y}_i = 0$ otherwise. Hence, vector $\bar{y}$ belongs to the polyhedron D, feasible set of the dual problem. In fact,

$$A^T \bar{y} = A_{I(\bar{x})}^T \bar{u} = c, \quad \bar{y} \geq 0.$$

Furthermore, the value of the objective function of the dual problem at $\bar{y}$ coincides with the value of the objective function of the primal problem. In fact:

$$\langle \bar{y}, b \rangle = \langle \bar{u}, b_{I(\bar{x})} \rangle = \langle \bar{u}, A_{I(\bar{x})}\bar{x} \rangle = \langle c, \bar{x} \rangle.$$

Thus, by virtue of the weak duality, $\bar{y}$ is an optimal solution of the dual problem, and the optimal values $v_{\mathcal{P}} = \langle c, \bar{x} \rangle$ and $v_{\mathcal{D}} = \langle \bar{y}, b \rangle$ coincide. $\square$

To wrap up what we have seen thus far, the content of Proposition (4.3) and Theorem (4.5) can now be summarised in the following table, where we still assume the same hypotheses and notations as in Definition (3.2).

	$D \neq \emptyset$	$D = \emptyset$
$P \neq \emptyset$	$v_{\mathcal{P}} = v_{\mathcal{D}} \in \mathbb{R}$	$v_{\mathcal{P}} = v_{\mathcal{D}} = +\infty$
$P = \emptyset$	$v_{\mathcal{P}} = v_{\mathcal{D}} = -\infty$	$v_{\mathcal{P}} = -\infty,\ v_{\mathcal{D}} = +\infty$

The most important takeaway from this table is the fact that, except in the case in which both the primal and the dual problems have an empty feasible region, the primal and the dual share the same optimal value, which, incidentally, is finite as soon as both the feasible regions of the primal and the dual are non-empty.

In the next example, let us see the case in which both polyhedra, primal and dual, are empty, but we shall save the more interesting cases (when both feasible regions are non-empty) when we have fully developed the tools to actually find the optimal solutions.

(4.6) Example. Consider the following linear programming problem in primal standard form:

$$\begin{array}{ll}
\text{maximise} & x_1 + 2x_2 \\
\text{subject to} & x_1 + x_2 \leq 0 \\
& -x_1 - x_2 \leq -1
\end{array}$$

Its dual problem is

$$\begin{array}{ll}
\text{minimise} & -y_2 \\
\text{subject to} & y_1 - y_2 = 1 \\
& y_1 - y_2 = 2 \\
& y_1, y_2 \geq 0
\end{array}$$

It is not difficult to see that the feasible regions

$$P = \left\{ (x_1, x_2) \in \mathbb{R}^2 : x_1 + x_2 \leq 0,\ -x_1 - x_2 \leq -1 \right\},$$

$$D = \left\{ (y_1, y_2) \in \mathbb{R}^2 : y_1 - y_2 = 1,\ y_1 - y_2 = 2,\ y_1 \geq 0,\ y_2 \geq 0 \right\}$$

are both empty. As a result, the optimal values of the primal and dual problems are $v_P = -\infty$ and $v_D = +\infty$, respectively.

So far, we have discussed the relationship between the optimal values of the primal and dual problem, and found that as soon as both the feasible regions of the primal and dual problems are non-empty, these two values coincide. This is however only half of the problem, as we also need to link the optimal solutions of the primal problems to the optimal solutions of the dual problems. This is just a little bit more complicated, because the feasible regions of the primal and dual space generally live in spaces with different dimensions. However, the following result shows a very strong link between optimal solutions of the primal and the dual problems, which is actually a simple consequence of the strong duality theorem.

(4.7) Theorem (Complementary slackness). *Under the hypotheses (4.1), assume that $\bar{x}$ and $\bar{y}$ are feasible solutions for $(\mathcal{P})$ and $(\mathcal{D})$, respectively. Then $\bar{x}$ and $\bar{y}$ are optimal solutions for the primal and dual problem, respectively, if and only if $\langle \bar{y}, b - A\bar{x} \rangle = 0$.*

Proof. By virtue of Theorem (4.5), $\bar{x}$ and $\bar{y}$ are optimal solutions for the primal and dual problems, respectively, if and only if $\langle c, \bar{x} \rangle = \langle \bar{y}, b \rangle$. On the other hand, the constraint of the dual problem ensures that $c = A^T \bar{y}$, so $\langle \bar{y}, b \rangle = \langle c, \bar{x} \rangle = \langle A^T \bar{y}, \bar{x} \rangle = \langle \bar{y}, A\bar{x} \rangle$. Rearranging this equality gives that $\bar{x}$ and $\bar{y}$ are optimal solutions for the primal and dual problems, respectively, if and only if $\langle \bar{y}, b - A\bar{x} \rangle = 0$. $\qquad\square$

(4.8) Remark. The theorem of complementary slackness takes a very expressive form in the special case of a primal/dual pair of linear programs in symmetrical form, that is to say in the form

$$
\begin{array}{ll}
\text{maximise} & \langle c, x \rangle \\
\text{subject to} & Ax \leq b \\
& x \geq 0
\end{array}
\qquad \text{and} \qquad
\begin{array}{ll}
\text{minimise} & \langle b, y \rangle \\
\text{subject to} & A^T y \geq c \\
& y \geq 0
\end{array}
$$

In this case, it is not difficult to prove that, if $\bar{x}$ and $\bar{y}$ are a feasible solution for the first and second problem, respectively, then $\bar{x}$ and $\bar{y}$ are optimal solutions if and only if

$$
\langle \bar{y}, b - A\bar{x} \rangle = 0
$$
$$
\langle A^T \bar{y} - c, \bar{x} \rangle = 0
$$

To this end, let us first show that these two problems are indeed one the dual of the other. In fact, augmenting the matrix A to include the constraint $x \geq 0$ in it, and using the transformation shown in Proposition III(2.7) in the dual problem, it is easy to check that the constraints of the two problems can be written in primal and dual standard form as

$$
\begin{bmatrix} A \\ -I \end{bmatrix} x \leq \begin{bmatrix} b \\ 0 \end{bmatrix}
\qquad \text{and} \qquad
\begin{cases} A^T y - z = c \\ y, z \geq 0 \end{cases}
$$

where z is a suitable slack variable. By means of Theorem (4.7), $\bar{x}$ and $(\bar{y}, \bar{z})$ are optimal solutions of the primal and dual problems, respectively, if and only if

$$
\begin{bmatrix} \bar{y} \\ \bar{z} \end{bmatrix} \cdot \left\{ \begin{bmatrix} b \\ 0 \end{bmatrix} - \begin{bmatrix} A \\ -I \end{bmatrix} \bar{x} \right\} = \begin{bmatrix} \bar{y} \\ \bar{z} \end{bmatrix} \cdot \begin{bmatrix} b - A\bar{x} \\ \bar{x} \end{bmatrix} = \langle \bar{y}, b - A\bar{x} \rangle + \langle \bar{z}, \bar{x} \rangle = 0.
$$

Since the two terms in the previous equality are both positive, this yields $\langle \bar{y}, b - A\bar{x} \rangle = \langle \bar{z}, \bar{x} \rangle = 0$, and from the fact that $z = A^T \bar{y} - c$, it finally follows that

$$
\langle \bar{y}, b - A\bar{x} \rangle = \langle A^T \bar{y} - \bar{c}, \bar{x} \rangle = 0.
$$

Theorem (4.7) may need some explanation. At first glance, it may even sound like a totally useless result, because, given a linear program, and calculated its dual, it does not give any clue on which feasible solution might be optimal. All it does is, if someone manages to get hold of a feasible solution, $\bar{x}$, for the primal linear program, and a feasible solution, $\bar{y}$, for the dual linear program, to establish whether these two solutions are optimal solutions for the two problems. Now, we already know that the one of the optimal solutions must lie on the vertices of the feasible region, so at least the number of possibilities is not infinite, but even in this case, for a general linear problem, a feasible region might have a huge number of vertices, and, more importantly, we do not have (yet), a procedure to find these vertices. This is where the algebra of linear programming really kicks, as the following section aims to unfold.

5. Algebra of linear programming

In this section we will take all the important results found thus far, and we shall put them together with the algebraic structure of a linear program, to develop an algorithm to find a vertex of the feasible regions of the primal and the dual problem (at the same time). This shall be done both for the primal and for the dual problem, because, depending on the type of practical problem at hand, it may be convenient to utilise the primal standard form, or the dual standard form. Practically, the algebra of linear programming consists in exploiting the properties of matrices, and systems of simultaneous equations, and we shall see how simple operations like the algebra of matrices becomes an extremely powerful tool, especially when equipped with the Complementary Slackness Theorem.

To this end, let us start with a simple algebraic observation.

(5.1) Remark. Let $A \in \mathfrak{M}(m, n, \mathbb{R})$ be a matrix, and observe that if the rank of A is less than n, then the equation $Ax = 0$ has infinite non-zero solutions. Hence, by Theorem I(5.16), each of these solutions is a direction of lineality. So, if the rank of A is less than n, then $\mathrm{lin}(P) \neq \{0\}$. Hence, by means of Corollary I(8.8), polyhedron P does not have vertices. On the other hand, if the rank of A is n, then the only solution of $Ax = 0$ is the trivial one: $x = 0$. This yields $\mathrm{lin}(P) = \{0\}$ and therefore, by means of Corollary I(8.8), polyhedron P does have vertices. Furthermore, in this case, the matrix A contains an $n \times n$ non-singular sub-matrix.

Furthermore, throughout this section, we shall always assume that the polyhedron with inequality $Ax \leq b$, for a certain $b \in \mathbb{R}^m$, is non-empty. We shall see, in Chapter VIII, section 6, how it is possible to check whether the polyhedron is non-empty.

This simple observation leads us to make the blanket assumption that the matrix A is of rank n. In order to describe the algebraic characterisation of the vertices of a polyhedron, it is convenient to introduce, one and for all, a notation that shall be used systematically in the sequel.

(5.2) Notation. If A is a $m \times n$ matrix and J is a subset of $\{1, \ldots, m\}$, we shall denote by A_J the matrix obtained selecting from A only the rows with indices in J. (This same notation can also be applied to vectors, for they are, by definition, special type of matrices with only one column.)

Now, consider a polyhedron P. We already know that such a polyhedron can be written in the form $P = \{x \in \mathbb{R}^n : Ax \leq b\}$ for some $A \in \mathfrak{M}(m, n, \mathbb{R})$ and $b \in \mathbb{R}^m$. The whole idea is that a vertex $\bar{x}$ of the polyhedron must lie on the boundary of P, hence the set $I(\bar{x}) = \{i : A_i\bar{x} = b_i\}$, called the set of the **active constraints** of $\bar{x}$, must be non-empty. In other words, this means that the point $\bar{x}$ can be found as the solution of a system of simultaneous equation $A_{I(\bar{x})} = b_{I(\bar{x})}$. This argument suggests that all vertices must be solutions of a system of simultaneous equations of the form $A_J x = b_J$, for some set of indices J (the 'active constraints' of that specific vertex). If we could prove such a statement, the problem of searching for the vertices would translate into the problem of identifying all possible sets J, such that $A_J x = b_J$ can be solved, and such that the solution is a vertex. This justifies the introduction of the following definition.

(5.3) Definition. Let $A \in \mathfrak{M}(m, n, \mathbb{R})$ be a matrix, where $m \geq n$ with rank n. A subset B of $\{1, \ldots, m\}$ with n elements shall be called a **basis** if the matrix A_B is non-singular. If B is a basis, the set, N of all indices in $\{1, \ldots, m\}$ that do not belong to B, that is to say, if $N = B^c$, the complementary set of B, shall be called the **non-basis** associated to B. Finally, the matrices A_B and A_N shall be called the **basic matrix** and **non-basic matrix**, respectively.

Let us see the consequences of this definition in the context of a linear program, or, more precisely, on a pair of primal/dual linear programs. To this end, consider a pair of primal and dual linear programs in standard form:

$$\begin{array}{ll} \text{maximise} & \langle c, x \rangle \\ \text{subject to} & Ax \leq b, \end{array} \quad \text{and} \quad \begin{array}{ll} \text{minimise} & \langle b, y \rangle \\ \text{subject to} & A^T y = c, \; y \geq 0, \end{array}$$

where $A \in \mathfrak{M}(m, n, \mathbb{R})$ is a matrix, $b \in \mathbb{R}^m$ is a m-vector and $c \in \mathbb{R}^n$ is a n-vector. A basis B induces a decomposition of A, y and b of the form

$$A = \begin{bmatrix} A_B \\ A_N \end{bmatrix}, \qquad y = \begin{bmatrix} y_B \\ y_N \end{bmatrix}, \qquad b = \begin{bmatrix} b_B \\ b_N \end{bmatrix}.$$

As a result of this, given a basis B, the vector inequality $Ax \leq b$ can be split as a system of the following two vector inequalities

$$A_B x \leq b_B, \qquad A_N x \leq b_N,$$

where A_B is a $n \times n$ non-singular matrix. Furthermore, the product $A^T y$ can be written as

$$A^T y = A_B^T y_B + A_N^T y_N.$$

As we can see, a basis naturally splits the m-space into two parts, one given by the basis, B, where the matrix A_B is non-singular, whence invertible, which is going to be utilised to find potential solutions, and a second part, N, which is going to be utilised to establish whether these potential solutions are indeed feasible or not. This idea is formalised in the definition below.

(5.4) Definition. Let $A \in \mathfrak{M}(m, n, \mathbb{R})$ is a matrix, $b \in \mathbb{R}^m$ is a m-vector and $c \in \mathbb{R}^n$ is a n-vector, and consider the pair of primal and dual linear programs

$$\begin{array}{ll} \text{maximise} & \langle c, x \rangle \\ \text{subject to} & Ax \leq b \end{array} \quad \text{and} \quad \begin{array}{ll} \text{minimise} & \langle b, y \rangle \\ \text{subject to} & A^T y = c, \; y \geq 0 \end{array}$$

The n-vector

$$\bar{x} = A_B^{-1} b_B,$$

solution of the system $A_B x = b_B$, shall called a **primal basic solution**. Similarly, the vector

$$\bar{y} = \begin{bmatrix} \bar{y}_B \\ \bar{y}_N \end{bmatrix} \qquad \text{where} \quad \bar{y}_B = (A_B^T)^{-1} c \quad \text{and} \quad \bar{y}_N = 0$$

shall be called a **dual basic solution**. Furthermore, the pair $\bar{x}$ and $\bar{y}$ of basic solutions relative to the same basis B is called a pair of **basic complementary solutions**.

For the primal basic solution, if, in addition, $b_N - A_N \bar{x} \geq 0$, $\bar{x}$ shall be called a **primal basic feasible solution**. Otherwise, it shall be called a **primal basic non-feasible solution**. If, regardless on whether or not $\bar{x}$ is feasible or non-feasible, there exists an index i in N, such that $A_i \bar{x} = b_i$, we shall say that the $\bar{x}$ is **degenerate**. If $\bar{x}$ is not degenerate, we shall say that it is **non-degenerate**.

Similarly, for the dual basic solution, if, in addition, $\bar{y}_B \geq 0$, $\bar{y}$ shall be called a **dual basic feasible solution**. Otherwise, it shall be called a **dual basic non-feasible solution**. If, regardless on whether or not $\bar{y}$ is feasible or non-feasible, there exists an index i in B, such that $\bar{y}_i = 0$, we shall say that the $\bar{y}$ is **degenerate**. If $\bar{y}$ is not degenerate, we shall say that it is **non-degenerate**.

In order to try to highlight the pattern behind these definitions, it may be convenient to summarise the content of the previous definition in a table:

	Primal solution, $\bar{x}$	**Dual solution, $\bar{y}$**
Feasible	$\forall i \in N,\ A_i \bar{x} \leq b_i$	$\forall i \in B,\ \bar{y}_i \geq 0$
Non-feasible	$\exists i \in N : A_i \bar{x} > b_i$	$\exists i \in B : \bar{y}_i < 0$
Non-degenerate	$\forall i \in N,\ A_i \bar{x} \neq b_i$	$\forall i \in B,\ \bar{y}_i > 0$
Degenerate	$\exists i \in N : A_i \bar{x} = b_i$	$\exists i \in B : \bar{y}_i = 0$

Even with the aid of the table, the whole definition may still look a bit obscure. Let's dive into a numerical example that ought to clarify the whole process.

(5.5) Example. Consider the constraints

$$\begin{aligned} \text{maximise} \quad & x_1 + 2x_2 \\ \text{subject to} \quad & -x_1 + x_2 \leq 2 \\ & 2x_1 + x_2 \leq 6 \\ & x_1,\ x_2 \geq 0 \end{aligned}$$

The corresponding feasible region is shown in Figure 5.1. In matrix form, the

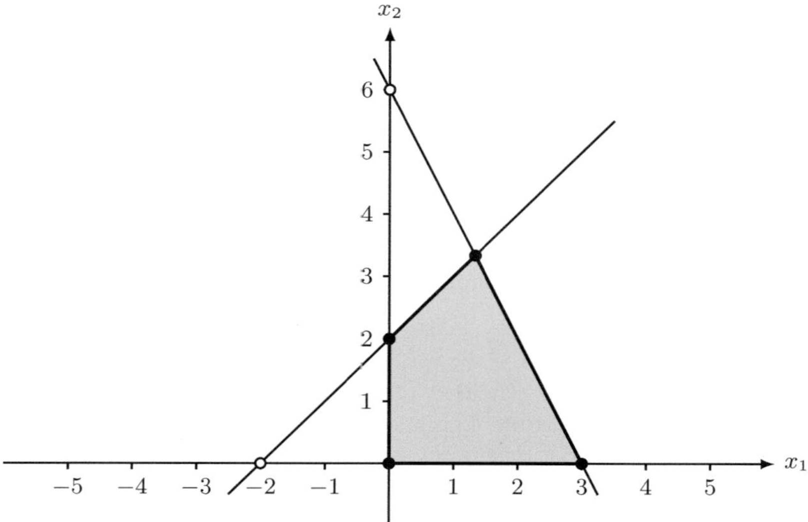

Figure 5.1. The diagram in figure shows the feasible region with constraints $-x_1 + x_2 \leq 2$, $2x_1 + x_2 \leq 6$ and $x_1, x_2 \geq 0$. The solid points denote the basic feasible solutions, and the hollow points denote the basic non-feasible solutions.

problem can be written as to maximise $\langle c, x \rangle$ subject to $Ax \leq b$, where

$$A = \begin{bmatrix} -1 & 1 \\ 2 & 1 \\ -1 & 0 \\ 0 & -1 \end{bmatrix}, \qquad b = \begin{bmatrix} 2 \\ 6 \\ 0 \\ 0 \end{bmatrix}, \qquad c = \begin{bmatrix} 1 \\ 2 \end{bmatrix}.$$

Let us first calculate the dual:

$$\begin{aligned} \text{minimise} \quad & 2y_1 + 6y_2 \\ \text{subject to} \quad & -y_1 + 2y_2 - y_3 = 1 \\ & y_1 + y_2 - y_4 = 2 \\ & y_1, y_2, y_3, y_4 \geq 0 \end{aligned}$$

Recall that a basis is a subset of the row indices such that the corresponding matrix is non-singular. Since a non-singular matrix is square, any basis has two elements. Let us therefore list all the possible combinations of two distinct indices, and for all pairs B such that A_B is non-singular, we can calculate the corresponding basic solutions. Let's take all possible bases and use them to find all possible basic solutions for both the primal and the dual programs.

Basis $B = \{1, 2\}$. Then

$$A_B = \begin{bmatrix} -1 & 1 \\ 2 & 1 \end{bmatrix}, \qquad A_N = \begin{bmatrix} -1 & 0 \\ 0 & -1 \end{bmatrix}, \qquad b_B = \begin{bmatrix} 2 \\ 6 \end{bmatrix}, \qquad b_N = \begin{bmatrix} 0 \\ 0 \end{bmatrix}.$$

Correspondingly, the primal basic solution is

$$\bar{x} = \begin{bmatrix} -1 & 1 \\ 2 & 1 \end{bmatrix}^{-1} \begin{bmatrix} 2 \\ 6 \end{bmatrix} = \begin{bmatrix} 4/3 \\ 10/3 \end{bmatrix}.$$

This means that a primal basic solution is the point $(4/3, 10/3)$. To see whether this solution is feasible or not, let us plug $\bar{x}$ in the expression $b_N - A_N\bar{x}$ and check whether all coordinates are positive.

$$\begin{bmatrix} 0 \\ 0 \end{bmatrix} - \begin{bmatrix} -1 & 0 \\ 0 & -1 \end{bmatrix} \begin{bmatrix} 4/3 \\ 10/3 \end{bmatrix} = \begin{bmatrix} 4/3 \\ 10/3 \end{bmatrix}$$

which shows that $(4/3, 10/3)$ is indeed a primal basic feasible solution.

Let us check the corresponding dual basic solution. We have:

$$\bar{y}_B^T = \begin{bmatrix} 1 & 2 \end{bmatrix} \begin{bmatrix} -1 & 1 \\ 2 & 1 \end{bmatrix}^{-1} = \begin{bmatrix} 1 & 1 \end{bmatrix}.$$

This means that the dual basic solution is the point $(1, 1, 0, 0)$. This is also a feasible solution, because all coefficients in the vector $\bar{y}_B = (1, 1)$ are positive.

Basis $B = \{1, 3\}$. Then

$$A_B = \begin{bmatrix} -1 & 1 \\ -1 & 0 \end{bmatrix}, \quad A_N = \begin{bmatrix} 2 & 1 \\ 0 & -1 \end{bmatrix}, \quad b_B = \begin{bmatrix} 2 \\ 0 \end{bmatrix}, \quad b_N = \begin{bmatrix} 6 \\ 0 \end{bmatrix}.$$

Correspondingly, the primal basic solution is

$$\bar{x} = \begin{bmatrix} -1 & 1 \\ -1 & 0 \end{bmatrix}^{-1} \begin{bmatrix} 2 \\ 0 \end{bmatrix} = \begin{bmatrix} 0 \\ 2 \end{bmatrix}.$$

This means that a primal basic solution is the point $(0, 2)$. Let us check whether this solution is feasible:

$$\begin{bmatrix} 6 \\ 0 \end{bmatrix} - \begin{bmatrix} 2 & 1 \\ 0 & -1 \end{bmatrix} \begin{bmatrix} 0 \\ 2 \end{bmatrix} = \begin{bmatrix} 4 \\ 2 \end{bmatrix}$$

which shows that $(0, 2)$ is indeed a primal basic feasible solution.

Let us check the corresponding dual basic solution. We have:

$$\bar{y}_B^T = \begin{bmatrix} 1 & 2 \end{bmatrix} \begin{bmatrix} -1 & 1 \\ -1 & 0 \end{bmatrix}^{-1} = \begin{bmatrix} 2 & -3 \end{bmatrix}.$$

This means that the dual basic solution is the point $(2, 0, -3, 0)$. This is not a feasible solution, because one of the coefficients of the vector $\bar{y}_B$ is negative.

Basis $B = \{1, 4\}$**.** Then

$$A_B = \begin{bmatrix} -1 & 1 \\ 0 & -1 \end{bmatrix}, \quad A_N = \begin{bmatrix} 2 & 1 \\ -1 & 0 \end{bmatrix}, \quad b_B = \begin{bmatrix} 2 \\ 0 \end{bmatrix}, \quad b_N = \begin{bmatrix} 6 \\ 0 \end{bmatrix}.$$

Correspondingly, the primal basic solution is

$$\bar{x} = \begin{bmatrix} -1 & 1 \\ 0 & -1 \end{bmatrix}^{-1} \begin{bmatrix} 2 \\ 0 \end{bmatrix} = \begin{bmatrix} -2 \\ 0 \end{bmatrix}.$$

This means that a primal basic solution is the point $(-2, 0)$. Let us check whether the solution is feasible:

$$\begin{bmatrix} 6 \\ 0 \end{bmatrix} - \begin{bmatrix} 2 & 1 \\ -1 & 0 \end{bmatrix} \begin{bmatrix} -2 \\ 0 \end{bmatrix} = \begin{bmatrix} 8 \\ -2 \end{bmatrix}$$

Since the second coordinate of this vector is negative, the primal basic solution $(-2, 0)$ is not feasible.

Let us check the corresponding dual basic solution. We have:

$$\bar{y}_B^T = \begin{bmatrix} 1 & 2 \end{bmatrix} \begin{bmatrix} -1 & 1 \\ 0 & -1 \end{bmatrix}^{-1} = \begin{bmatrix} -1 & -3 \end{bmatrix}.$$

This means that a dual basic solution is $(-1, 0, 0, -3)$. This is not a feasible solution, because the coefficients of the vector $\bar{y}_B$ are both negative.

Basis $B = \{2, 3\}$**.** Then

$$A_B = \begin{bmatrix} 2 & 1 \\ -1 & 0 \end{bmatrix}, \quad A_N = \begin{bmatrix} -1 & 1 \\ 0 & -1 \end{bmatrix}, \quad b_B = \begin{bmatrix} 6 \\ 0 \end{bmatrix}, \quad b_N = \begin{bmatrix} 2 \\ 0 \end{bmatrix}.$$

Correspondingly, the primal basic solution is

$$\bar{x} = \begin{bmatrix} 2 & 1 \\ -1 & 0 \end{bmatrix}^{-1} \begin{bmatrix} 6 \\ 0 \end{bmatrix} = \begin{bmatrix} 0 \\ 6 \end{bmatrix}.$$

This means that a primal basic solution is the point $(0, 6)$. Let us check whether the solution is feasible:

$$\begin{bmatrix} 2 \\ 0 \end{bmatrix} - \begin{bmatrix} -1 & 1 \\ 0 & -1 \end{bmatrix} \begin{bmatrix} 0 \\ 6 \end{bmatrix} = \begin{bmatrix} -4 \\ 6 \end{bmatrix}$$

Since the first coordinate of this vector is negative, the basic solution $(0, 6)$ is not feasible.

Let us check the corresponding dual basic solution. We have:

$$\bar{y}_B^T = \begin{bmatrix} 1 & 2 \end{bmatrix} \begin{bmatrix} 2 & 1 \\ -1 & 0 \end{bmatrix}^{-1} = \begin{bmatrix} 2 & 3 \end{bmatrix}.$$

This means that a dual basic solution is $(0, 2, 3, 0)$. This is indeed a dual basic feasible solution, because all coefficients of vector $\bar{y}_B$ are positive.

Basis $B = \{2, 4\}$**.** Then

$$A_B = \begin{bmatrix} 2 & 1 \\ 0 & -1 \end{bmatrix}, \quad A_N = \begin{bmatrix} -1 & 1 \\ -1 & 0 \end{bmatrix}, \quad b_B = \begin{bmatrix} 6 \\ 0 \end{bmatrix}, \quad b_N = \begin{bmatrix} 2 \\ 0 \end{bmatrix}.$$

Correspondingly, the primal basic solution is

$$\bar{x} = \begin{bmatrix} 2 & 1 \\ 0 & -1 \end{bmatrix}^{-1} \begin{bmatrix} 6 \\ 0 \end{bmatrix} = \begin{bmatrix} 3 \\ 0 \end{bmatrix}.$$

This means that a primal basic solution is the point $(3, 0)$. Let us check whether the solution is feasible:

$$\begin{bmatrix} 2 \\ 0 \end{bmatrix} - \begin{bmatrix} -1 & 1 \\ -1 & 0 \end{bmatrix} \begin{bmatrix} 3 \\ 0 \end{bmatrix} = \begin{bmatrix} 5 \\ 3 \end{bmatrix}$$

which shows that $(3, 0)$ is indeed a basic feasible solution.

Let us check the corresponding dual basic solution. We have:

$$\bar{y}_B^T = \begin{bmatrix} 1 & 2 \end{bmatrix} \begin{bmatrix} 2 & 1 \\ 0 & -1 \end{bmatrix}^{-1} = \begin{bmatrix} 1/2 & -3/2 \end{bmatrix}.$$

This means that a dual basic solution is $(0, 1/2, 0, -3/2)$. This is not a feasible solution, because the second coefficients of the vector $\bar{y}_B$ is negative.

Basis $B = \{3, 4\}$**.** This case is trivial, because b_B is zero, and therefore the basic solution is $(0, 0)$, which is clearly feasible. As regard to the corresponding dual basic solution, we have

$$\bar{y}_B^T = \begin{bmatrix} 1 & 2 \end{bmatrix} \begin{bmatrix} -1 & 0 \\ 0 & -1 \end{bmatrix}^{-1} = \begin{bmatrix} -1 & -2 \end{bmatrix}.$$

This shows that the corresponding dual basic solution is $(0, 0, -1, -2)$ which is not feasible, because both coefficients of vector $\bar{y}_B$ are negative.

To sum up this long example, we have shown that the primal basic feasible solutions are $(0, 0)$, $(3, 0)$, $(4/3, 10/3)$ and $(0, 2)$, that is exactly the vertices of the feasible region, as we can see from the diagram in Figure 5.1. In order to get a graphical picture of the dual basis, let us observe that $c = A_B^T \bar{y}_B = \sum_{i \in B} A_i \bar{y}_i$, that is, vector c is a linear combination of the rows of matrix A whose indices belong to the basis B, and the coefficients of this linear combination are the components of vector $\bar{y}_B$. The dual solution is feasible if and only if the components of this vector are all positive, i.e. if and only if vector c belongs to the cone generated by the rows of matrix A with index in the basis B. This gives a graphical representation of the dual basis as shown in Figure 5.2.

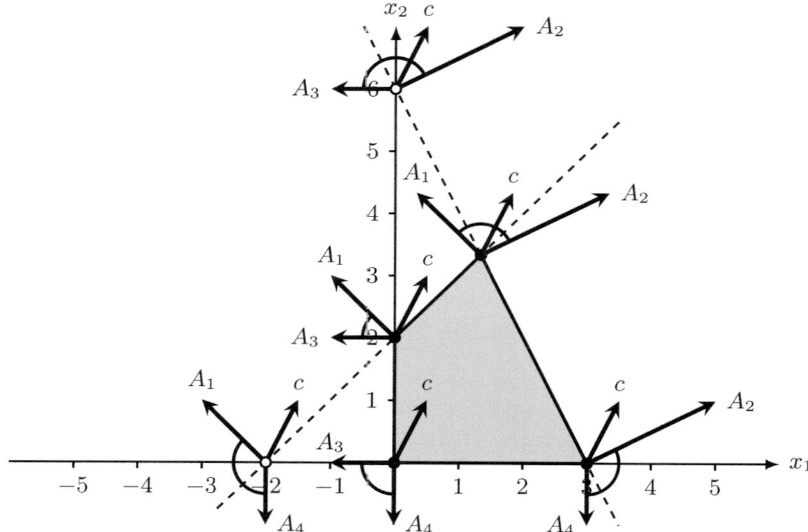

Figure 5.2. The diagram shows the region defined by the constraints $-x_1 + x_2 \leq 2$, $2x_1 + x_2 \leq 6$ and $x_1, x_2 \geq 0$. Each primal basic feasible solution is represented with a solid circle, and each primal basic non-feasible solution is represented with a hollow circle. At each primal basic solution the vector $c = [1,2]^T$ is represented and the row of the corresponding matrix A_B. The arc shows the region of the cone generated by these vectors. Notice that there are only two points, $(0,6)$ and $(4/3, 10/3)$ where vector c lies in the cone generated by the rows of A_B. Indeed, these are the only two dual basic feasible solutions.

The previous example contains a wealth of geometrical insights. In particular, the first thing that should stand out is that the primal basic feasible solutions correspond to the vertices of the feasible region. This is also quite intuitive, because the vertices of the feasible region coincide with the intersection of the faces of the polyhedron, which are reached when a constraint gets to its extreme values. This intuitive idea is formalised in the next two theorems, which offers an algebraic characterisation of the vertices of a polyhedron expressed in primal and dual standard form, respectively, as the set of all primal and dual basic feasible solutions, respectively.

(5.6) Theorem. *Let $P = \{x \in \mathbb{R}^n : Ax \leq b\}$ be a polyhedron. Then $\bar{x} \in P$ is a vertex if and only if $\bar{x}$ is a primal basic feasible solution.*

Proof. ($\Rightarrow$): Assume that $\bar{x}$ is a vertex of P and consider the set of all 'active constraints' at $\bar{x}$, which means the set

$$I(\bar{x}) = \left\{ i : A_i \bar{x} = b_i \right\}.$$

To prove that $\bar{x}$ is a primal basic feasible solution, it suffices to prove that the rank of $A_{I(\bar{x})}$ is n. By contradiction, let us suppose that the rank of this matrix is less than n. Then, the columns of matrix $A_{I(\bar{x})}$ are linearly dependent, i.e. there exists a vector $d \neq 0$ such that $A_{I(\bar{x})}d = 0$. Consider the two vectors

$$z^+ = \bar{x} + \lambda d, \qquad\qquad z^- = \bar{x} - \lambda d,$$

where λ is a positive scalar. We have that $z^+ \neq z^-$ and we can also suppose that $z^+, z^- \in P$, providing to take λ sufficiently small. In fact, if $i \in I(\bar{x})$, $A_i z^\pm = A_i \bar{x} \pm \lambda A_i d = A_i \bar{x} = b_i$, and if $i \notin I(\bar{x})$, then $A_i \bar{x} < b_i$, hence $A_i z^\pm = A_i \bar{x} \pm \lambda A_i d < b_i \pm \lambda A_i d < b_i$, for λ sufficiently small. On the other hand, if $z^+, z^- \in P$, then $\bar{x} = \frac{1}{2}z^+ + \frac{1}{2}z^-$ contradicts the hypothesis that $\bar{x}$ is a vertex of P.

($\Leftarrow$): Let us now assume that $\bar{x}$ is a primal basic feasible solution relative to the basis B. In other words, suppose that $\bar{x} = A_B^{-1}b_B$ and that $b_N - A_N \bar{x} \geq 0$. By contradiction, let us assume that $\bar{x}$ is not a vertex, i.e. that there exist in P two distinct points z and w such that $\bar{x} = \frac{1}{2}z + \frac{1}{2}w$. For every index $i \in B$, we have $b_i = A_i \bar{x} = \frac{1}{2}A_i z + \frac{1}{2}A_i w \leq \frac{1}{2}b_i + \frac{1}{2}b_i = b_i$, so $A_i z = A_i w = b_i$. In other words, it is proved that $A_B z = A_B w = b_B$ and therefore, by means of the fact that A_B is non-singular, $z = w = A_B^{-1}b_B = \bar{x}$, which contradicts the fact that $z \neq w$. $\qquad\square$

A similar result holds for the dual problem, where the dual basic feasible solutions correspond to the vertices of the dual feasible region.

(5.7) Theorem. *Let $D = \{y \in \mathbb{R}^m : A^T y = b, \, y \geq 0\}$ be a polyhedron. Then $\bar{y} \in D$ is a vertex if and only if $\bar{y}$ is a dual basic feasible solution.*

Proof. ($\Rightarrow$): Assume that $\bar{y}$ is a vertex of D and consider the set

$$J(\bar{y}) = \left\{ j : \bar{y}_j > 0 \right\}.$$

To prove that $\bar{y}$ is a dual basic feasible solution, it suffices to prove that the the rows A_j with index j in $J(\bar{y})$ are linearly independent. By contradiction, let us suppose that these rows are linearly dependent, whence there exists a vector $d \neq 0$ such that $A_{J(\bar{y})}^T d = 0$. Consider the two vectors z^+ and z^- whose components are, for every index $j = 1, \ldots, m$,

$$z_j^\pm = \begin{cases} \bar{y}_j \pm \lambda d_j & \text{if } j \in J(\bar{y}), \\ 0 & \text{if } j \notin J(\bar{y}), \end{cases}$$

where λ is a positive scalar. We have that $z^+ \neq z^-$. Thanks to the fact that $\bar{y} \geq 0$, for λ sufficiently small, one has $z^+ \geq 0$ and $z^- \geq 0$. In addition,

$$\begin{aligned} A^T z^\pm = A_{J(\bar{y})}^T z_{J(\bar{y})}^\pm &= A_{J(\bar{y})}^T \left(\bar{y}_{J(\bar{y})} \pm \lambda d_{J(\bar{y})} \right) \\ &= A_{J(\bar{y})}^T \bar{y}_{J(\bar{y})} \pm \lambda A_{J(\bar{y})}^T \bar{y} \\ &= A_{J(\bar{y})}^T \bar{y}_{J(\bar{y})} \\ &= c. \end{aligned}$$

Thus $z^+, z^- \in D$, whence $\bar{y} = \frac{1}{2}z^+ + \frac{1}{2}z^-$, which contradicts the hypothesis that $\bar{y}$ is a vertex of D.

($\Leftarrow$): Let us now assume that $\bar{y}$ is a dual basic feasible solution relative to the basis B. In other words, suppose that $\bar{y}_B = (A_B^{-1})^T c \geq 0$ and $\bar{y}_N = 0$. By contradiction, let us assume that $\bar{y}$ is not a vertex, that is to say, that there exist in D two distinct points z and w such that $\bar{y} = \frac{1}{2}z + \frac{1}{2}w$. Since $z_N \geq 0$ and $w_N \geq 0$, and $\frac{1}{2}z_N + \frac{1}{2}w_N = 0$, we have that $z_N = w_N = 0$. Furthermore, since $z, w \in D$, then $c = A^T z = A_B^T z_B$ and $c = A^T w = A_B^T w_B$. In other words, it is proved that $A_B^T z_B = A_B^T w_B = c$ and therefore, by means of the fact that A_B is non-singular, $z_B = w_B = (A_B^T)^{-1} c = \bar{y}_B$, which contradicts the fact that $z \neq w$. $\qquad\square$

To fully unveil this connection between the primal and dual basic solutions, let us prove the following important result.

(5.8) Proposition. *Let $A \in \mathfrak{M}(m, n, \mathbb{R})$ be a matrix, $b \in \mathbb{R}^m$ a m-vector and $c \in \mathbb{R}^n$ a n-vector, consider the pair of primal and dual linear programming problems*

$$(\mathcal{P}) : \begin{cases} \text{maximise} & \langle c, x \rangle \\ \text{subject to} & Ax \leq b \end{cases} \qquad (\mathcal{D}) : \begin{cases} \text{minimise} & \langle b, y \rangle \\ \text{subject to} & A^T y = c, \ y \geq 0 \end{cases}$$

and let $\bar{x}$ and $\bar{y}$ be a pair of basic complementary solutions. Then, $\bar{x}$ and $\bar{y}$ are in complementary slackness, which means that $\langle \bar{y}, b - A\bar{x} \rangle = 0$.

Proof. Let B be a basis, and let $N = B^c$ be the complement of the basis. Using the decomposition

$$A = \begin{bmatrix} A_B \\ A_N \end{bmatrix}, \qquad y = \begin{bmatrix} y_B \\ y_N \end{bmatrix}, \qquad b = \begin{bmatrix} b_B \\ b_N \end{bmatrix},$$

and by means of Definition (5.4), we have

$$\begin{aligned} \langle \bar{y}, b - A\bar{x} \rangle &= \langle \bar{y}_B, b_B - A_B \bar{x} \rangle + \langle \bar{y}_N, b_N - A_N \bar{x} \rangle \\ &= \langle c, A_B^{-1}(b_B - A_B A_B^{-1} b_B) \rangle + \langle 0, b_N - A_N \bar{x} \rangle \\ &= \langle c, A_B^{-1}(b_B - b_B) \rangle \\ &= 0 \end{aligned}$$

and the result is fully proved. $\qquad\square$

We have now all the ingredients to prove the desired criterion of optimality.

(5.9) Theorem (Optimality criterion for basic solutions). *In the same hypotheses as in Proposition (5.8), if $\bar{x}$ and $\bar{y}$ are both feasible, then $\bar{x}$ is an optimal solution for $(\mathcal{P})$, and $\bar{y}$ is an optimal solution for $(\mathcal{D})$.*

Proof. By virtue of Proposition (5.8), the two solutions $\bar{x}$ and $\bar{y}$ satisfy the relation $\langle \bar{y}, b - A\bar{x} \rangle = 0$. Thus, by virtue of Theorem (4.7), this means that $\bar{x}$ is an optimal solution for $(\mathcal{P})$, and $\bar{y}$ is an optimal solution for $(\mathcal{D})$. $\hspace{1cm}$ $\square$

Utilising this criterion, we now know that, given a linear program (that can always be written in primal standard form) to maximise the function $\langle c, x \rangle$ subject to the constraint $Ax \leq b$, we now know that an optimal solution lies on one of its vertices, and each vertex is a primal basic feasible solution. To find the basic solutions, we can take a basis, B, and calculate the corresponding basic complementary solutions, $\bar{x}$ and $\bar{y}$. If both these basic solutions are feasible, then we have reached the optimum, otherwise, we can change the basis, and repeat the same operation. This algorithm to find the optimal value of a linear program is pretty simple and neat, but still it leaves an important question, that will be answered in Chapter VIII: is there an efficient way to select the basis?

Before completing the theory of linear programming, it is important to notice that Theorem (5.9) gives a sufficient condition for optimality. This condition is necessary only if both the primal and dual basic solutions are non-degenerate. Indeed, if one of the two solutions is degenerate, it may happen that different bases give the same vertex and, depending on the basis chosen, one may find a pair of basic complementary solutions such that the primal basic solution is feasible, the dual basic solution is non-feasible, but the primal basic solution is optimal. The following example should clarify this crucial point.

(5.10) Example. Consider the linear program

$$
\begin{aligned}
\text{maximise} \quad & x_1 + 2x_2 \\
\text{subject to} \quad & x_1 + x_2 \leq 2 \\
& x_1 \leq 1 \\
& x_2 \leq 1 \\
& x_1,\, x_2 \geq 0
\end{aligned}
$$

The corresponding feasible region is shown in Figure 5.3. This problem can be written in standard form, setting

$$
A = \begin{bmatrix} 1 & 1 \\ 1 & 0 \\ 0 & 1 \\ -1 & 0 \\ 0 & -1 \end{bmatrix}, \qquad b = \begin{bmatrix} 2 \\ 1 \\ 1 \\ 0 \\ 0 \end{bmatrix}, \qquad c = \begin{bmatrix} 1 \\ 2 \end{bmatrix}.
$$

Let us start by observing that there are a few subsets of the set of all row indices, $\{1, 2, 3, 4, 5\}$, who do not form a basis. Precisely, $B = \{2, 4\}$ and $B = \{3, 5\}$ are not bases, because the corresponding matrices A_B are singular. Taking a look at Figure 5.3 again, we can easily verify that these indices

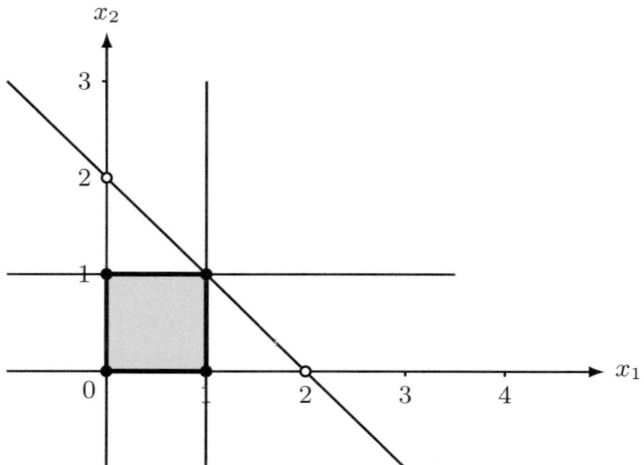

Figure 5.3. The diagram shows the feasible region of the linear program in Example (5.10). As we can see from the diagram, the vertex with coordinates $(1, 1)$ can be described as the intersection of two out of three constrains: $x_1 = 1$, $x_2 = 1$ and $x_1 + x_2 = 2$. Using the geometric argument, it is easy to check that this is the optimal solution of the linear program.

correspond to constraints that are parallel, and therefore have no intersections. Overall, looking at the diagram in Figure 5.3, we see that there are four primal basic feasible solutions, $(0, 0)$, $(1, 0)$, $(0, 1)$ and $(1, 1)$, and two primal basic non-feasible solutions, $(2, 0)$ and $(0, 2)$.

Let us consider the three bases $B_1 = \{1, 2\}$, $B_2 = \{1, 3\}$ and $B_3 = \{2, 3\}$ and let us calculate the corresponding pair of basic complementary solutions.

Basis $B_1 = \{1, 2\}$. We have

$$A_{B_1} = \begin{bmatrix} 1 & 1 \\ 1 & 0 \end{bmatrix}, \quad A_{N_1} = \begin{bmatrix} 0 & 1 \\ -1 & 0 \\ 0 & -1 \end{bmatrix}, \quad b_{B_1} = \begin{bmatrix} 2 \\ 1 \end{bmatrix}, \quad b_{N_1} = \begin{bmatrix} 1 \\ 0 \\ 0 \end{bmatrix}.$$

Correspondingly, the primal basic solution is:

$$\bar{x} = \begin{bmatrix} 1 & 1 \\ 1 & 0 \end{bmatrix}^{-1} \begin{bmatrix} 2 \\ 1 \end{bmatrix} = \begin{bmatrix} 1 \\ 1 \end{bmatrix}$$

This means that $(1, 1)$ is the primal basic solution corresponding to the basis B_1. Let us check whether this solution is feasible:

$$\begin{bmatrix} 1 \\ 0 \\ 0 \end{bmatrix} - \begin{bmatrix} 0 & 1 \\ -1 & 0 \\ 0 & -1 \end{bmatrix} \begin{bmatrix} 1 \\ 1 \end{bmatrix} = \begin{bmatrix} 0 \\ 1 \\ 1 \end{bmatrix}.$$

Thus, $(1, 1)$ is a primal degenerate basic feasible solution.

Let us now check the complementary dual basic solution. We have

$$\bar{y}_{B_1}^T = \begin{bmatrix} 1 & 2 \end{bmatrix} \begin{bmatrix} 1 & 1 \\ 1 & 0 \end{bmatrix}^{-1} = \begin{bmatrix} 2 & -1 \end{bmatrix}$$

and this shows that complementary dual basic solution $(2, -1, 0, 0)$ is non-feasible.

Basis $B_2 = \{1, 3\}$. We have

$$A_{B_2} = \begin{bmatrix} 1 & 1 \\ 0 & 1 \end{bmatrix}, \quad A_{N_2} = \begin{bmatrix} 1 & 0 \\ -1 & 0 \\ 0 & -1 \end{bmatrix}, \quad b_{B_2} = \begin{bmatrix} 2 \\ 1 \end{bmatrix}, \quad b_{N_2} = \begin{bmatrix} 1 \\ 0 \\ 0 \end{bmatrix}.$$

Correspondingly, the primal basic solution is:

$$\bar{x} = \begin{bmatrix} 1 & 1 \\ 0 & 1 \end{bmatrix}^{-1} \begin{bmatrix} 2 \\ 1 \end{bmatrix} = \begin{bmatrix} 1 \\ 1 \end{bmatrix}$$

This means that $(1, 1)$ is the primal basic solution corresponding to the basis B_2. Let us check whether this solution is feasible:

$$\begin{bmatrix} 1 \\ 0 \\ 0 \end{bmatrix} - \begin{bmatrix} 1 & 0 \\ -1 & 0 \\ 0 & -1 \end{bmatrix} \begin{bmatrix} 1 \\ 1 \end{bmatrix} = \begin{bmatrix} 0 \\ 1 \\ 1 \end{bmatrix}.$$

Thus, $(1, 1)$ is a primal degenerate basic feasible solution.

Let us now check the complementary dual basic solution. We have

$$\bar{y}_{B_2}^T = \begin{bmatrix} 1 & 2 \end{bmatrix} \begin{bmatrix} 1 & 1 \\ 0 & 1 \end{bmatrix}^{-1} = \begin{bmatrix} 1 & 1 \end{bmatrix}$$

and this shows that complementary dual basic solution $(1, 0, 1, 0)$ is feasible.

Basis $B_3 = \{2, 3\}$. We have

$$A_{B_3} = \begin{bmatrix} 1 & 0 \\ 0 & 1 \end{bmatrix}, \quad A_{N_3} = \begin{bmatrix} 1 & 1 \\ -1 & 0 \\ 0 & -1 \end{bmatrix}, \quad b_{B_3} = \begin{bmatrix} 1 \\ 1 \end{bmatrix}, \quad b_{N_3} = \begin{bmatrix} 2 \\ 0 \\ 0 \end{bmatrix}.$$

Correspondingly, the primal basic solution is:

$$\bar{x} = \begin{bmatrix} 1 & 0 \\ 0 & 1 \end{bmatrix}^{-1} \begin{bmatrix} 1 \\ 1 \end{bmatrix} = \begin{bmatrix} 1 \\ 1 \end{bmatrix}$$

This means that $(1, 1)$ is the primal basic solution corresponding to the basis B_2. Let us check whether this solution is feasible:

$$\begin{bmatrix} 2 \\ 0 \\ 0 \end{bmatrix} - \begin{bmatrix} 1 & 1 \\ -1 & 0 \\ 0 & -1 \end{bmatrix} \begin{bmatrix} 1 \\ 1 \end{bmatrix} = \begin{bmatrix} 0 \\ 1 \\ 1 \end{bmatrix}.$$

Thus, $(1, 1)$ is a primal degenerate basic feasible solution.

Let us now check the complementary dual basic solution. We have

$$\bar{y}_{B_3}^T = \begin{bmatrix} 1 & 2 \end{bmatrix} \begin{bmatrix} 1 & 0 \\ 0 & 1 \end{bmatrix}^{-1} = \begin{bmatrix} 1 & 2 \end{bmatrix}$$

and this shows that complementary dual basic solution $(1, 0, 1, 0)$ is feasible.

As we can see, when a primal basic feasible solution is degenerate, the same vertex can be obtained with different bases. Furthermore, even though the point $(1, 1)$ is clearly the optimal solution, as we can see using the geometric argument, or applying the optimality criterion with the basis B_2 or B_3, it may also happen that one of these basis, in this case B_1, gives a dual basic solution that is not feasible. Incidentally, this also shows that the hypotheses of the Optimality Criterion for Basic Solutions (Theorem (5.9)) cannot be reversed.

Chapter V
Non-Linear Programming

In this chapter, we are going to explore the field of non-linear programming, that is to say, problems consisting in minimising or maximising an objective function subject to a finite number of equality and/or inequality constraints. We begin by introducing the foundational tools required for the analysis of non-linear programming problems, such as feasible, descent and tangent directions. Building on these, we will derive and examine the essential first-order and second-order optimality conditions that form the core of non-linear optimisation theory. Specifically, we will study the Fritz-John conditions and the Karush-Kuhn-Tucker (KKT) conditions, as well as constraint qualifications Additionally, we will investigate sufficient conditions for optimality and second-order conditions for optimality.

1. Feasible, descent and tangent directions

The main goal of non-linear programming is to efficiently locate points that minimise (or maximise) the objective function over a feasible region. Unlike linear programming, where the feasible region is just a polyhedron, the most general feasible region of a non-linear programming problem exhibits a complex structure, and the fact that the objective function need not be linear yields to potentially numerous local optimal solutions. To systematically guide the search towards improved solutions, we shall need a quantity that represents a direction along which the objective function's value is guaranteed to decrease (or increase). This is the notion of descent (or ascent) direction. As we shall see in detail in Chapter IX and Chapter X, by iteratively moving along such directions, well designed optimisation algorithms will then be able to navigate the feasible region, gradually approaching a local optimum.

To simplify future statements, in the rest of this chapter, unless otherwise specified, we shall make the blanket assumption that all functions involved are at least of class $\mathcal{C}^1$ on a non-empty open set containing the feasible region.

A. Carpignani and M. Pappalardo, *Theory and Methods of Optimisation*,
UNITEXT 175, https://doi.org/10.1007/978-3-032-01514-3_5

(1.1) Definition. Let D be a non-empty set in $\mathbb{R}^n$, $f(x)$ a real function defined on D and let x be a point in D. A vector d is said to be a **descent direction** of f at x if there exists a strictly positive real number τ such that $f(x + td) < f(x)$ for every real number t, with $0 \le t \le \tau$. The set of such vectors d, called the **cone of descent directions** of f at x, shall be denoted by $\mathcal{D}_f(x)$. In other words, the set of descent directions of f at x is

$$\mathcal{D}_f(x) = \left\{ d \in \mathbb{R}^n : f(x + td) < f(x) \text{ for all } t \in (0, \tau) \text{ for some } \tau > 0 \right\}.$$

Let us observe immediately that $\mathcal{D}_f(x)$ is a cone. Indeed, is d is a descent direction, and c is a strictly positive real number, then there exists a strictly positive real number τ, such that, $f(x + td) < f(x)$ for all $t \in (0, \tau)$, hence taking $\tau' = c^{-1}\tau$, one gets that

$$f\big(x + t'(cd)\big) = f\big(x + (t'c)d\big) < f(x) \qquad \text{for all } t' \in (0, \tau').$$

Descent directions are quite simple and intuitive, but only take into account the objective function. Thus, they tend to be particularly useful when dealing with unconstrained problems. Indeed, when we are dealing with constrained problems, it becomes essential to focus, as well as on the objective function, also on the constraint. The simplest way to achieve this is by studying feasible directions. Intuitively, these are directions that allow movement within the feasible region without violating any constraints. The collection of all such directions at a given point forms a cone, usually referred to as the 'cone of feasible directions', which provides insight into potential paths for improvement that adhere to the problem's constraints.

(1.2) Definition. Let D be a non-empty set in $\mathbb{R}^n$, and x a point in D. A vector d is said to be a **feasible direction** to D at x if there exists a strictly positive real number τ such that

$$x + td \in D \qquad \text{for every real number } t, \text{ with } 0 \le t \le \tau.$$

The set of such vectors d is called the **cone of feasible directions** of D at x and shall be denoted by $\mathcal{A}_D(x)$. In other words, the cone of feasible directions of D at x is the set

$$\mathcal{A}_D(x) = \left\{ d \in \mathbb{R}^n : x + td \in D \text{ for all } t \in (0, \tau) \text{ for some } \tau > 0 \right\}.$$

It is easy to show that $\mathcal{A}_D(x)$ is indeed a cone. In fact, if d is a feasible direction, then, for every strictly positive real number λ, choosing $\lambda^{-1}\tau$, it is straightforward to recognise that the vector λd is also a feasible direction.

The two definitions of feasible direction and descent direction already allow us to state a first very simple necessary condition for optimality, whose proof is just a direct consequence of the definitions of descent direction, feasible direction, and of local minimum.

(1.3) Proposition. *Let D be a non-empty set in $\mathbb{R}^n$, $f(x)$ a real function defined on D, and x^* be a point in D. If x^* is a local minimum of $f(x)$, then there exists no feasible direction at x^* which is also a descent direction for $f(x)$ at x^*.*

Proof. Assume that x^* is a local minimum of $f(x)$. If there were a feasible direction d which is also a descent direction, there would exist a strictly positive real number τ such that, for every strictly positive real number t smaller than τ,

$$x^* + td \in D, \qquad f(x^* + td) < f(x^*) \leq f(x^* + td)$$

where the last inequality follows from the fact that x^* is a local minimum. Thus, no feasible direction at x^* is also a descent direction for $f(x)$ at x^*. $\quad\square$

(1.4) Remark. The condition presented in the previous proposition leads to an interesting consequence. Specifically, let x^* be a point in D, and consider an arbitrary direction d. Define the real function

$$\phi(t) = f(x^* + td) \qquad \text{for all } t > 0.$$

Now, if we further assume that d is a descent direction of $f(x)$ at the point x^*, it follows from the definition of a descent direction that the function $\phi(t)$ must be strictly decreasing at $t = 0$. As a result of this simple observation, evaluating the derivative $\phi'(t) = \langle \nabla f(x^* + td), d \rangle$ at $t = 0$, gives

$$\phi'(0) = \langle \nabla f(x^*), d \rangle < 0.$$

However, if we assume additionally that x^* is a local minimum of f over the set D, Proposition (1.3) implies that such an inequality cannot hold for any feasible direction d. Consequently, we obtain the condition

$$\langle \nabla f(x^*), d \rangle \geq 0 \qquad \text{for all } d \in \mathcal{A}_D(x^*),$$

where $\mathcal{A}_D(x^*)$ denotes the set of feasible directions at x^* with respect to D.

Furthermore, if the set D is convex and contains at least two distinct points, any vector of the form $x - x^*$ with $x \in D$ is a feasible direction. In this case, the optimality condition above can be reformulated as

$$\langle \nabla f(x^*), x - x^* \rangle \geq 0 \qquad \text{for all } x \in D.$$

Finally, if the function f is convex in addition to the aforementioned assumptions, this condition becomes both necessary and sufficient for x^* to be not merely a local minimum but a global minimum of f over D thanks to the Theorem III(3.1).

It is worth noticing that, if one happens to have the cone $\mathcal{A}_D(x^*)$ of all feasible directions, Proposition (1.3) and even more so Remark (1.4) give a very simple criterion for optimality. However, there is quite a significant limit to this. Indeed, feasible directions need not exist. To see this, it suffices to consider the problem of minimising the objective function $f(x_1, x_2) = x_1$ subject to the constraint $x_1^2 + x_2^2 = 1$. The point $(-1, 0)$ is clearly the optimal solution, but no feasible directions can be identified at this point due to the curvature of the constraint.

Since feasible directions may or may not exist, depending on the local geometry of the feasible region, we must consider a different type of direction, which is more robust to the geometry and to the topology of the feasible region. Precisely, let us give the following definition.

(1.5) Definition. Let D be a non-empty set in $\mathbb{R}^n$, and x a point in D. A direction $d \in \mathbb{R}^n$ is said to be a **tangent** vector to D at x if there exists a sequence $\{z^k\}$ of points in D that converges towards x and a sequence $\{t_k\}$ of strictly positive real numbers, which converges to 0, such that

$$\lim_{k \to \infty} \frac{z^k - x}{t_k} = d.$$

The set of all tangent vectors to D at x is called the **cone of tangents** of D at x, and shall be denoted by $\mathcal{T}_D(x)$.

The phrase 'cone of tangents' tacitly assumes that this set be indeed a cone. To show that the set $\mathcal{T}_D(x)$ is a cone is quite a straightforward thing to do: in fact, if d is a tangent vector, and if $\{z^k\}$ and $\{t_k\}$ are the corresponding sequences, the first convergent towards x and the second convergent towards 0, then, by replacing t_k with $\alpha^{-1} t_k$, we obtain that the vector αd is also tangent to D at x, for every $\alpha > 0$. Furthermore, one can always consider the sequence $\{z^k\}$ to be the constant sequence $z^k = x$, which shows that 0 is a tangent vector at x.

The following result shows an equivalent description of a tangent vector.

(1.6) Proposition. *Let D be a non-empty set in $\mathbb{R}^n$, and x a point in D. A vector d is tangent to D at x if and only if there exists a sequence $\{d^k\}$ of points of D, convergent towards d, and a sequence $\{t_k\}$ of strictly positive real numbers, convergent towards 0, such that $x + t_k d^k \in D$.*

Proof. $(\Rightarrow)$: Let d be an element of $\mathcal{T}_D(x)$. By hypothesis, there exist a sequence $\{z^k\}$ in D convergent towards x, and a sequence $\{t_k\}$ of strictly positive real numbers convergent towards 0, such that $(z^k - x)/t_k \to d$ as $k \to \infty$. In other words, this limit can be rewritten saying that there exists a sequence $\{u^k\}$, convergent towards 0, such that $(z^k - x)/t_k = d + u^k$. Rearranging this expression gives $z^k = x + t_k(d + u^k) = x + t_k d^k$, where we have set $d^k = d + u^k$. Since D is assumed to be open, we may also assume that the vectors d^k lie in D for k sufficiently large. Extracting from $\{d^k\}$ a subsequence, we may as well assume that all points in $\{d^k\}$ lie in D.

($\Leftarrow$): Assume that there exists a sequence $\{d^k\}$ in D, and a sequence $\{t_k\}$ of strictly positive real numbers, the first convergent towards d and the second convergent towards 0, such that $x + t_k d^k \in D$ for every index k, and let us set $z^k = x + t_k d^k$. Then $\{z^k\}$ is a sequence of points of D, converges towards x, and we have $(x - z^k)/t_k = d^k \to d$. Thus, d is a tangent vector to D at the point x. $\qquad\square$

An interesting relationship between the cone of feasible direction and the cone of tangents is that the former is always included in the latter. In other words, every feasible direction is also a tangent direction. This is the content of the following result.

(1.7) Proposition. *Let D a non-empty set in $\mathbb{R}^n$, and let $x \in D$. Then,*

$$\mathcal{A}_D(x) \subseteq \mathcal{T}_D(x).$$

Proof. Let $d \in \mathcal{A}_D(x)$. By definition, there exists a strictly positive real number τ such that $x + td \in D$ for $t \in (0, \tau)$. Take any sequence $\{t_k\}$ in $(0, \tau)$ that converges towards 0 and set $z^k = x + t_k d$. This ensures, by construction, that $z^k \in D$ for every k, and it is also obvious that

$$\frac{z^k - x}{t_k} \to d \qquad \text{as } k \to \infty.$$

This proves that $d \in \mathcal{T}_D(x)$, whence $\mathcal{A}_D(x) \subseteq \mathcal{T}_D(x)$. $\qquad\square$

The previous preposition indicates that tangent directions generalise feasible directions. Furthermore, we have already mentioned that that the cone of tangents is never empty, because at least 0 always is a tangent vector. All these clues seem to indicate that the cone of tangents might be the most suited set to study criteria of optimality. While this is to some extent the truth, as the criteria of optimality that we shall study in the sequel heavily rely on the cone of tangents, in the general case of a non-empty set D of $\mathbb{R}^n$ this set of direction may present all sorts of complications. For instance, sometimes the cone of tangents may be only the vector 0. For example, take $D = \{x\}$, with $x \in \mathbb{R}^n$. Then $\mathcal{T}_D(x) = \{0\}$. Conversely, some other times, the cone of tangents may coincide with the entire $\mathbb{R}^n$, as shown in details below.

(1.8) Example. Let D be a non-empty subset of $\mathbb{R}^n$, and suppose that its interior, $\text{int}(D)$, is non-empty. If $x \in \text{int}(D)$, then the cone of tangents $\mathcal{T}_D(x)$ coincides with the entire space $\mathbb{R}^n$.

To see this, let us start by observing that the assumption $x \in \text{int}(D)$ implies the existence of an open ball $B(x, \varrho)$, centred at the point x, with radius $\varrho > 0$, such that $B(x, \varrho) \subseteq \text{int}(D)$. Now, consider any unit vector $u \in \mathbb{R}^n$. For any real number t satisfying $|t| < \varrho$, the point $d = x + tu$ clearly lies in the open ball $B(x, \varrho)$, whence $d \in D$.

Let $\{t_k\}$ be a sequence of strictly positive real numbers, such that, for every index k, $t_k < \varrho$, and such that $t_k \to 0$, as $k \to \infty$. Let us define the sequence $\{d^k\}$ by setting $d^k = x + t_k u$. For this choice of $\{t_k\}$ and $\{d^k\}$, Proposition (1.6) ensures that the vector u belongs to the cone of tangents $\mathcal{T}_D(x)$. Since the choice of the vector $u \in \mathbb{R}^n$ was arbitrary and $\mathcal{T}_D(x)$ is a cone, it follows that $\mathcal{T}_D(x)$ must include all of $\mathbb{R}^n$.

From the previous example we understand in particular that the only points where the cone of tangents may be non-trivial (i.e. neither reduces to the singleton $\{0\}$, nor to the entire space), are those that lie on the topological boundary, ∂D, of the set D. This topological restriction alone, however, does not ensure that the cone of tangents is anyhow informative. For this to happen, we must require some sort of structure for the set D.

Even though the cone of tangent may be extremely complicated to explicitly compute, it does indeed lead to a very simple optimality condition, which turns out to be the basis of all other optimality conditions that we are going to discuss in the sequel. In fact, in spite of its simplicity, the following result may be considered the core of optimality in non-linear programming.

(1.9) Lemma. *Let D be a non-empty set in $\mathbb{R}^n$, $f(x)$ a real differentiable function defined on D and let x^* be a point in D. If x^* is a local minimum of f on D, then*

$$\langle \nabla f(x^*), d \rangle \geq 0 \qquad \text{for all } d \in \mathcal{T}_D(x^*).$$

Proof. The proof is done by contradiction. To this end, let us assume that there exists a tangent vector d such that $\langle \nabla f(x^*), d \rangle < 0$. Since d is a tangent direction, there exist a sequence $\{z^k\}$ of points in D and a sequence $\{t_k\}$ of strictly positive real numbers, the first convergent towards x^* and the second convergent towards 0, such that:

$$\lim_{k \to \infty} \frac{z^k - x^*}{t_k} = d.$$

This means that $z^k = x^* + t_k d + o(t_k)$, where $o(t_k)$ represents an infinitesimal with order higher than t_k. Let us observe that Taylor's formula of the function $f(x)$ at x^* yields

$$f(z^k) = f(x^*) + t_k \langle \nabla f(x^*), d \rangle + o(t_k).$$

Rearranging this expression, from the fact that $f(x)$ is assumed to be of class $\mathcal{C}^1$, and that we had assumed, by contradiction, that $\langle \nabla f(x^*), d \rangle < 0$, for k sufficiently large, we shall have

$$\frac{f(z^k) - f(x^*)}{t_k} = \langle \nabla f(x^*), d \rangle + \frac{o(t_k)}{t_k} < 0.$$

In particular, this inequality shows that, for k sufficiently large, we must have $f(z^k) < f(x^*)$, which is a contradiction, because x^* is assumed to be a local minimum of f on D. It is thereby proved that that for all tangent directions d, we have $\langle \nabla f(x^*), d \rangle \geq 0$. $\qquad\square$

We have also a sufficient optimality condition of the same type:

(1.10) Lemma. *Let D be a non-empty set in $\mathbb{R}^n$, $f(x)$ a real differentiable function defined on D and let x^* be a point in D. If*

$$\langle \nabla f(x^*), d \rangle > 0 \qquad \text{for all } d \in \mathcal{T}_D(x^*), \quad d \neq 0,$$

then x^* is a local minimum of f on D.

Proof. The proof is done by contradiction. Let $d \neq 0$ be a tangent direction; then there exists a sequence $\{z^k\}$ of points in D and a sequence $\{t_k\}$ of strictly positive real numbers, the first convergent towards x^* and the second convergent towards 0, such that:

$$\lim_{k \to \infty} \frac{z^k - x^*}{t_k} = d,$$

and that $f(z^k) < f(x^*)$. We have that $z^k = x^* + t_k d + o(t_k)$, where $o(t_k)$ represents an infinitesimal with order higher than t_k. Let us observe that Taylor's formula of the function $f(x)$ at x^* yields

$$0 > f(z^k) - f(x^*) = t_k \langle \nabla f(x^*), d^k \rangle + o(t_k).$$

and therefore

$$0 > \frac{f(z^k) - f(x^*)}{t_k} = \langle \nabla f(x^*), d \rangle + \frac{o(t_k)}{t_k};$$

passing to the limit when $\{t_k\}$ tends to 0 we have

$$\langle \nabla f(x^*), d \rangle \leq 0,$$

in contradiction with the assumption. $\qquad\square$

These lemmata already mark a significant achievement, which at this stage remains, however, more theoretical than practical. In fact, as we have mentioned before, for a general set D, the cone $\mathcal{T}_D(x^*)$ may be trivial, and, even when we can guarantee that this set is not trivial, computing explicitly the cone of tangents may be an extremely complicated task to perform. Fortunately, however, in the special framework of a non-linear programming problem, things are way smoother. Indeed, the fact that the feasible region is described in terms of equalities and/or inequalities will play a significant role in simplifying the computation of the cone of tangent or, at least, in finding a suitable other cone, included in the cone of tangent, that is determinant for describing criteria of optimality.

Before being able to do so, however, essentially with the aim to clarify the notations adopted in the sequel, we shall have to recall an important theorem of differential calculus that will be used to fully exploit the particular structure of the feasible regions that are used in non-linear programming. This is the content of the next section.

2. The Implicit Function Theorem

Before delving into the study of the optimality conditions for a non-linear programming problem, we shall need to recall a definition that generalises the one of differentiable function, and state an important theorem, due to the Italian mathematician ULISSE DINI (1845–1918), which is usually referred to as the 'Implicit Function Theorem'. To this end, let us start by recalling that, if D is a subset of $\mathbb{R}^n$, a function $f : D \to \mathbb{R}^m$, from D to $\mathbb{R}^m$ can be seen as an m-tuple of real function $f_1, \ldots, f_m$ defined on D, such that,

$$f(x) = \big(f_1(x), \ldots, f_m(x)\big) \qquad \text{for all } x \in D.$$

The function $f_1, \ldots, f_m$ are called the **components** of f. In other words, a function with values in $\mathbb{R}^m$ may always be seen as an m-tuple of real functions all defined over the same domain. With this representation in mind, let us give the following definition.

(2.1) Definition. Let D be an open set in $\mathbb{R}^n$, let $f : D \to \mathbb{R}^m$ be a function defined on the set D with values in $\mathbb{R}^m$, and let x^* be a point in D. We shall say that f is **differentiable at** x^* if there exists a (necessarily unique) linear transformation from $\mathbb{R}^n$ to $\mathbb{R}^m$, which is denoted by $Jf(x^*)$ and is called the **Jacobian** of f at x^*, such that

$$f(x) = f(x^*) + Jf(x^*)(x - x^*) + o(x - x^*) \qquad \text{for all } x \in D.$$

If, in addition, the function f is differentiable at every point $x \in D$, we shall simply say that f is a **differentiable function**.

(2.2) Remark. If D is an open set in $\mathbb{R}^n$ and $f : D \to \mathbb{R}^m$ is a differentiable function at $x^* \in D$, the Jacobian, $Jf(x^*)$, can be expressed as a $m \times n$ matrix whose rows are the gradients $\nabla f_1(x^*), \ldots, \nabla f_m(x^*)$ of the components of f. More explicitly, we can say that the Jacobian, $Jf(x^*)$, of f at x^* can be represented by the matrix

$$Jf(x^*) = \begin{bmatrix} \dfrac{\partial f_1}{\partial x_1}(x^*) & \dfrac{\partial f_1}{\partial x_2}(x^*) & \cdots & \dfrac{\partial f_1}{\partial x_n}(x^*) \\[2ex] \dfrac{\partial f_2}{\partial x_1}(x^*) & \dfrac{\partial f_2}{\partial x_2}(x^*) & \cdots & \dfrac{\partial f_2}{\partial x_n}(x^*) \\[2ex] \vdots & \vdots & \ddots & \vdots \\[2ex] \dfrac{\partial f_m}{\partial x_1}(x^*) & \dfrac{\partial f_m}{\partial x_2}(x^*) & \cdots & \dfrac{\partial f_m}{\partial x_n}(x^*) \end{bmatrix}.$$

Also, observe that the Jacobian of f at x^* becomes the gradient of f at x^* when $m = 1$, that is to say when the function f takes values on the real line, ensuring that the two definitions are consistent.

The following proposition will be useful to compute the Jacobian matrix of a composite function. Its proof is just a mere application of the definition of differentiable function.

(2.3) Proposition (Chain rule). *If D is an open set in $\mathbb{R}^n$, E is an open set of $\mathbb{R}^m$, $f : D \to E$ and $g : E \to \mathbb{R}^p$ are two differentiable functions, then the composite function $h = g \circ f : D \to \mathbb{R}^p$ is also differentiable, and the Jacobian matrix of h at a point $x \in D$ can be expressed in terms of the Jacobian matrices of f and g as follows:*

$$Jh(x) = Jg\big(f(x)\big)Jf(x).$$

Proof. Set $y = f(x)$, $A = Jf(x)$ and $B = Jg(y)$. We have

$$f(x + \delta x) = f(x) + Jf(x)\delta x + o(\delta x) = y + A\delta x + o(\delta x)$$

and

$$\begin{aligned}
h(x + \delta x) &= g\big(f(x + \delta x)\big) \\
&= g\big(y + Ah + o(\delta x)\big) \\
&= g(y) + Jg(y)\big(A\delta x + o(\delta x)\big) + o\big(A\delta x + o(\delta x)\big) \\
&= g\big(f(x)\big) + BA\delta x + Bo(\delta x) + o\big(A\delta x + o(\delta x)\big) \\
&= h(x) + BA\delta x + o(\delta x).
\end{aligned}$$

Comparing the above equality with

$$h(x + \delta x) = h(x) + Jh(x)\delta x + o(\delta x)$$

we conclude immediately that

$$Jh(x) = BA = Jg\big(f(x)\big)Jf(x).$$

This proves the proposition. $\qquad\square$

When a function is defined over (a subset of) $\mathbb{R}^n$, we can sometimes think of it as a function of several variables. This can be done in many ways, and even if this would not change the function *per se*, it would sometimes make an awful lot of difference in how the function is used, as we shall see in the statement of Dini's Implicit Function Theorem.

(2.4) Remark. If D is the product $A \times B$, with $A \subseteq \mathbb{R}^p$ and $B \subseteq \mathbb{R}^{n-p}$, a function $f : D \to \mathbb{R}^m$ may be seen as a function of two vector variables, the first, x, defined over A and the second, y, defined over B. In particular, if (x^*, y^*) is a point in D, we can then consider the two functions

$$x \mapsto f(x, y^*) \qquad \text{and} \qquad y \mapsto f(x^*, y),$$

the first defined over A, and the second defined over B. If the first function is differentiable at x^*, its Jacobian shall be denoted by $J_x f(x^*, y^*)$. Similarly, if the second is differentiable at y^*, its Jacobian shall be denoted by $J_y(x^*, y^*)$. Let us also observe that, in the special case when $m = 1$, this notations become $\nabla_x f(x^*, y^*)$ and $\nabla_y f(x^*, y^*)$, respectively, and, for the Hessian matrix, we shall write $H_x f(x^*, y^*)$ and $H_y f(x^*, y^*)$.

With the definitions and the notations introduced above, we are now finally ready to state and prove the aforementioned Dini's Implicit Function Theorem. The proof that we are going to give is due to the Greek mathematician CONSTANTIN CARATHÉODORY (1873–1950), and it is particularly remarkable because it is at the same time clever and elementary, showing the deep foundations of the theorem itself.

(2.5) Theorem (Dini). *Let D be an open set in $\mathbb{R}^n \times \mathbb{R}^m$, and $f : D \to \mathbb{R}^m$ be a differentiable function. Let (x^*, y^*) be a point in D and suppose that*

$$f(x^*, y^*) = 0 \qquad and \qquad J_y f(x^*, y^*) : \mathbb{R}^m \to \mathbb{R}^m \text{ is non-singular.}$$

Then, there exist a neighbourhood U of x^* in $\mathbb{R}^n$, a neighbourhood V of y^* in $\mathbb{R}^m$, and a mapping $\varphi : U \to V$, such that $U \times V \subseteq D$ and

$$f\big(x, \varphi(x)\big) = 0 \qquad \text{for all } x \in U.$$

Furthermore, the function φ is differentiable and

$$J\varphi(x) = -J_y f\big(x, \varphi(x)\big)^{-1} J_x f\big(x, \varphi(x)\big) \qquad \text{for all } x \in U.$$

Proof. In the hypotheses of the theorem, with no loss of generality, let us assume that $x^* = 0$ and $y^* = 0$, by replacing, if necessary, the function f with $(x, y) \mapsto f(x + x^*, y + y^*) - f(x^*, y^*)$. Let us write

$$f(x, y) = \big(f_1(x, y), \dots, f_m(x, y)\big),$$

where $f_1, \dots, f_m$ are the components of the function f. Since the function Jf is linear, hence continuous, there exist neighbourhoods U_0 and V_0 of the origins in $\mathbb{R}^n$ and $\mathbb{R}^m$, respectively, such that the matrix, M, with rows $\nabla_y f_1(x, y_1), \dots, \nabla_y f_m(x, y_m)$ is invertible for every $(x, y_i) \in U_0 \times V_0$.

Let us now prove that, for every $x \in U_0$, there exists at most one $y \in V_0$ such that $f(x, y) = 0$. By contradiction, assume that there existed two points $y, z \in V_0$, with $y \neq z$, such that $f(x, y) = f(x, z) = 0$. By virtue of the mean value theorem, this would imply the existence of a point y_i, lying in the line segment between y and z, such that

$$f_i(x, z) - f_i(x, y) = \langle \nabla_y f(x, y_i), z - y \rangle = 0 \qquad \text{for all } i = 1, \dots, m.$$

Since the matrix M is non-singular, this gives $y = z$, which is a contradiction.

Let $B_r = \overline{B(0,r)}$ be a closed ball with centre 0 and a suitable radius r to ensure that $B_r \subseteq V_0$. Since $f(0,0) = 0$, we have

$$f(0,y) \neq 0 \qquad \text{for } y \in \partial B_r = \{y \in \mathbb{R}^m : \|y\| = r\},$$

and since the function f is by hypothesis continuous on $U_0 \times V_0$, there exists a strictly positive real number ε such that $\|f(0,y)\| \geq \varepsilon$, for every $y \in \partial B_r$. Hence, the function $F(x,y) = \|f(x,y)\|^2$ satisfies the properties

$$F(0,y) \geq \varepsilon > 0 \quad \text{for all } y \in \partial B_r \quad \text{and } F(0,0) = 0.$$

Furthermore, since the function F is continuous, there exists a neighbourhood $U \subseteq U_0$ of 0 in $\mathbb{R}^n$ such that

$$F(x,y) \geq \tfrac{1}{2}\varepsilon \quad \text{and } F(x,0) \leq \tfrac{1}{2}\varepsilon \quad \text{for all } x \in U \quad \text{and } y \in \partial B(0,r).$$

This, for a fixed $x \in U$, the function $y \mapsto F(x,y)$ achieves its minimum on B_r at the point $\varphi(x)$ in the interior of B_r, and we have

$$J_y F\big(x,\varphi(x)\big) = 2 J_y\big(x,\varphi(x)\big) f\big(x,\varphi(x)\big) = 0,$$

since the matrix $J_y f\big(x,\varphi(x)\big)$ is non-singular, we conclude that

$$f\big(x,\varphi(x)\big) = 0.$$

It remains to prove that the function φ is differentiable. To this end, let us start by writing $\Delta y = \varphi(x + \Delta x) - \varphi(x)$. By virtue of the mean value theorem, this gives

$$0 = J_x f(\tilde{x},\tilde{y})\Delta x + J_y f(\tilde{x},\tilde{y})\Delta y$$

for some point $(\tilde{x},\tilde{y})$ lying on the line segment between the point with coordinates $\big(x,\varphi(x)\big)$ and the point with coordinates $\big(x+\Delta x, \varphi(x+\Delta x)\big)$. This implies that, as $\|\Delta x\|$ goes to zero, so does $\|\Delta y\|$, proving that the function $\varphi(x)$ is continuous. In fact, the function $\varphi(x)$ is actually differentiable, since by means of Taylor's expansion,

$$\begin{aligned}
0 &= f\big(x + \Delta x, \varphi(x + \Delta x)\big) - f\big(x,\varphi(x)\big) \\
&= J_x f\big(x,\varphi(x)\big)\Delta x + J_y f\big(x,\varphi(x)\big)\Delta y + o\big((\Delta x, \Delta y)\big),
\end{aligned}$$

and since $o\big((\Delta x, \Delta y)\big) = o(\Delta x)$, by continuity of $\varphi(x)$, we have

$$\Delta y = -J_y f\big(x,\varphi(x)\big)^{-1} J_x f\big(x,\varphi(x)\big) + o(\Delta x).$$

This proves that $\varphi(x)$ is differentiable at x with

$$J\varphi(x) = -J_y f\big(x,\varphi(x)\big)^{-1} J_x\big(x,\varphi(x)\big)$$

and the theorem is completely proved. $\qquad\square$

3. First order optimality conditions

As we have mentioned in Chapter III, the general non-linear program can be
stated in the form:

$$\begin{array}{ll}
\text{minimise} & f(x) \\
\text{subject to} & g_i(x) \leq 0 \quad \text{for every } i = 1, \ldots, m, \\
& h_j(x) = 0 \quad \text{for every } j = 1, \ldots, p,
\end{array}$$

where f, g_i $(i = 1, \ldots, m)$, and h_j $(j = 1, \ldots, p)$ are real differentiable
functions defined on a suitable common domain, and at least one of them is
non-linear. Thanks to this hypothesis, it turns out that, as we shall see in
the sequel, finding tangent directions is not only a rather simpler task, but
in most cases this cone can also be calculated algorithmically. Consequently,
all necessary conditions for optimality based on Lemma (1.9) will assume
a much simpler form that can be used to construct algorithms for finding
optimal solutions.

To illustrate this important link between the structure of the feasible region
and the structure of the cone of tangents, let us focus on the structure of the
feasible region. In other words, assume we are in the following hypotheses.

(3.1) Hypothesis. Let E be a non-empty open set in $\mathbb{R}^n$, and assume that
the domain of the real functions g_i (for $i = 1, \ldots, m$) and h_j (for $j = 1, \ldots, p$)
are all contained in E. Let us set

$$D = \left\{ x \in \mathbb{R}^n : \begin{array}{ll} g_i(x) \leq 0 & \text{for every } i = 1, \ldots, m \\ h_j(x) = 0 & \text{for every } j = 1, \ldots, p \end{array} \right\}.$$

Given a point $x^* \in D$, in the sequel we shall always denote by $I(x^*)$ the set
of all indices representing the 'active' inequality constraints at x^*, that is to
say all the indices for which the value of $g_i(x)$ at x^* is exactly 0. In other
words, let us set $I(x^*) = \{i : g_i(x^*) = 0\}$.

We can now give the following definition.

(3.2) Definition. In the hypotheses and with the notations introduced
in (3.1), let $x^* \in D$. The set of all those directions that satisfy the linearised
approximations of the problem's constraints at x^*, that is to say the set

$$\mathcal{L}_D(x^*) = \left\{ d \in \mathbb{R}^n : \begin{array}{ll} \langle \nabla g_i(x^*), d \rangle \leq 0 & \text{for all } i \in I(x^*) \\ \langle \nabla h_j(x^*), d \rangle = 0 & \text{for all } j = 1, \ldots, p \end{array} \right\}$$

shall be called the **cone of linearised feasible directions** to D at x^*.

It is straightforward to verify that the cone of linearised feasible direction
is indeed a cone. It is important to note, however, that in the definition of
the linearised feasible direction cone, $\mathcal{L}_D(x^*)$, all the equality constraints h_j

are included, while only the inequality constraints g_i that are 'active' at the point x^* are considered. Furthermore, the cone of linearised feasible directions to D at x^* offers a linear approximation of the feasible region's behaviour near the point of interest. In other words, this cone consists of all vectors that represent directions for infinitesimal movements towards x^* remaining within the feasible region.

Coupled with the cone of linearised feasible directions we will consider also the following other cone. Again, we should verify that this is actually a cone, but this is once again straightforward.

(3.3) Definition. In the hypothesis and with the notations of (3.1), let x^* be a point of D. The set of all those directions that satisfy the strict linearised approximations of the problem's constraints at x^*, that is to say the set

$$\mathcal{S}_D(x^*) = \left\{ d \in \mathbb{R}^n : \begin{array}{ll} \langle \nabla g_i(x^*), d \rangle < 0 & \text{for all } i \in I(x^*) \\ \langle \nabla h_j(x^*), d \rangle = 0, & \text{for all } j = 1, \ldots, p \end{array} \right\}$$

shall be called **cone of strict linearised feasible directions** at x^*.

It is clear, from the definition, that $\mathcal{S}_D(x^*) \subseteq \mathcal{L}_D(x^*)$. A more subtle relationship connects these cones to the cone of tangents. This is discussed in detail in the next theorem, whose proof will require to apply the Implicit Function Theorem.

(3.4) Theorem. *In the hypothesis and with the notations defined in (3.1), let x^* be a point in D. Then,*

$$\mathcal{T}_D(x^*) \subseteq \mathcal{L}_D(x^*).$$

If, in addition, $\nabla h_1(x^*), \ldots, \nabla h_p(x^*)$ are linearly independent, then

$$\mathcal{S}_D(x^*) \subseteq \mathcal{T}_D(x^*).$$

Proof. Let us start by proving that $\mathcal{T}_D(x^*) \subseteq \mathcal{L}_D(x^*)$. To this end, let us take $d \in \mathcal{T}_D(x^*)$ and consider the sequences $\{z^k\}$, of elements of D, and $\{t_k\}$, of strictly positive real numbers, the first convergent towards x^* and the second convergent towards 0, such that $(z^k - x)/t_k \to 0$ as $k \to \infty$. Observe that this limit can be rephrased by saying that the sequence $\{z^k\}$ may be defined for every index k as $z^k = x + t_k d + o(t_k)$, where $o(t_k)$ denotes an infinitesimal with order higher than t_k. Using Taylor's expansions of the functions g_i and h_j, with $i \in I(x^*)$ and $j = 1, \ldots, p$, and observing that z^k lies in D by hypothesis, we get

$$g_i(z^k) = t_k \langle \nabla g_i(x^*), d \rangle + o(t_k) \leq 0,$$
$$h_j(z^k) = t_k \langle \nabla h_j(x^*), d \rangle + o(t_k) = 0,$$

Dividing throughout by t_k and taking the limit as $k \to \infty$, this proves that $\langle \nabla g_i(x^*), d \rangle \leq 0$ and $\langle \nabla h_j(x^*), d \rangle = 0$, that is to say $d \in \mathcal{L}_D(x^*)$.

It remains to prove that, if $\nabla h_1(x^*),\dots,\nabla h_p(x^*)$ are linearly independent, then $\mathcal{S}_D(x^*) \subseteq \mathcal{T}_D(x^*)$. To this end, take a direction d in $\mathcal{S}_D(x^*)$ and set

$$h(x) = \big(h_1(x),\dots,h_p(x)\big).$$

By definition, $h(x^*) = 0$, and the Jacobian, $Jh(x^*)$, of h at x^*, being the matrix whose rows are the vectors $\nabla h_1(x^*),\dots,\nabla h_p(x^*)$, has maximum rank p. Let us take a direction $d \in \mathcal{S}_D(x^*)$. By definition of the cone of strict linearised feasible directions, we have $\langle \nabla h_j(x^*), d \rangle = 0$ for every $j = 1,\dots,p$ and this yields in particular $Jh(x^*)d = 0$. Consider the set

$$S = \Big\{(t,u) \in \mathbb{R}^{p+1} : x^* + td + Jh(x^*)^T u \in \Omega\Big\}$$

and define the function $K : S \to \mathbb{R}^p$ by

$$K(t,u) = h\big(x^* + td + Jh(x^*)^T u\big) \qquad \text{for all } (t,u) \in S.$$

Observe that $K(0,0) = h(x^*) = 0$. In addition, observe that by chain rule we have $J_u K(0,0) = Jh(x^*)Jh(x^*)^T$, which yields that the Jacobian of K with respect to the second variable at $(0,0)$ is positive definite, hence is non-singular. To see this, observe that, if $v \neq 0$, we have

$$\langle v, J_u K(0,0)v \rangle = \langle Jh(x^*)^T v, Jh(x^*)^T v \rangle = \|Jh(x^*)v\|^2 > 0$$

because $Jh(x^*)$ has maximum rank, and therefore $Jh(x^*)Jh(x^*)^T$ is non-singular. This, we may now apply Dini's Implicit Function Theorem to infer that there exist a strictly positive real number τ, and a differentiable function $\varphi : (-\tau, \tau) \to \mathbb{R}^p$ such that $\varphi(0) = 0$ and

$$K\big(t, \varphi(t)\big) = 0 \qquad \text{for all } t \in (-\tau, \tau),$$

Furthermore, observe that

$$h\big(x^* + td + Jh(x^*)^T \varphi(t)\big) = K\big(t, \varphi(t)\big) = 0 \qquad \text{for all } t \in (-\tau, \tau),$$

which guarantees that

$$0 = \left[\frac{\mathrm{d}}{\mathrm{d}t}h\big(x^* + td + Jh(x^*)^T \varphi(t)\big)\right]_{t=0} = Jh(x^*)\big[d + Jh(x^*)^T \varphi'(0)\big]$$
$$= Jh(x^*)Jh(x^*)^T \varphi'(0)$$

hence $\varphi'(0) = 0$. Let us finally consider any sequence $\{t_k\}$ of strictly positive real numbers, bounded from above by τ, and convergent towards 0, and let us set $d^k = d + Jh(x^*)^T (t_k^{-1}\varphi(t_k))$ and $z^k = x^* + t_k d^k$. By constructions, it is easy to check that $(z^k - x^*)/t_k = d^k \to d$. To complete the proof, it suffices to show that $z^k \in D$ for every index k. Indeed, $h_j(z^k) = h_j(x^*v + t_k d^k) = 0$ for every $j = 1,\dots,p$. To see that the inequality constraints are also satisfied, let us start by taking $i \notin I(x^*)$. Then $g_i(x^* + t_k d^k) \to g_i(x^*) < 0$ as $k \to \infty$, whence $g_i(z^k) = g_i(x^* + t_k d^k) < 0$ for k sufficiently large, if necessary. On the other hand, if $i \in I(x^*)$, by means of Taylor's expansion, we get

$$\frac{g_i(z^k)}{t_k} = \frac{g_i(x^*)}{t_k} + \langle \nabla g_i(x^*), d^k \rangle + \frac{o(t_k d^k)}{t_k} = \langle \nabla g_i(x^*), d^k \rangle + \frac{o(t_k d^k)}{t_k}$$

and the last term tends towards $\langle \nabla g_i(x^*), d \rangle$, which is negative, because by hypothesis $d \in \mathcal{S}_D(x^*)$. Thus $g_i(z^k) = g_i(x^* + t_k d^k) < 0$ for k sufficiently large. In conclusion, up to extracting a sub-sequence from $\{z^k\}$, we have that $z^k \in D$ for every index k, and this completes the proof. $\qquad\square$

We are ready to prove one of the main results of this chapter.

(3.5) Theorem (Fritz-John conditions). *In the hypothesis and with the notations defined in (3.1), let x^* be a local minimum of the non-linear programming problem*

$$\begin{aligned} \text{minimise} \quad & f(x) \\ \text{subject to} \quad & g_i(x) \le 0 \text{ for all } i = 1, \ldots, m \\ & h_j(x) = 0 \text{ for all } j = 1, \ldots, p. \end{aligned}$$

Assume that the vectors $\nabla h_j(x^*)$ are linearly independent. Then, there exist a real number $\theta^* \in \mathbb{R}$, a vector $\lambda^* \in \mathbb{R}^m$, with $(\theta^*, \lambda^*) \ge 0$ and not all zero, and a vector $\mu^* \in \mathbb{R}^p$ such that the quadruple $(x^*, \theta^*, \lambda^*, \mu^*)$ satisfies the following system of simultaneous equalities and inequalities, which is referred to as the **Fritz-John (FJ) system**:

(3.6)
$$\begin{cases} \theta^* \nabla f(x^*) + \sum_{i=1}^{m} \lambda_i^* \nabla g_i(x^*) + \sum_{j=1}^{p} \mu_j^* \nabla h_j(x^*) = 0, \\ \lambda_i^* g_i(x^*) = 0 \quad \text{for all } i = 1, \ldots, m, \\ g_i(x^*) \le 0 \quad \text{for all } i = 1, \ldots, m, \\ h_j(x^*) = 0 \quad \text{for all } j = 1, \ldots, p. \end{cases}$$

Proof. Since, by hypothesis, the vectors $\nabla h_j(x^*)$ with $j = 1, \ldots, p$ are linearly independent, Theorem (3.4) ensures that $\mathcal{S}_D(x^*) \subseteq \mathcal{T}_D(x^*)$. Thus, from Lemma (1.9), it follows that

$$\langle \nabla f(x^*), d \rangle \ge 0 \qquad \text{for all } d \in \mathcal{S}_D(x^*).$$

In other words, the previous inequality shows that the following system has no solutions

$$\begin{cases} \langle \nabla f(x^*), d \rangle < 0, \\ \langle \nabla g_i(x^*), d \rangle < 0 \quad \text{for all } i \in I(x^*), \\ \langle \nabla h_j(x^*), d \rangle = 0 \quad \text{for all } j = 1, \ldots, p. \end{cases}$$

By virtue of Motzkin's Theorem I(9.1), this means that there must therefore be a solution $(\theta^*, \alpha^*, \mu^*)$ of the following system:

$$\begin{cases} \theta^* \nabla f(x^*) + \sum_{i \in I(x^*)} \alpha_i^* \nabla g_i(x^*) + \sum_{j=1}^{p} \mu_j^* \nabla h_j(x^*) = 0 \\ \alpha_i \ge 0 \quad \text{for all } i \in I(x^*). \end{cases}$$

If we define the vector λ^* as:

$$\lambda_i^* = \begin{cases} \alpha_i^* & \text{if } i \in I(x^*) \\ 0 & \text{if } i \notin I(x^*) \end{cases}$$

then $(x^*, \theta^*, \lambda^*, \mu^*)$ solves the FJ system. $\qquad\qquad\qquad\square$

4. Regular constraints and KKT conditions

There is an important class of non-linear programming problems which are called regular. For this class of problems, the Fritz-John theorem holds in a more elegant form, and this will be useful when it comes to develop algorithms to approximate optimal solutions.

(4.1) Definition. In the hypothesis and with the notations defined in (3.1), the constraints are said to be **regular** at a point $x \in D$ if

$$\mathcal{T}_D(x) = \mathcal{L}_D(x).$$

When D is regular at every point $x \in D$ we say that the domain D is **regular**.

The following examples aim to clarify the notion of regularity.

(4.2) Example. Let us consider the non-linear programming problem consisting in minimising the function $f(x_1, x_2) = x_1$ over the feasible region

$$D = \left\{ x \in \mathbb{R}^2 : x_1^2 + x_2^2 - 1 = 0 \right\},$$

which is the circumference of centre $(0,0)$ and radius 1. It is straightforward to recognise that the global maximum and global minimum are the points $(1,0)$ and $(-1,0)$, respectively. Let us study the regularity of the constraint at the point $x^* = (-1,0)$. To this end, let us start by computing the cone of tangents of D at the point $(-1,0)$. A sequence of feasible points converging to $(-1,0)$ can be defined as

$$z^k = \left(-\sqrt{1 - \frac{1}{k^2}}, \frac{1}{k} \right).$$

If we choose the sequence of scalars $t_k = \|z^k - x^*\|$, then we obtain the tangent vector $d = (0,1)$. Similarly, choosing the sequences

$$z^k = \left(-\sqrt{1 - \frac{1}{k^2}}, -\frac{1}{k} \right).$$

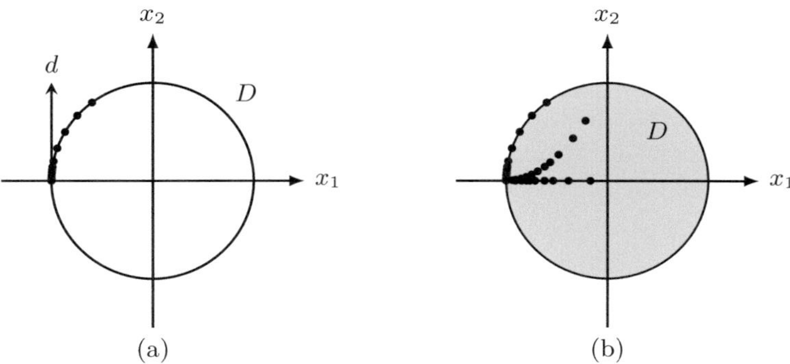

Figure 4.1. Diagram (a) refers to Example (4.2) and shows a sequence $\{z^k\}$ of point on the constraint $D = \{(x_1, x_2) : x_1^2 + x_2^2 = 1\}$ converging towards the point $(-1, 0)$, the minimum point of $f(x_1, x_2) = x_1$. Diagram (b), instead, refers to Example (4.3), and shows that, in this case, there are several directions from which a sequence $\{z^k\}$ of points in D can approach $(-1, 0)$.

and setting again $t_k = \|z^k - x^*\|$, gives the tangent vector $d = (0, -1)$. Therefore, the cone of tangents of D in $(-1, 0)$ is

$$\mathcal{T}_D(-1, 0) = \left\{(0, d_2) : d_2 \in \mathbb{R}\right\}.$$

Let us move to the cone of linearised feasible directions. To this end observe that, since $\nabla h(-1, 0) = (-2, 0)$, we have to find the vectors $d \in \mathbb{R}^2$ such that

$$0 = \langle \nabla h(x^*), d \rangle = -2d_1,$$

and this gives $\mathcal{L}_D(-1, 0) = \{d \in \mathbb{R}^2 : d_1 = 0\}$, which, in turn, coincides with the cone of tangents, $\mathcal{T}_D(-1, 0)$, thereby proving that the constraint is regular at the point $(-1, 0)$.

(4.3) Example. Let us modify the previous example by considering the region $D = \{x \in \mathbb{R}^2 : x_1^2 + x_2^2 - 1 \leq 0\}$, but let us take the same objective function, $f(x_1, x_2) = x_1$, and observe that the global maximum and the global minimum remain the points $(1, 0)$ e $(-1, 0)$, respectively. Let us start by the regularity of the point $x^* = (-1, 0)$ by computing the cone of tangents of D at the point $(-1, 0)$. In this case in addition to the two sequences $\{z^k\}$ described in Example (4.2), we may also take any sequence of points belonging to the interior of the circle and converge towards $(-1, 0)$. For instance, a sequence of this type may be the following:

$$z^k = (-1, 0) + k^{-1}d,$$

where $d = (d_1, d_2)$ is a vector such that the component d_1 is positive. The points z^k are feasible so long as $\|z^k\| \leq 1$, that is to say, if

$$\left(-1 + \frac{d_1}{k}\right)^2 + \left(\frac{d_2}{k}\right)^2 \leq 1,$$

which yields $k \geq \|d\|^2/(2d_1)$. By choosing $t_k = \|z^k - x^*\|/\|d\|$ we obtain that d is a tangent vector. Moreover we should also consider the sequences converging to $(-1, 0)$ following a curve which lies in the interior part of our circle; but only the already computed tangent vectors would be found. In summary, the cone of tangents to D at $(-1, 0)$ is

$$\mathcal{T}_D(-1, 0) = \left\{ d \in \mathbb{R}^2 : d_1 \geq 0 \right\}.$$

Let us now compute the cone of linearised feasible directions. Since the constraint g is active in $(-1, 0)$ and $\nabla g(-1, 0) = (-2, 0)$, we have to find all vectors $d = (d_1, d_2)$ such that

$$\langle \nabla g(-1, 0), d \rangle = -2d_1 \leq 0,$$

therefore we obtain that $\mathcal{L}_D(-1, 0) = \{d \in \mathbb{R}^2 : d_1 \geq 0\}$, which, in turn, coincides with the cone of tangents, $\mathcal{T}_D(-1, 0)$, at $(-1, 0)$, thereby proving that the constraints are regular at the point $(-1, 0)$.

(4.4) Example. Consider the non-linear programming problem to find the maximum and minimum of $f(x_1, x_2) = x_1 + x_2$ over the feasible region

$$D = \left\{ x \in \mathbb{R}^2 : (x_1 - 1)^2 + (x_2 - 1)^2 - 1 \leq 0, \ x_2 \leq 0 \right\},$$

which coincides with the singleton $\{(1, 0)\}$. In such a point, it is easy to see that the cone of tangents is $\mathcal{T}_D(1, 0) = \{(0, 0)\}$, because the only feasible sequence convergent towards $(1, 0)$ is the constant sequence $\{z^k\}$ defined by $z^k = (1, 0)$ for every index k.

On the other hand, both constraints that define the region D are active at $(1, 0)$, and since we have $\nabla g_1(1, 0) = (0, -2)$ and $\nabla g_2(1, 0) = (0, 1)$, the cone of linearised feasible directions is $\mathcal{L}_D(1, 0) = \{d \in \mathbb{R}^2 : d_2 = 0\}$ which does not coincide with $\mathcal{T}_D(1, 0)$. This shows that, in this case, the constraints are not regular at the point $(1, 0)$.

If the constraints of a non-linear programming problem are regular, we have mentioned before that Fritz-John conditions can be rewritten in a more elegant and effective form. Even though we have not discussed yet whether regular constraints are a common or uncommon condition, we can immediately say that, as a consequence of the following theorem, these become a 'desirable' condition, because the KKT conditions that are described in the theorem are considered amongst the most useful conditions to solve non-linear programming problems.

(4.5) Theorem (Karush-Kuhn-Tucker conditions). *In the hypothesis and with the notations defined in (3.1), let x^* be a local minimum of the non-linear programming problem*

$$\begin{aligned}
\text{minimise} \quad & f(x) \\
\text{subject to} \quad & g_i(x) \leq 0 \text{ for all } i = 1, \ldots, m \\
& h_j(x) = 0 \text{ for all } j = 1, \ldots, p.
\end{aligned}$$

*Assume that the constraints are regular at x^*. Then, there exist two vectors $\lambda^* \in \mathbb{R}^m$ and $\mu^* \in \mathbb{R}^p$, with $\lambda^* \geq 0$, such that (x^*, λ^*, μ^*) satisfies the following system of simultaneous equalities and inequalities, which is referred to as the **Karush-Kuhn-Tucker (KKT) system**:*

$$(4.6) \quad \begin{cases} \nabla f(x^*) + \sum_{i=1}^m \lambda_i^* \nabla g_i(x^*) + \sum_{j=1}^p \mu_j^* \nabla h_j(x^*) = 0, \\ \lambda_i^* g_i(x^*) = 0 & \text{for all } i = 1, \ldots, m, \\ g_i(x^*) \leq 0 & \text{for all } i = 1, \ldots, m, \\ h_j(x^*) = 0 & \text{for all } j = 1, \ldots, p. \end{cases}$$

Proof. Since, by hypothesis, the constraints are regular in x^*, by definition, we have that $\mathcal{T}_D(x^*) = \mathcal{L}_D(x^*)$. Thus, Lemma (1.9) entails

$$\langle \nabla f(x^*), d \rangle \geq 0 \qquad \text{for all } d \in \mathcal{L}_D(x^*).$$

This inequality implies the following system of equalities and inequalities have no solution:

$$\begin{cases} \langle \nabla f(x^*), d \rangle < 0 \\ \langle \nabla g_i(x^*), d \rangle \leq 0, & \text{for all } i \in I(x^*), \\ \langle \nabla h_j(x^*), d \rangle = 0, & \text{for all } j = 1, \ldots, p. \end{cases}$$

Thus, by virtue of Motzkin's Theorem I(9.1), there is a solution (α^*, μ^*) of the following system of simultaneous inequalities and equalities:

$$\begin{cases} \nabla f(x^*) + \sum_{i \in I(x^*)} \alpha_i^* \nabla g_i(x^*) + \sum_{j=1}^p \mu_j^* \nabla h_j(x^*) = 0 \\ \alpha_i \geq 0 & \text{for all } i \in I(x^*). \end{cases}$$

If we define the vector λ^* as:

$$\lambda_i^* = \begin{cases} \alpha_i^* & \text{if } i \in I(x^*) \\ 0 & \text{if } i \notin I(x^*) \end{cases}$$

then the triple (x^*, λ^*, μ^*) solves the KKT system. In fact, if $i \in I(x^*)$ then, by definition, $g_i(x^*) = 0$, while if $i \notin I(x^*)$ then $\lambda_i^* = 0$, therefore

$$\lambda_i^* \, g_i(x^*) = 0 \qquad \text{for all } i = 1, \ldots, m.$$

So, (x^*, λ^*, μ^*) is a solution of the KKT system, and the theorem is thereby fully proved. $\qquad \square$

(4.7) Remark. In the hypothesis and with the notations defined in (3.1), if we define the Lagrangian function of the problem as:

$$L(x, \lambda, \mu) = f(x) + \sum_{i=1}^m \lambda_i \, g_i(x) + \sum_{j=1}^p \mu_j \, h_j(x),$$

with $x \in \mathbb{R}^n$, $\lambda \in \mathbb{R}^m$ e $\mu \in \mathbb{R}^p$, then the necessary conditions of optimality now established can be written, in an equivalent fashion, as follows:

$$\nabla_x L(x^*, \lambda^*, \mu^*) = 0$$

where, as we have already mentioned, the symbol $\nabla_x L$ means the gradient of the Lagrangian with respect the set of variables x. This aspect shall be discussed in further detail in Chapter VI.

(4.8) Remark. Similar results as Fritz-John's and Karush-Kuhn-Tucker's conditions can be obtained for a non-linear programming problem consisting of maximising an objective function, by only replacing the condition $\lambda_i \geq 0$, with the condition $\lambda_i \leq 0$.

Theorem (3.5) and Theorem (4.5) provide us only necessary optimality conditions as the following example shows.

(4.9) Example. Let us consider the problem of minimising the objective function $f(x_1, x_2) = x_1 + x_2$ over the feasible region

$$D = \left\{ x \in \mathbb{R}^2 : -x_1^2 - x_2^2 + 2 \leq 0 \right\}.$$

It is easy to check that the KKT solutions of this problem are:

$$\begin{cases} x^1 = (1, 1), & \lambda_1 = \frac{1}{2}, \\ x^2 = (-1, -1), & \lambda_2 = -\frac{1}{2}. \end{cases}$$

Let us prove that x^1, while satisfying necessary optimality condition for a local minimum, is not an actual local minimum of f on D, because the sequence of points

$$x^k = \left(1 - \frac{1}{k}, \sqrt{2 - \left(1 - \frac{1}{k} \right)^2} \right),$$

belongs to D, converges to x^1 e $f(x^k) < f(x^1)$ for every index k. Thus, x^1 is not a local minimum. Similarly, it is also possible to show that the point x^2 is not a local maximum of f on D.

The previous example justifies the necessity for the following definition.

(4.10) Definition. In the same hypothesis and with the same notations as in Theorem (4.5), any solution of the KKT system shall be called a **stationary point**. A stationary point which is neither a local minimum nor a local maximum shall be called a **saddle point**.

(4.11) Remark. In the same hypothesis and with the same notations as in Theorem (4.5), if (x^*, λ^*, μ^*) is a solution of the KKT system, then:

- if $\lambda^* \geq 0$, then x^* is a candidate local minimum of f on D;
- if $\lambda^* \leq 0$, then x^* is a candidate local maximum of f on D;
- if there exist two indices i_1 and i_2 such that $\lambda_{i_1} > 0$ and $\lambda_{i_2} < 0$, then x^* is a saddle point.

5. Constraint qualifications

In the previous section, we have mentioned that for a non-linear programming problem, to have regular constraints is a 'desirable' property, because it entails, by means of Theorem (4.5), the KKT conditions. However, thus far we have only shown numerical examples of regular and non-regular constraints, without discussing under which circumstances the constraints are regular. This is the aim of this section, where we are going to show some tests that guarantees the regularity of the domain, that is to say $\mathcal{T}_D(x) = \mathcal{L}_D(x)$. These conditions are generally referred to as the 'constraint qualifications'.

(5.1) Remark. The definition of the cone of tangents of a subset D of $\mathbb{R}^n$ is based on the geometry of the region D and not on its algebraic description. On the other hand, the cone of linearised feasible directions, which can only be defined when the set D is described as

$$D = \left\{ x \in \mathbb{R}^n : \begin{array}{ll} g_i(x) \leq 0 & \text{for all } i = 1, \ldots, m \\ h_j(x) = 0 & \text{for all } j = 1, \ldots, p \end{array} \right\},$$

where g_i $(i = 1, \ldots, m)$ and h_j $(j = 1, \ldots, p)$ are real functions defined on a common domain Ω, is based on its algebraic representation, and in particular on the function g_i and h_j. Let us consider, for example, the region D introduced in Example (4.2). Such a set can be defined, in an equivalent fashion, as follows:

$$D = \left\{ x \in \mathbb{R}^2 : (x_1^2 + x_2^2 - 1)^2 = 0 \right\}.$$

The gradient of the constraint is

$$\nabla h(x) = \begin{bmatrix} 4x_1(x_1^2 + x_2^2 - 1) \\ 4x_2(x_1^2 + x_2^2 - 1) \end{bmatrix} = \begin{bmatrix} 0 \\ 0 \end{bmatrix} \qquad \text{for all } x \in D,$$

and therefore the cone of linearised feasible directions is $\mathcal{L}_D(x) = \mathbb{R}^2$ for every point $x \in D$. Notice that, with this description algebraic description of the feasible region, the constraints are not regular at $(-1, 0)$ because the cones $\mathcal{L}_D(-1, 0)$ and $\mathcal{T}_D(-1, 0) = \{(0, d_2) : d_2 \in \mathbb{R}\}$ do not coincide.

The previous remark shows that the choice of the expressions used to describe the constraints in a non-linear programming problem is important, because a certain choice may lead to regular constraints, while another, in spite of describing the same set, may lead to constraints that are not regular. This fact becomes even clearer by looking at some of the conditions that guarantee the regularity of constraints. The following are amid the most well-known and widely used ones, but far from being the only possible ones.

(5.2) Theorem. *In the same hypotheses as in (3.1), the following is true:*

1. *(Linear constraints) If g_i and h_j are linear functions for every $i = 1, \ldots, m$ and for every $j = 1, \ldots, p$, the constraints are regular at each point $x^* \in D$.*

2. (Slater condition) *If the functions g_i are convex for each $i = 1, \ldots, m$, the functions h_j are linear for every $j = 1, \ldots, p$ and there exists a point $\bar{x}$ in D such that $g(\bar{x}) < 0$ and $h(\bar{x}) = 0$, then the constraints are regular at each point $x^* \in D$.*

3. (Linear independence of the active constraints' gradients) *If $x^* \in D$ and the vectors*
$$\begin{cases} \nabla g_i(x^*) & \text{for all } i \in I(x^*), \\ \nabla h_j(x^*) & \text{for all } j = 1, \ldots, p \end{cases}$$
are linearly independent, then the constraints are regular at the point x^*.

Proof. Let us start by proving the first statement. Let us start by observing that, by means of Theorem (3.4), it suffices to prove that $\mathcal{L}_D(x^*) \subseteq \mathcal{T}_D(x^*)$ for every $x^* \in D$. To this end, take a point $x^* \in D$, and let us consider a direction $d \in \mathcal{L}_D(x^*)$. Since the constraints are linear, we can write:

$$g_i(x) = \langle a_i, x \rangle - b_i, \quad \text{for all } i = 1, \ldots, m$$
$$h_j(x) = \langle c_j, x \rangle - d_j, \quad \text{for all } j = 1, \ldots, p$$

Since $d \in \mathcal{L}_D(x)$, we have

$$\langle a_i, d \rangle \leq 0, \quad \text{for all } i \in I(x^*)$$
$$\langle c_j, d \rangle = 0, \quad \text{for all } j = 1, \ldots, p.$$

Consider any sequence $\{t_k\}$ of strictly positive real numbers which converges towards 0, and define the sequence $\{z^k\}$ by setting $z^k = x^* + t_k d$. Observe that $\{x^k\}$ obviously converges towards x^*. To prove the statement, it suffices to prove that the sequence $\{z^k\}$ is included in D, at least when k is sufficiently large. In fact, if $i \in I(x)$, then

$$g_i(z^k) = \langle a_i, z^k \rangle - b_i = \langle a_i, x^* \rangle - b_i + t_k \langle a_i, d \rangle = t_k \langle a_i, d \rangle \leq 0.$$

On the other hand, if $i \notin I(x^*)$, then

$$g_i(z^k) = \langle a_i, z^k \rangle - b_i = \langle a_i, x^* \rangle - b_i + t_k \langle a_i, d \rangle \leq 0$$

where the last inequality follows from the fact that, since by hypothesis we have $i \notin I(x^*)$, then $\langle a_i, x \rangle - b_i < 0$, and from the fact that, since $\{t_k\}$ is a sequence of strictly positive real numbers convergent towards 0, then, for sufficiently large values of k, we have $t_k \langle a_i, d \rangle \leq \langle a_i, x^* \rangle - b_i$. Furthermore, for every index j, we also have

$$h_j(z^k) = \langle c_j, z^k \rangle - d_j = \langle c_j, x \rangle - d_j + t_k \langle c_j, d \rangle = 0.$$

This proves that, up to extracting from $\{z^k\}$ a suitable sub-sequence, we may assume that $z^k \in D$ for every index k. Since $(z^k - x^*)/t_k$ converges towards d by construction, it follows that $d \in \mathcal{T}_D(x)$, as required.

Let us now move onto the second statement. By means of Theorem (3.4), it is going to be sufficient to prove that $\mathcal{L}_D(x^*) \subseteq \mathcal{T}_D(x^*)$ for every point $x^* \in D$. To this end, take a point $x^* \in D$, and consider a direction $d \in \mathcal{L}_D(x)$. By definition, this means that

$$\langle \nabla g_i(x^*), d \rangle \le 0, \quad \text{for all } i \in I(x^*)$$
$$\langle \nabla h_j(x^*), d \rangle = 0, \quad \text{for all } j = 1, \ldots, p.$$

Since, for every index $i \in I(x^*)$, the functions g_i are convex, and since, by hypothesis, $g_i(x^*) = 0$, we have

$$\langle \nabla g_i(x^*), x - x^* \rangle = g_i(x^*) + \langle \nabla g_i(x^*), x - x^* \rangle \le g_i(x) < 0.$$

Furthermore, since, for each $j = 1, \ldots, p$, the functions h_j are linear, and since, by hypothesis, $h_j(x^*) = 0$, we have

$$\langle \nabla h_j(x^*), x - x^* \rangle = h_j(x^*) + \langle \nabla h_j(x^*), x - x^* \rangle = h_j(x) = 0.$$

Let us now define the sequence $\{d^k\}$ by setting $d^k = \frac{1}{k}(x^* - x) + (1 - \frac{1}{k})d$. It is easy to check that

$$\langle \nabla g_i(x), d^k \rangle < 0 \quad \text{for all } i \in I(x^*)$$
$$\langle \nabla h_j(x), d^k \rangle = 0 \quad \text{for all } j = 1, \ldots, p.$$

For every index k, let us show that $x^* + td^k \in D$ for every sufficiently small strictly positive real number t. In fact, if $i \in I(x^*)$, then

$$g_i(x^* + td^k) = g_i(x^*) + t\langle \nabla g_i(x^*), d^k \rangle + o(t) = t\langle \nabla g_i(x^*), d^k \rangle + o(t),$$

where $o(t)$ is an infinitesimal of order higher than t. Hence, $g_i(x^* + td^k) \le 0$ for every sufficiently small strictly positive real number t. On the other hand, if $i \notin I(x^*)$ then of course $g_i(x^*) < 0$ and therefore, by continuity of g_i, it follows that $g_i(x^* + td^k) \le 0$, for every sufficiently small strictly positive real number t. Moreover, for every index $j = 1, \ldots, p$ we also have that

$$h_j(x^* + td^k) = h_j(x^*) + t\langle \nabla h_j(x^*), d^k \rangle = h_j(x^*) = 0.$$

For every index k, let us now choose a real number t_k, with $0 < t_k < 1/k$, and such that $x^* + t_k d^k \in D$. This allows us to define the sequence $\{z^k\}$, by setting $z^k = x^* + t_k d^k$. By construction, we have $z^k \in D$, the sequence $\{z^k\}$ converges towards x^*, and

$$\lim_{k \to \infty} \frac{z^k - x}{t_k} = \lim_{k \to \infty} d_k = d.$$

This proves that $d \in \mathcal{T}_D(x)$, and the second statement is now proved.

Finally, let us prove the third and last statement. This is slightly more complicated by, once again, by means of Theorem (3.4), it is going to be sufficient to prove the inclusion $\mathcal{L}_D(x^*) \subseteq \mathcal{T}_D(x^*)$. To this end, consider a direction $d \in \mathcal{L}_D(x^*)$. In order to prove that $d \in \mathcal{T}_D(x^*)$ we are going to need to utilise Dini's Implicit Function Theorem. To this end, let us denote by V the $r \times n$ matrix whose rows are the gradients of all the active constraints at x^*. In other words, let us set

$$V = \begin{bmatrix} \nabla g_i(x^*)^T \\ \nabla h_j(x^*)^T \end{bmatrix} \qquad \text{where } i \in I(x^*) \text{ and } j = 1, \ldots, p.$$

By assumption, V has rank r. In addition, denote by Z the $(n-r) \times n$ matrix having as rows an orthonormal basis for the kernel of V. Then, by definition, Z has rank $n - r$, and $VZ = 0$. Consider the function $F : \Omega \times \mathbb{R} \to \mathbb{R}^n$ defined as:

$$F(z, t) = \begin{bmatrix} g_i(z) - t\langle V_i, d\rangle \\ h_j(z) - t\langle V_j, d\rangle \\ Z(z - x^* - td) \end{bmatrix} \qquad \text{where } i \in I(x^*) \text{ and } j = 1, \ldots, p.$$

It is not difficult to recognise that this function is of class $\mathcal{C}^1$, (because such are g_i and h_j, and because such is the scalar product), that $F(x^*, 0) = 0$, and that the Jacobian matrix of F at $(x^*, 0)$ is the $n \times n$ matrix

$$J_z F(x^*, 0) = \begin{bmatrix} V \\ Z \end{bmatrix}.$$

Furthermore, by hypothesis, $J_z F(x^*, 0)$ is non-singular, because the rows of V and Z are linearly independent by construction.

By virtue of Dini's Implicit Function Theorem (see Theorem (2.5)), there exists a real function $\varphi(t)$ of class $\mathcal{C}^1$, defined in a neighbourhood I of 0, such that $\varphi(0) = x^*$, $F(\varphi(t), t) = 0$ for every $t \in I$, and

$$\varphi'(t) = -J_z F\big(\varphi(t), t\big)^{-1} J_t F\big(\varphi(t), t\big) \qquad \text{for all } t \in I.$$

To complete the proof, let us now consider a sequence $\{t_k\}$ of strictly positive real numbers, that converges towards 0 and such that $t_k \in I$, for every index k, and define the sequence $\{z^k\}$ by setting $z^k = \varphi(t_k)$. By the properties of the function $\varphi(t)$, and by the fact that $d \in D(x^*)$, it follows that

$$\begin{aligned} g_i(z^k) &= t_k\langle \nabla g_i(x^*), d\rangle \leq 0 \qquad \text{for all } i \in I(x^*), \\ h_j(z^k) &= t_k\langle \nabla h_j(x^*), d\rangle = 0 \qquad \text{for all } j = 1, \ldots, p. \end{aligned}$$

Hence, the sequence $\{z^k\}$ is all included in D, and, in addition, thanks to the continuity of the function $\varphi(t)$, it follows that the sequence $\{z^k\}$ converges towards x^*. Now, from Taylor's expansion of $\varphi(t)$ at the point $t = 0$, we obtain

$$\varphi(t) = \varphi(0) + t\varphi'(0) + o(t) \qquad \text{for all } t \in I.$$

Since we have

$$J_t F\big(\varphi(0),0\big) = - \begin{bmatrix} V \\ Z \end{bmatrix} d,$$

it follows that

$$\varphi'(0) = -J_z F\big(\varphi(0),0\big)^{-1} J_t F\big(\varphi(0),0\big) = \begin{bmatrix} V \\ Z \end{bmatrix}^{-1} \begin{bmatrix} V \\ Z \end{bmatrix} d = d,$$

and this gives, in particular, $z^k = x^* + t_k d + o(t_k)$. Thus, ultimately,

$$\lim_{k\to\infty} \frac{z^k - x^*}{t_k} = d.$$

This proves that $d \in \mathcal{T}_D(x^*)$, as required. $\qquad\square$

(5.3) Example. We wish to find the maximum and minimum points of the objective function $f(x_1, x_2) = x_1 + x_2$ over the feasible region

$$D = \left\{ (x_1, x_2) \in \mathbb{R}^2 : x_1^2 + x_2^2 - 2 \le 0 \right\}.$$

Let us start by observing that the function f is continuous, and that D is a closed and bounded set, whose boundary is the circle with centre $(0,0)$ and radius $\sqrt{2}$. Thus, by virtue of Weierstrass's Theorem, both the global minimum and the global maximum exist.

The Slater condition is also satisfied, because $g(x_1, x_2) = x_1^2 + x_2^2 - 2$ is convex and $g(0,0) < 0$. Hence, the feasible region D is regular at every point. To find the global maximum and minimum, let us start by solving the KKT system, that is to say the system of simultaneous inequalities:

$$\begin{cases} 1 + 2\lambda x_1 = 0 \\ 1 + 2\lambda x_2 = 0 \\ \lambda(x_1^2 + x_2^2 - 2) = 0 \\ x_1^2 + x_2^2 - 2 \le 0 \end{cases}$$

Let us observe that, if $\lambda = 0$, the first equation is impossible. This means that it must be $\lambda \ne 0$. In this case, the first two equations can be solved for x_1 and x_2, respectively, giving $x_1 = x_2 = -1/(2\lambda)$, while the third equation becomes $x_1^2 + x_2^2 = 2$. This gives finally $x_1 = \pm 1$, whence the solutions of the KKT system are:

$$\begin{aligned} x' &= (1,1) & \lambda' &= -\tfrac{1}{2} \\ x'' &= (-1,-1) & \lambda'' &= \tfrac{1}{2} \end{aligned}$$

Now, taking into account the simple criterion described in Remark (4.11), observe that, since $\lambda' = -\tfrac{1}{2} \le 0$, x' turns out to be the only candidate maximum point; hence it must be the global maximum of f on D. With the same argument, we can also infer that x'' is the global minimum of f on D.

(5.4) Remark. Without the regularity assumption on domain D, Theorem (4.5) may be false. To see this, let us consider, for example, the problem described in the Example (4.4). The domain, D, consists of a single point $x^* = (1,0)$, which is global maximum and global minimum of f on D. On the other hand, we have

$$\nabla f(x^*) = (1,1), \quad \nabla g_1(x^*) = (0,-2), \quad \nabla g_2(x^*) = (0,1),$$

whence it cannot exist a vector $\lambda^* \in \mathbb{R}^2$ such that the pair (x^*, λ^*) solves the KKT system, that is to say such that $\nabla f(x^*) + \lambda_1^* \nabla g_1(x^*) + \lambda_2^* \nabla g_2(x^*) = 0$.

6. First order sufficient conditions

As it happens for functions of a single variable, beside necessary conditions, we can also find sufficient conditions for optimality. Let us start by stating and proving those involving only the first derivatives, i.e. gradient and directional derivatives.

(6.1) Theorem (First order sufficient conditions). *In the hypotheses and with the same notations described in (3.1), assume that we are given the non-linear programming problem to minimise a function $f : E \to \mathbb{R}$ over the feasible region D. Assume, in addition, that the functions g_i are convex (for every $i = 1, \ldots, m$), h_j are linear (for every $j = 1, \ldots, p$), and the triple (x^*, λ^*, μ^*) is a solution of the KKT system. If f is convex and $\lambda^* \geq 0$, then x^* is a global minimum of f on D.*

Proof. Assume that the two vectors λ^* and μ^* are fixed, and observe that the function $L(x, \lambda^*, \mu^*)$ is convex with respect to x, being the sum of convex functions. Furthermore, the triple (x^*, λ^*, μ^*) is a solution of KKT system, which means that $\nabla_x L(x^*, \lambda^*, \mu^*) = 0$. Thus,

$$L(x, \lambda^*, \mu^*) \geq L(x^*, \lambda^*, \mu^*) \qquad \text{for all } x \in \mathbb{R}^n.$$

On the other hand, we have that $L(x^*, \lambda^*, \mu^*) = f(x^*)$ and, for every $x \in D$,

$$L(x, \lambda^*, \mu^*) = f(x) + \sum_{i=1}^{m} \lambda_i^* \, g_i(x) + \sum_{j=1}^{p} \mu_j^* \, h_j(x) \leq f(x),$$

so

$$f(x) \geq f(x^*) \qquad \text{for all } x \in D,$$

that is x^* is a global minimum of the function f on D. $\qquad\qquad\square$

(6.2) Example. Let us apply Theorem (6.1) to find the global maximum and minimum of the objective function $f(x) = x_1 - x_2$ over the feasible region D defined by

$$D = \left\{ (x_1, x_2) \in \mathbb{R}^2 : -x_1 + x_2^2 - 1 \leq 0 \quad x_1 + x_2^2 - 2 \leq 0 \right\}.$$

Let us start by observing that f is a linear function, hence it is at the same time both convex and concave. In addition, the constraints are both convex. Let us solve the KKT system:

$$\begin{cases} 1 - \lambda_1 + \lambda_2 = 0 \\ -1 + 2\lambda_1 x_2 + 2\lambda_2 x_2 = 0 \\ \lambda_1(-x_1 + x_2^2 - 1) = 0 \\ \lambda_2(x_1 + x_2^2 - 2) = 0 \\ -x_1 + x_2^2 - 1 \le 0 \\ x_1 + x_2^2 - 2 \le 0 \end{cases}$$

To this end, it is necessary to distinguish 4 cases. If $\lambda_1 = \lambda_2 = 0$, then, from the first equation, we have $\lambda_1 = 1$, and from the second equation we get $x_2 = \frac{1}{2}$. Furthermore, the third equation gives $x_1 = -\frac{3}{4}$. Since the point with coordinates $\left(-\frac{3}{4}, \frac{1}{2}\right)$ is feasible, a solution of the KKT system is:

$$x' = \left(-\tfrac{3}{4}, \tfrac{1}{2}\right) \qquad \lambda' = (1, 0).$$

Since $\lambda' \ge 0$, x' is a global minimum of f on D, thanks to Theorem (6.1).

If $\lambda_1 = 0$ and $\lambda_2 \ne 0$, then from the first equation, we have $\lambda_2 = -1$, and from the second one $x_2 = -\frac{1}{2}$. Furthermore, from the fourth equation we have $x_1 = \frac{7}{4}$. Since the point with coordinates $\left(\frac{7}{4}, -\frac{1}{2}\right)$ is feasible, another solution of the KKT system is

$$x'' = \left(\tfrac{7}{4}, -\tfrac{1}{2}\right) \qquad \lambda'' = (0, -1).$$

Since $\lambda'' \le 0$, x'' is a global maximum of f on D, thanks to Theorem (6.1).

Finally, if $\lambda_1 \ne 0$ and $\lambda_2 \ne 0$, then from the third and the fourth equations, we have $2x_2^2 - 3 = 0$, whence $x_2 = \pm\frac{1}{2}\sqrt{6}$. From the first equation we then have $\lambda_1 = 1 + \lambda_2$, while from the third equation we get $x_1 = \frac{1}{2}$. If $x_2 = \frac{1}{2}\sqrt{6}$, then from the second equation we also have

$$\lambda_2 = \tfrac{1}{12}\sqrt{6} - \tfrac{1}{2} < 0 \quad \text{and} \quad \lambda_1 = \tfrac{1}{12}\sqrt{6} + \tfrac{1}{2} > 0.$$

Thanks to Theorem (4.5), the point $\left(\frac{1}{2}, \frac{1}{2}\sqrt{6}\right)$ is a saddle point, because the conditions λ_1 and λ_2 have opposite sign. On the other hand, if $x_2 = -\frac{1}{2}\sqrt{6}$, then

$$\lambda_2 = -\tfrac{1}{12}\sqrt{6} - \tfrac{1}{2} < 0 \quad \text{and} \quad \lambda_1 = -\tfrac{1}{12}\sqrt{6} + \tfrac{1}{2} > 0.$$

So, for this reason, we obtain that $\left(\frac{1}{2}, -\frac{1}{2}\sqrt{6}\right)$ is also a saddle point.

7. Second order optimality conditions

In the previous sections, we established that when the feasible region D is regular, a stationary point x^* with $\lambda^* \ge 0$ is a potential local minimum of f

on D. Additionally, we know that $\langle \nabla f(x^*), d \rangle \geq 0$ for every $d \in \mathcal{S}_D(x^*)$. However, if the first-order sufficient conditions do not hold, we cannot determine the nature of the stationary points in a direction $d \in \mathcal{S}_D(x^*)$ such that $\langle \nabla f(x^*), d \rangle = 0$. In such cases, relying solely on first-order derivative information becomes insufficient to fully establish whether f is increasing or decreasing along that direction. To overcome this difficulty, we may wish to consider the second-order derivatives of f, g, and h.

(7.1) Definition.   (Critical cone) In the hypotheses and with the same notations described in (3.1), assume we are given a solution (x^*, λ^*, μ^*) of KKT system. The set

$$C(x^*, \lambda^*, \mu^*) = \left\{ d \in \mathbb{R}^n : \begin{array}{ll} \langle \nabla g_i(x^*), d \rangle = 0 & \text{for all } i \in I(x^*) \text{ with } \lambda_i^* \neq 0 \\ \langle \nabla g_i(x^*), d \rangle \leq 0 & \text{for all } i \in I(x^*) \text{ with } \lambda_i^* = 0 \\ \langle \nabla h_j(x^*), d \rangle = 0 & \text{for all } j = 1, \ldots, p \end{array} \right\}$$

is called the **critical cone** at (x^*, λ^*, μ^*).

As usual, it is easy to recognise that the critical cone is actually a cone.

(7.2) Remark.   If $\lambda^* \geq 0$ (respectively, $\lambda^* \leq 0$), it is not difficult to show that the critical cone, $C(x^*, \lambda^*, \mu^*)$, coincides with the set

$$\left\{ d \in \mathcal{S}_D(x^*) : \langle \nabla f(x^*), d \rangle = 0 \right\}.$$

In fact, if $d \in C(x^*, \lambda^*, \mu^*)$, then, by definition, we have $d \in \mathcal{S}_D(x^*)$. Furthermore, from the KKT system it follows that

$$\langle \nabla f(x^*), d \rangle = -\sum_{i=1}^m \lambda_i^* \langle \nabla g_i(x^*), d \rangle - \sum_{j=1}^p \mu_j^* \langle \nabla h_j(x^*), d \rangle = 0.$$

Conversely, let us assume that $d \in \mathcal{S}_D(x^*)$ and that $\langle \nabla f(x^*), d \rangle = 0$. Then,

$$\begin{array}{ll} \langle \nabla h_j(x^*), d \rangle = 0 & \text{for all } j = 1, \ldots, p, \\ \langle \nabla g_i(x^*), d \rangle \leq 0 & \text{for all } i = 1, \ldots, m \end{array}$$

Moreover, from the aforementioned KKT system we obtain that

$$\sum_{i=1}^m \lambda_i^* \langle \nabla g_i(x^*), d \rangle = 0 \qquad \text{for all } i \in I(x^*) \text{ with } \lambda_i^* \neq 0.$$

Since the terms in the summation all have the same sign, the only possibility would be that $\langle \nabla g_i(x^*), d \rangle = 0$ for every $i \in I(x^*)$, with $\lambda_i^* \neq 0$, which, in turn, proves that $d \in C(x^*, \lambda^*, \mu^*)$.

We are now ready to introduce the second-order necessary conditions.

(7.3) Theorem (Second order necessary conditions). *In the hypotheses and with the notations described in (3.1), assume we are given a non-linear programming problem to minimise an objective function $f : E \to \mathbb{R}$ over the feasible region D. In addition, suppose that f, g_i and h_j are all of class C^2, and suppose that (x^*, λ^*, μ^*) be a solution of the KKT system, and that the gradients of the active constraints are linearly independent. If x^* is a local minimum of f on D, then*

$$\langle d, H_x L(x^*, \lambda^*, \mu^*) d \rangle \geq 0 \qquad \text{for all } d \in C(x^*, \lambda^*, \mu^*).$$

Proof. The proof is done by contradiction. To this end, let us suppose that there exists a direction $d \in C(x^*, \lambda^*, \mu^*)$ such that

$$\langle d, H_x L(x^*, \lambda^*, \mu^*) d \rangle < 0.$$

Since $d \in \mathcal{S}_D(x^*)$, we can use an argument similar to the one described in the proof of Theorem (5.2) point 3. Precisely, we can choose a sequence $\{t_k\}$ of strictly positive real numbers that converges towards 0, and a sequence $\{z^k\}$ of feasible points that converges towards x^*, such that

$$\lim_{k \to \infty} \frac{z^k - x^*}{t_k} = d,$$

$$g_i(z^k) = t_k \langle \nabla g_i(x^*), d \rangle \qquad \text{for all } i \in I(x^*),$$

$$h_j(z^k) = t_k \langle \nabla h_j(x^*), d \rangle \quad \text{for all } j = 1, \ldots, p.$$

As a result, we obtain

$$
\begin{aligned}
L(z^k, \lambda^*, \mu^*) &= f(z^k) + \sum_{i=1}^m \lambda_i^* g_i(z^k) + \sum_{j=1}^p \mu_j^* h_j(z^k) \\
&= f(z^k) + \sum_{i \in I(x^*)} \lambda_i^* t_k \langle \nabla g_i(x^*), d \rangle + \sum_{j=1}^p \mu_j^* t_k \langle \nabla h_j(x^*), d \rangle \\
&= f(z^k)
\end{aligned}
$$

On the other hand, by means of the second order Taylor's expansion of the Lagrangian function, we obtain:

$$
\begin{aligned}
L(z^k, \lambda^*, \mu^*) = L(x^*, \lambda^*, \mu^*) &+ \langle \nabla_x L(x^*, \lambda^*, \mu^*), z^k - x^* \rangle \\
&+ \tfrac{1}{2} \langle z^k - x^*, H_x L(x^*, \lambda^*, \mu^*)(z^k - x^*) \rangle + o\big(\|z^k - x^*\|^2\big).
\end{aligned}
$$

Since $L(x^*, \lambda^*, \mu^*) = f(x^*)$ and $\nabla_x L(x^*, \lambda^*, \mu^*) = 0$, we have

$$
\begin{aligned}
(7.4) \qquad L(z^k, \lambda^*, \mu^*) = f(x^*) &+ \tfrac{1}{2} \langle z^k - x^*, H_x L(x^*, \lambda^*, \mu^*)(z^k - x^*) \rangle \\
&+ o\big(\|z^k - x^*\|^2\big),
\end{aligned}
$$

and therefore

$$f(z^k) - f(x^*) = \tfrac{1}{2} \langle z^k - x^*, H_x L(x^*, \lambda^*, \mu^*)(z^k - x^*) \rangle + o\big(\|z^k - x^*\|^2\big).$$

By dividing both members by t_k^2 and taking the limit as $k \to \infty$ we obtain

$$\lim_{k \to \infty} \frac{f(z^k) - f(x^*)}{t_k^2} < 0,$$

and this is a contradiction, because x^* is a local minimum of f on D. $\qquad \square$

(7.5) Remark. Theorem (7.3) only gives a necessary optimality conditions for local minimum. To see that this condition is, generally, not sufficient, let us consider the non-linear programming problem of minimising the function $f(x) = x_1^3 + x_2$ over the feasible region $D = \{x \in \mathbb{R}^2 : -x_2 \leq 0\}$. The unique solution of the KKT system is $x^* = (0,0)$, $\lambda^* = 1$. The linear constraint is active at x^* and $\nabla g(x^*) = (0, -1)$. Since the matrix $H_x L(x^*, \lambda^*)$ is null, the point x^* satisfies the necessary optimality second order conditions both for minimum and for maximum, but x^* is neither a local minimum, because $f(t, 0) < f(0, 0)$ for every $t < 0$, nor a local maximum because $\lambda^* > 0$.

(7.6) Remark. Let us notice that, when the constraints g e h are absent, the second order necessary optimality condition takes a simpler form. Precisely, if x^* is a local minimum of f, then $Hf(x^*)$ is semidefinite positive on the critical cone, which, in absence of the constraints g and h reduces to the whole space. Observe that, even in this case, the condition is necessary but not sufficient.

(7.7) Theorem (Second order sufficient condition). *Let us suppose that the same hypotheses and with the same notations as in Theorem (7.3) hold.*

1. *If $\lambda^* \geq 0$ and*

$$\langle d, H_x L(x^*, \lambda^*, \mu^*)d \rangle > 0 \qquad \text{for all } d \in C(x^*, \lambda^*, \mu^*), \ d \neq 0.$$

   Then, x^* is a local minimum of f on D.

2. *If $\lambda^* \leq 0$ and*

$$\langle d, H_x L(x^*, \lambda^*, \mu^*)d \rangle < 0 \qquad \text{for all } d \in C(x^*, \lambda^*, \mu^*), \ d \neq 0.$$

   Then, x^* is a local maximum of f on D.

Proof. With no loss of generality, we can limit ourselves to proving the first statement, because the second statement can be transformed in the first by the simple transformation $L \mapsto -L$. In the hypotheses of the first assumption, let us start by observing that the set $\bar{C} = \{d \in C(x^*, \lambda^*, \mu^*) : \|d\| = 1\}$ is closed and bounded, hence compact. Thus, by means of Weierstrass's Theorem, the quadratic function

$$d \mapsto \langle d, H_x L(x^*, \lambda^*, \mu^*)d \rangle$$

has a minimum value over $\bar{C}$. Let us denote it by σ. By hypothesis, we also have $\sigma > 0$. Since $C(x^*, \lambda^*, \mu^*)$ is a cone, we have that $\|d\|^{-1}d \in \bar{C}$ for every $d \in C(x^*, \lambda^*, \mu^*)$. This means that

$$\langle d, H_x L(x^*, \lambda^*, \mu^*)d\rangle \geq \sigma\|d\|^2 \qquad \text{for all } d \in C(x^*, \lambda^*, \mu^*).$$

To prove that x^* is a local minimum of f on D, let us start by checking that there exists a neighbourhood U of x^* such that

$$f(x) \geq f(x^*) + \tfrac{1}{4}\sigma\|x - x^*\|^2 \qquad \text{for all } x \in D \cap U.$$

By contradiction, let us suppose that this latter condition is not true. In other words, let us assume that there exists a sequence $\{z^k\}$ of point in D that converges towards x^* and such that, for every index k, we have

(7.8) $$f(z^k) < f(x^*) + \tfrac{1}{4}\sigma\|z^k - x^*\|^2.$$

Then, u to passing to a convergent subsequence, if necessary, we know that there exists a vector $v \in \mathbb{R}^n$ such that

$$\lim_{k \to \infty} \frac{z^k - x^*}{\|z^k - x^*\|} = v.$$

By definition of cone of tangents, this means that $v \in \mathcal{T}_D(x^*)$. Moreover, from the first part of Theorem (5.2), we have that $\mathcal{T}_D(x^*) \subseteq \mathcal{S}_D(x^*)$ and therefore $v \in \mathcal{S}_D(x^*)$.

Let us now prove that neither $v \in C(x^*, \lambda^*, \mu^*)$, nor $v \notin C(x^*, \lambda^*, \mu^*)$. To this end, let us start by assuming the first of these inclusions, that is to say, let us assume that $v \in C(x^*, \lambda^*, \mu^*)$. This implies $\langle v, H_x L(x^*, \lambda^*, \mu^*)v\rangle \geq \sigma$. Since $\lambda^* \geq 0$ and the sequence $\{z^k\}$ is feasible we have that

$$L(z^k, \lambda^*, \mu^*) = f(z^k) + \sum_{i=1}^{m} \lambda_i^* g_i(z^k) + \sum_{j=1}^{p} \mu_j^* h_j(z^k) \leq f(z^k).$$

Then from equation (7.4), we obtain

$$f(z^k) \geq f(x^*) + \tfrac{1}{2}\langle z^k - x^*, H_x L(x^*, \lambda^*, \mu^*)(z^k - x^*)\rangle + o\big(\|z^k - x^*\|^2\big).$$

By using (7.8), we then get

$$\tfrac{1}{4}\sigma\|z^k - x^*\|^2 > \tfrac{1}{2}\langle z^k - x^*, H_x L(x^*, \lambda^*, \mu^*)(z^k - x^*)\rangle + o\big(\|z^k - x^*\|^2\big).$$

By dividing both sides by $\|z^k - x^*\|^2$ gives

$$\frac{1}{4}\sigma > \frac{1}{2}\left\langle \frac{z^k - x^*}{\|z^k - x^*\|}, H_x L(x^*, \lambda^*, \mu^*)\frac{z^k - x^*}{\|z^k - x^*\|}\right\rangle + \frac{o\big(\|z^k - x^*\|^2\big)}{\|z^k - x^*\|^2}.$$

Then, taking the limit as $k \to \infty$, we obtain

$$\tfrac{1}{4}\sigma \geq \tfrac{1}{2}\langle v, H_x L(x^*, \lambda^*, \mu^*)v\rangle \geq \tfrac{1}{2}\sigma,$$

which is impossible because $\sigma > 0$.

Conversely, let us now suppose that $v \notin C(x^*, \lambda^*, \mu^*)$. This means that there exists an index $i_0 \in I(x^*)$ such that

$$(7.9) \qquad \lambda_{i_0}^* \langle \nabla g_{i_0}(x^*), v \rangle < 0.$$

Using the first order Taylor's expansion of g_{i_0} we have

$$\lambda_{i_0}^* g_{i_0}(z^k) = \lambda_{i_0}^* g_{i_0}(x^*) + \lambda_{i_0}^* \langle \nabla g_{i_0}(x^*), z^k - x^* \rangle + o(\|z^k - x^*\|)$$
$$= \lambda_{i_0}^* \langle \nabla g_{i_0}(x^*), z^k - x^* \rangle + o(\|z^k - x^*\|).$$

and therefore we deduce that

$$L(z^k, \lambda^*, \mu^*) = f(z^k) + \sum_{i=1}^m \lambda_i^* g_i(z^k) + \sum_{j=1}^p \mu_j^* h_j(z^k)$$
$$\leq f(z^k) + \lambda_{i_0}^* g_{i_0}(z^k)$$
$$= f(z^k) + \lambda_{i_0}^* \langle \nabla g_{i_0}(x^*), z^k - x^* \rangle + o(\|z^k - x^*\|).$$

Moreover, the equation (7.4) can be written as

$$L(z^k, \lambda^*, \mu^*) = f(x^*) + o(\|z^k - x^*\|),$$

thus

$$(7.10) \qquad f(z^k) \geq f(x^*) - \lambda_{i_0}^* \langle \nabla g_{i_0}(x^*), z^k - x^* \rangle + o(\|z^k - x^*\|).$$

From the relationships (7.8) and (7.10) we obtain that

$$\tfrac{1}{4}\sigma \|z^k - x^*\|^2 > -\lambda_{i_0}^* \langle \nabla g_{i_0}(x^*), z^k - x^* \rangle + o(\|z^k - x^*\|).$$

Dividing both members by $\|z^k - x^*\|$ and taking the limit as $k \to \infty$ gives

$$0 \geq -\lambda_{i_0}^* \langle \nabla g_{i_0}(x^*), v \rangle,$$

which is impossible, because it contradicts (7.9). In conclusion, we have proved that there exists a neighbourhood U of x^* such that

$$f(x) \geq f(x^*) + \tfrac{1}{4}\sigma \|x - x^*\|^2 \qquad \text{for all } x \in D \cap U,$$

and therefore x^* is a local minimum of f on D. $\qquad\qquad\square$

8. Some worked examples

In this final section, we are going to show a few worked numerical examples of application of the theory developed thus far to solve non-linear programming problems. These examples aim to show how the theory developed in this chapter can be applied, in practice, to solve simple optimisation problems by solving the corresponding KKT system of simultaneous equations to determine the stationary point, and then using second-order conditions to establish the nature of these points.

(8.1) Example. Suppose we are seeking the maximum and minimum of the objective function $f(x) = x_1 x_2$ over the feasible region

$$D = \left\{ x \in \mathbb{R}^2 : x_1^2 + 2x_2^2 - 2 \leq 0 \right\}.$$

The set D is closed and bounded; therefore, a global maximum and global minimum exist. Furthermore, the constraint satisfies the Slater condition because the function $g(x) = x_1^2 + 2x_2^2 - 2$ is convex and $g(0,0) < 0$; thus, the constraint is regular at every point of D. Let us solve the KKT system:

$$\begin{cases} x_2 + 2\lambda x_1 = 0 \\ x_1 + 4\lambda x_2 = 0 \\ \lambda(x_1^2 + 2x_2^2 - 2) = 0 \\ x_1^2 + 2x_2^2 - 2 \leq 0 \end{cases}$$

If $\lambda = 0$, then from the first two equations we deduce that $x_1 = x_2 = 0$, and therefore the first solution of the KKT system is $x^1 = (0,0)$, $\lambda_1 = 0$. Through straightforward calculations, we find the solutions of the KKT system:

$$x^2 = \left(1, \tfrac{1}{2}\sqrt{2}\right), \qquad \lambda_2 = -\tfrac{1}{4}\sqrt{2}, \qquad x^3 = \left(-1, -\tfrac{1}{2}\sqrt{2}\right), \qquad \lambda_3 = -\tfrac{1}{4}\sqrt{2},$$

$$x^4 = \left(1, -\tfrac{1}{2}\sqrt{2}\right), \qquad \lambda_4 = \tfrac{1}{4}\sqrt{2}, \qquad x^5 = \left(-1, \tfrac{1}{2}\sqrt{2}\right), \qquad \lambda_5 = \tfrac{1}{4}\sqrt{2}.$$

The points x^1, x^2, and x^3 are candidates for the global maximum. Since $f(x^1) = 0$ and $f(x^2) = f(x^3) = \tfrac{1}{2}\sqrt{2}$, the points x^2 and x^3 are the global maxima of f on D.

Similarly, the points x^1, x^4, and x^5 are candidates for the global minimum. Since $f(x^1) = 0$ and $f(x^4) = f(x^5) = -\tfrac{1}{2}\sqrt{2}$, the global minima of f on D are x^4 and x^5. The point x^1, which is neither a global maximum nor a global minimum, requires further analysis. We cannot use the first-order sufficient optimality conditions because f is neither convex nor concave, given that the matrix

$$Hf(x^1) = \begin{bmatrix} 0 & 1 \\ 1 & 0 \end{bmatrix}$$

is indefinite. Now, we consider the second-order conditions. The Hessian matrix of the Lagrangian is

$$H_x L(x, \lambda) = \begin{bmatrix} 2\lambda & 1 \\ 1 & 4\lambda \end{bmatrix}$$

Since the constraint is not active at x^1, the critical cone is $C(x^1, \lambda_1) = \mathbb{R}^2$. The matrix

$$H_x L(x^1, \lambda_1) = \begin{bmatrix} 0 & 1 \\ 1 & 0 \end{bmatrix}$$

is indefinite, and therefore the necessary second-order optimality conditions are not satisfied for either a local maximum or a local minimum. In conclusion, x^1 is a saddle point.

(8.2) Example. Suppose we are seeking the maximum and minimum of the objective function $f(x) = (x_1 + 1)^2 + (x_2 + 1)^2$ over the feasible region

$$D = \left\{ x \in \mathbb{R}^2 : x_1 + x_2 - 2 \leq 0, \, -x_1 + x_2 - 2 \leq 0 \right\}.$$

which is unbounded. The function f is coercive (being quadratic with a positive definite Hessian matrix) and therefore admits a global minimum on D, but a global maximum does not exist. The constraints are linear and thus regular at every point of D. Let us solve the KKT system:

$$\begin{cases} 2(x_1 + 1) + \lambda_1 - \lambda_2 = 0 \\ 2(x_2 + 1) + \lambda_1 + \lambda_2 = 0 \\ \lambda_1(x_1 + x_2 - 2) = 0 \\ \lambda_2(-x_1 + x_2 - 2) = 0 \\ x_1 + x_2 - 2 \leq 0 \\ -x_1 + x_2 - 2 \leq 0 \end{cases}$$

- If $\lambda_1 = \lambda_2 = 0$, we get $x^1 = (-1, -1)$, $\lambda_1 = (0, 0)$.
- If $\lambda_1 = 0$ and $\lambda_2 \neq 0$, we get $x^2 = (-2, 0)$, $\lambda_2 = (0, -2)$.
- If $\lambda_1 \neq 0$ and $\lambda_2 = 0$, we get $x^3 = (1, 1)$, $\lambda_3 = (-4, 0)$.
- Finally, if $\lambda_1 \neq 0$ and $\lambda_2 \neq 0$, we get $x^4 = (0, 2)$, $\lambda_4 = (-4, -2)$.

Since x^1 is the only possible minimum, x^1 is the global minimum. Points x^2, x^3, and x^4 are candidates for local maxima. Using the second-order necessary optimality conditions, we examine the Hessian matrix of the Lagrangian, which does not depend on (x, λ):

$$H_x L(x, \lambda) = \begin{bmatrix} 2 & 0 \\ 0 & 2 \end{bmatrix}$$

For each candidate, we check the conditions and conclude that x^2 and x^3 are saddle points, whereas x^4 satisfies the second-order sufficient optimality conditions for a local maximum. Thus, x^4 is a local maximum.

(8.3) Example. Suppose we are seeking the maximum and minimum of the objective function $f(x) = x_1 - x_2^2$ over the feasible region

$$D = \left\{ x \in \mathbb{R}^2 : 4 - x_1^2 - x_2^2 \leq 0 \right\},$$

which is unbounded. We note that neither f nor $-f$ are coercive because they vanish on the parabola $x_1 = x_2^2$. Moreover, neither the sub-level sets

$$L_\alpha^- = \left\{ x \in D : x_1 - x_2^2 \leq \alpha \right\}$$

nor the super-level sets, that is to say,

$$L_\alpha^+ = \left\{ x \in D : x_1 - x_2^2 \geq \alpha \right\}$$

are bounded, so we cannot use any of the sufficient conditions we have seen. However, we observe that f is neither bounded above nor below on D since

$$\lim_{t\to+\infty} f(t,0) = +\infty, \qquad \lim_{t\to-\infty} f(t,0) = -\infty.$$

Thus, there are neither global maxima nor global minima of f on D.

We now study the local extrema. The constraint is regular at all points in D, as the linear independence condition for active constraints holds. Solving the KKT system:

$$\begin{cases} 1 - 2\lambda x_1 = 0, \\ -2x_2 - 2\lambda x_2 = 0, \\ \lambda(4 - x_1^2 - x_2^2) = 0, \\ 4 - x_1^2 - x_2^2 \le 0. \end{cases}$$

If $\lambda = 0$, the first equation is impossible. If $\lambda \ne 0$, the system reduces to:

$$\begin{cases} 1 - 2\lambda x_1 = 0, \\ x_2(1 + \lambda) = 0, \\ 4 - x_1^2 - x_2^2 = 0. \end{cases}$$

We distinguish the cases $x_2 = 0$ and $x_2 \ne 0$. If $x_2 = 0$, from the third equation we get $x_1^2 = 4$, leading to two solutions:

$$x^1 = (2,0), \quad \lambda_1 = \tfrac{1}{4}, \qquad x^2 = (-2,0), \quad \lambda_2 = -\tfrac{1}{4}.$$

If $x_2 \ne 0$, from the second equation we obtain $\lambda = -1$, yielding two more solutions:

$$x^3 = \left(-\tfrac{1}{2}, \tfrac{1}{2}\sqrt{15}\right), \quad \lambda_3 = -1, \qquad x^4 = \left(-\tfrac{1}{2}, -\tfrac{1}{2}\sqrt{15}\right), \quad \lambda_4 = -1.$$

Applying the second-order conditions, the Hessian of the Lagrangian is:

$$H_x L(x, \lambda) = \begin{bmatrix} -2\lambda & 0 \\ 0 & -2 - 2\lambda \end{bmatrix}.$$

- At $x^1 = (2,0)$, $\lambda_1 = \tfrac{1}{4}$:

$$H_x L(x^1, \lambda_1) = \begin{bmatrix} -\tfrac{1}{2} & 0 \\ 0 & -\tfrac{5}{2} \end{bmatrix}.$$

A direction d of the critical cone satisfies:

$$\langle \nabla g(x^1), d \rangle = \langle (-4,0), d \rangle = -4d_1 = 0 \Rightarrow d_1 = 0.$$

Thus, $d = (0, d_2)$ with d_2 arbitrary. The second-order condition requires:

$$\langle d, H_x L(x^1, \lambda_1)d \rangle = -\tfrac{5}{2}d_2^2 \ge 0,$$

which is false for $d_2 \ne 0$, so x^1 is a saddle point.

- At $x^2 = (-2, 0)$, $\lambda_2 = -\frac{1}{4}$:

$$H_x L(x^2, \lambda_2) = \begin{bmatrix} \frac{1}{2} & 0 \\ 0 & -\frac{3}{2} \end{bmatrix}.$$

The same critical cone condition gives $d_1 = 0$, so $d = (0, d_2)$. Evaluating:

$$\langle d, H_x L(x^2, \lambda_2)d \rangle = -\tfrac{3}{2}d_2^2 \leq 0.$$

This confirms x^2 is a local maximum.

- At x^3 and x^4: wee substitute $\lambda = -1$:

$$H_x L(x^3, \lambda_3) = \begin{bmatrix} 2 & 0 \\ 0 & 0 \end{bmatrix}.$$

The quadratic form is:

$$\langle d, H_x L(x^3, \lambda_3)d \rangle = 2d_1^2.$$

Since d_1 is free and positive semi-definiteness is not guaranteed, x^3 and x^4 are saddle points.

So, to sum up, x^2 is a local maximum, and x^1, x^3, x^4 are saddle points.

Chapter VI
Lagrangian Duality

The notion of duality, in non-linear programming, extends the concept of duality from linear programming that we have developed in Chapter IV.

In full analogy with that of linear programming, in the context of non-linear programming, duality provides a framework for understanding the relationship between the original optimisation problem, which is more naturally called in this context the *primal problem*, and a new problem, derived by the primal problem, which is referred to as the *dual problem*. The dual problem often offers insights into the properties of the primal problem and can sometimes be easier to solve.

As we have previously seen in Chapter IV, the primal problem is the name that we give to the original optimisation problem, to avoid to always refer to it as the 'original problem', while the dual problem is a new optimisation problem which is derived from the primal problem's constraints and objective function in such a way that the solutions to the primal and the dual problems are related, and, under certain conditions, solving the dual problem can provide a mean to find the solutions to the primal problem.

1. Lagrangian duality

The aim of Lagrangian duality is to transform a non-linear programming problem into a dual problem by means of the KKT conditions.

(1.1) Definition. Assume that the domain of the real functions f, g_i (for $i = 1, \ldots, m$), and h_j (for $j = 1, \ldots, p$), is a non-empty open set E in $\mathbb{R}^n$. The non-linear programming problem

$$(\mathcal{P}) : \quad \begin{cases} \text{minimise} & f(x) \\ \text{subject to} & g_i(x) \leq 0 \quad \text{for all } i = 1, \ldots, m, \\ & h_j(x) = 0 \quad \text{for all } j = 1, \ldots, p, \end{cases}$$

shall be referred to as a **primal problem**, and its optimal value shall be denoted by $v_\mathcal{P}$. For every pair of vectors $\lambda \in \mathbb{R}^m$, with $\lambda \geq 0$, and $\mu \in \mathbb{R}^p$, the function L defined as

$$L(x, \lambda, \mu) = f(x) + \sum_{i=1}^{m} \lambda_i g_i(x) + \sum_{j=1}^{p} \mu_j h_j(x)$$

for $x \in E$, shall be called the **Lagrangian function** associated to the primal problem $(\mathcal{P})$. Furthermore, the real function φ defined by

$$\varphi(\lambda, \mu) = \inf\left\{ L(x, \lambda, \mu) : x \in E \right\} \qquad \text{for every } \lambda \in \mathbb{R}^m \text{ and } \mu \in \mathbb{R}^p,$$

shall be called the **dual function** associated to the primal problem $(\mathcal{P})$.

The rest of this section focuses on the study of some important analytical properties of this function, and on the usage of this function to define a new problem, the dual problem, which is related to the original problem, the primal problem, in such a way (in analogy with the case of linear programming) that the pair of these two problems have related optimal solutions.

To this end, let us observe that, by definition, the function $\varphi(\lambda, \mu)$ may be $-\infty$, and even when the function $\varphi(\lambda, \mu)$ does not take the value $-\infty$, it is, in general, non-differentiable. Nonetheless, we always have the following simple result.

(1.2) Proposition. *In the same hypotheses as in Definition (1.1), the dual function φ is concave.*

Proof. Observe that, by definition,

$$L(x, \lambda, \mu) = f(x) + \sum_{i=1}^{m} \lambda_i g_i(x) + \sum_{j=1}^{p} \mu_j h_j(x)$$

is a linear function for every x. Since a linear function is both concave and convex, hence it is, in particular, concave, the results follows because φ is the infimum of a family of concave functions (see Proposition II(1.10)). $\square$

So, even though we cannot assure that the function φ is differentiable, at least we can study problems related to this function using results from convex analysis. In particular, we now wish to study a non-linear programming problem related to the function φ. Precisely, let us give the following definition.

(1.3) Definition. In the same hypotheses and with the same notations as in Definition (1.1), the non-linear programming problem

$$(\mathcal{D}) : \qquad \begin{cases} \text{maximise} & \varphi(\lambda, \mu) \\ \text{subject to} & \lambda \geq 0 \end{cases}$$

shall be called the **dual** of the non-linear programming problem $(\mathcal{P})$, and its optimal value shall be denoted by $v_\mathcal{D}$.

In the sequel, let us assume that we are assigned a non-empty open set E in $\mathbb{R}^n$, and that the real functions $f, g_1, \ldots, g_m, h_1, \ldots, h_p$ are all defined in a domain that is contained in E. We shall also assume we are given a primal non-linear programming problem

$$(\mathcal{P}): \quad \begin{cases} \text{minimise} & f(x) \\ \text{subject to} & g_i(x) \le 0 \quad \text{for all } i = 1, \ldots, m, \\ & h_j(x) = 0 \quad \text{for all } j = 1, \ldots, p, \end{cases}$$

we shall denote by $L(x, \lambda, \mu) = f(x) + \sum_{i=1}^{m} \lambda_i g_i(x) + \sum_{j=1}^{p} \mu_j h_j(x)$ its Lagrangian function, by D the feasible region and by $\varphi = \inf\{L(x, \lambda, \mu) : x \in D\}$ its dual function, and we shall assume that its dual problem

$$(\mathcal{D}): \quad \begin{cases} \text{maximise} & \varphi(\lambda, \mu) \\ \text{subject to} & \lambda \ge 0 \end{cases}$$

is also assigned. Finally, let us denote by $v_{\mathcal{P}}$ and $v_{\mathcal{Q}}$ the optimal values of $(\mathcal{P})$ and $(\mathcal{D})$, respectively.

Let us notice that problem $(\mathcal{D})$ is a more general type of non-linear programming problem, as we have thus far assumed that the functions involved in a non-linear program were differentiable. Nevertheless, problem $(\mathcal{D})$ is a problem of maximising a concave function over a convex set, hence it has a wealth of interesting properties. Furthermore, its optimal value is linked to that of the primal problem, $(\mathcal{P})$, as shown in the following proposition.

(1.4) Proposition (Weak Duality). *Let E be a a non-empty open set in $\mathbb{R}^n$, f, g_i (for $i = 1, \ldots, m$), and h_j (for $j = 1, \ldots, p$), real functions whose domain contains E, and consider the optimisation problem*

$$(\mathcal{P}): \quad \begin{cases} \text{minimise} & f(x) \\ \text{subject to} & g_i(x) \le 0 \quad \text{for all } i = 1, \ldots, m, \\ & h_j(x) = 0 \quad \text{for all } j = 1, \ldots, p. \end{cases}$$

Denote by L the Lagrangian function of this problem, by φ the dual function and by $(\mathcal{D})$ the corresponding dual problem. If $v_{\mathcal{P}}$ and $v_{\mathcal{D}}$ are the optimal values of problems $(\mathcal{P})$ and $(\mathcal{D})$, respectively, one always has $v_{\mathcal{D}} \le v_{\mathcal{P}}$.

Proof. Let x be a feasible solution; in other words, assume that x is a point in the feasible region $D = \{x \in \mathbb{R}^n : g(x) \le 0, h(x) = 0\}$. Then,

$$L(x, \lambda, \mu) = f(x) + \sum_{i=1}^{m} \lambda_i\, g_i(x) \le f(x),$$

Hence, for every $\lambda \ge 0$ and $\mu \in \mathbb{R}^p$, we have

$$\varphi(\lambda, \mu) = \inf_{x \in \Omega} L(x, \lambda, \mu) \le \inf_{x \in D} L(x, \lambda, \mu) \le \inf_{x \in D} f(x) = v_{\mathcal{P}},$$

which in turn yields $v_{\mathcal{D}} \le v_{\mathcal{P}}$, as required. $\qquad\square$

(1.5) Remark. The weak duality condition expressed by this result can be rewritten, in terms of the Lagrangian function, in the following way:

$$\sup_{\lambda \geq 0,\, \mu \in \mathbb{R}^p} \inf_{x \in D} L(x, \lambda, \mu) \leq \inf_{x \in D} \sup_{\lambda \geq 0,\, \mu \in \mathbb{R}^p} L(x, \lambda, \mu).$$

As a consequence, note that, the optimal value $v_\mathcal{D}$ of the problem $(\mathcal{D})$, that is also referred to as a **Lagrangian relaxation** of $(\mathcal{P})$, always provides a 'lower bound' to the value $v_\mathcal{P}$.

The difference between the objective values of the dual and the primal problems is known as the **duality gap**. We shall prove that, for convex problems, the duality gap is zero, which means that the optimal values of the primal problem and that of the dual problem turn out to be equal. With this information, we can see that:

- Evaluating the function φ at a point means to find a lower bound of $(\mathcal{P})$.

- Solving $(\mathcal{D})$ means to find the best lower bound for $v_\mathcal{P}$.

- The fact that φ is concave, and that the domain of $(\mathcal{D})$ is a convex set, means that $(\mathcal{D})$ is an, a priori non-differentiable, convex problem, and this allows us to apply sub-gradient methods to solve it.

- Knowing that a problem has zero duality gap is going to help finding a solution numerically.

The Strong Duality Theorem states that, under certain conditions, the optimal value of the primal problem is equal to the optimal value of the dual problem. However, unlike in linear programming, where strong duality is relatively straightforward, in non-linear programming, the conditions for strong duality are somewhat more intricate, and the proof is correspondingly more involved. As we have seen—and similarly to linear programming—the primal problem refers to the original optimisation problem, while the dual problem is derived from the primal using KKT multipliers. For strong duality to hold in non-linear programming, the primal problem typically must be a convex optimisation problem. This means that both the objective function, the constraints g must be convex functions and the constraints h must be linear functions. Convexity ensures that any local minimum is also a global minimum, which is a crucial property for establishing strong duality.

In addition to convexity, another common condition that ensures strong duality in non-linear programming is the 'regularity condition', also known (as already seen) as 'constraint qualification'. This condition ensures that the feasible region of the primal problem is well-behaved, preventing pathological cases where duality might fail. The most frequently used regularity conditions is the Slater's condition, which requires, as already seen, the convexity of feasible region and that there exists at least one feasible point in the relative interior of the feasible region.

Thus, the Strong Duality Theorem in non-linear programming guarantees that, provided these conditions are met, solving the dual problem will yield the same optimal value as solving the primal problem. This equivalence is not only theoretically significant but also practically useful, as it often allows us to choose the simpler of the two problems to solve, depending on the context.

(1.6) Theorem (Strong duality). *In the same hypotheses as in (1.4), let us further assume that the function f and the functions $g_1, \ldots, g_m$ are convex, and that the functions $h_1, \ldots, h_p$ are linear. Let us further assume that there exists a global minimum x^* for the primal problem, and that a constraint qualification holds at x^*. Then:*

1. *The optimal values of the primal problem and the dual problem coincide, that is to say, $v_{\mathcal{D}} = v_{\mathcal{P}}$*

2. *The pair (λ^*, μ^*) is an optimal solution for the dual problem, $(\mathcal{D})$, if and only if the triple (x^*, λ^*, μ^*) satisfies the KKT conditions associated to the primal problem, $(\mathcal{P})$.*

Proof. Observe that the real function $L(x, \lambda, \mu)$ is convex with respect to x, because the functions f, $g_1, \ldots, g_m$, and $h_1, \ldots, h_p$, are all convex. Then, assume that the triple (x^*, λ^*, μ^*) is a solution of the KKT system. Since $v_{\mathcal{D}}$ is the optimal value of the problem $\max \varphi(\lambda, \mu)$, it follows that $v_{\mathcal{D}} \geq \varphi(\lambda^*, \mu^*)$. On the other side, observe that

$$
\begin{aligned}
\varphi(\lambda^*, \mu^*) &= \min\{L(x, \lambda^*, \mu^*) : x \in D\} \\
&= L(x^*, \lambda^*, \mu^*) \\
&= f(x^*) + \sum_{i=1}^{m} \lambda_i^* g_i(x^*) + \sum_{j=1}^{p} \mu_j^* h_j(x^*) \\
&= f(x^*) = v_{\mathcal{P}},
\end{aligned}
$$

where the first equality follows from the fact that L is convex and the penultimate equality follows from the fact that x^* is a KKT solution. Since, from the weak duality, we have $v_{\mathcal{P}} \leq v_{\mathcal{D}}$, it follows that $v_{\mathcal{P}} = v_{\mathcal{D}}$, and (λ^*, μ^*) is an optimal solution for the dual problem $(\mathcal{D})$.

Let us now prove the converse. To this end, let us start by assuming that (λ^*, μ^*) is an optimal solution for the dual problem $(\mathcal{D})$. Then

$$
\begin{aligned}
v_{\mathcal{P}} = v_{\mathcal{D}} = f(x^*) &= \min\{L(x, \lambda^*, \mu^*) : x \in D\} \\
&\leq L(x^*, \lambda^*, \mu^*) \\
&= f(x^*) + \sum_{i=1}^{m} \lambda_i^* g_i(x^*) + \sum_{j=1}^{p} \mu_j^* h_j(x^*) \\
&= f(x^*) + \sum_{i=1}^{m} \lambda_i^* g_i(x^*) \\
&\leq f(x^*).
\end{aligned}
$$

As a result, $\sum_{i=1}^{m} \lambda_i^* g_i(x^*) = 0$ and

$$
\min\{L(x, \lambda^*, \mu^*) : x \in D\} = L(x^*, \lambda^*, \mu^*).
$$

Hence, $\nabla_x L(x^*, \lambda^*, \mu^*) = 0$. This means that (x^*, λ^*, μ^*) is a solution of the KKT conditions associated to the primal problem $(\mathcal{P})$. $\qquad\qquad\square$

(1.7) Example. Consider the (non-convex) problem

$$\begin{aligned} \text{minimise} \quad & -x_1^2 - x_2^2 \\ \text{subject to} \quad & x_1^2 + x_2^2 - 1 \le 0. \end{aligned}$$

We have $v_{\mathcal{P}} = -1$. The Lagrangian function is

$$\begin{aligned} L(x, \lambda) &= -x_1^2 - x_2^2 + \lambda(x_1^2 + x_2^2 - 1) \\ &= (\lambda - 1)x_1^2 + (\lambda - 1)x_2^2 - \lambda \end{aligned}$$

and the corresponding dual function is

$$\varphi(\lambda) = \begin{cases} -\infty & \text{if } \lambda < 1 \\ -\lambda & \text{if } \lambda \ge 1 \end{cases}$$

As a result, the dual problem is

$$\begin{aligned} \text{maximise} \quad & -\lambda \\ \text{subject to} \quad & \lambda \ge 1 \end{aligned}$$

hence its optimal solution is $\lambda^* = 1$ and $v_{\mathcal{D}} = -1$.

In the general case of a non-convex function, the strong duality does not necessarily hold, as the following example shows.

(1.8) Example. Let us consider the following non-convex problem:

$$\begin{aligned} \text{minimise} \quad & -x^2 \\ \text{subject to} \quad & x - 1 \le 0 \\ & -x \le 0. \end{aligned}$$

It is simple to recognise that $v_{\mathcal{P}} = -1$. On the other hand, the Lagrangian associated to this problem is is $L(x, \lambda_1, \lambda_2) = -x^2 + \lambda_1(x - 1) - \lambda_2 x$, hence the dual function is

$$\varphi(\lambda_1, \lambda_2) = \min\big\{ L(x, \lambda_1, \lambda_2) : x \in \mathbb{R} \big\} = -\infty \qquad \text{for all } (\lambda_1, \lambda_2) \in \mathbb{R}^2,$$

which obviously gives $v_{\mathcal{D}} = -\infty$.

2. Duality and saddle points

In non-linear programming, saddle points play a crucial role in understanding the behaviour of optimisation problems. Roughly speaking, in this context, a 'saddle point' is a point where the Lagrangian function, $L(x, \lambda, \mu)$, achieves a minimum with respect to x, and a maximum with respect to (λ, μ). Saddle points are significant because they can provide insights into the nature of the optimal solutions. Identifying saddle points helps in understanding the landscape of the objective function, which is essential for developing efficient algorithms. In fact, as we will see, there are strict relationships both between saddle points and KKT points and between saddle points and duality.

This section delves into the mathematical foundations of saddle points, KKT points and duality, exploring their theoretical underpinnings and practical applications. By understanding these concepts, readers will gain a deeper appreciation of the role of these concepts in non-linear programming.

To this end, let us start by recalling and fixing the notations and the working hypotheses that are going to be assumed throughout the rest of this section. Given a non-empty open set E in $\mathbb{R}^n$, assume that the domain D of the real functions f, g_i (for $i = 1, \ldots, m$), and h_j (for $j = 1, \ldots, p$), is contained in E. In the sequel we shall also use the vector notation, hence we shall set

$$g(x) = \big(g_1(x), \ldots, g_m(x)\big), \qquad h(x) = \big(h_1(x), \ldots, h_p(x)\big).$$

Let us consider the primal non-linear programming problem

$$(\mathcal{P}): \quad \begin{cases} \text{minimise} & f(x) \\ \text{subject to} & g(x) \leq 0, \\ & h(x) = 0 \end{cases}$$

and its corresponding dual problem

$$(\mathcal{D}): \quad \begin{cases} \text{maximise} & \varphi(\lambda, \mu) \\ \text{subject to} & \lambda \geq 0 \end{cases}$$

where $\varphi(\lambda, \mu) = \inf\big\{L(x, \lambda, \mu) : x \in D\big\}$ is the dual function, and

$$L(x, \lambda, \mu) = f(x) + \langle \lambda, g(x) \rangle + \langle \mu, h(x) \rangle$$

is the corresponding Lagrangian function.

(2.1) Definition. A triple (x^*, λ^*, μ^*) shall be called a **saddle point** of the Lagrangian function if, for every $x \in D$, every $\lambda \in \mathbb{R}^m$, with $\lambda \geq 0$, and every $\mu \in \mathbb{R}^p$, we have:

$$L(x^*, \lambda, \mu) \leq L(x^*, \lambda^*, \mu^*) \leq L(x, \lambda^*, \mu^*).$$

The following results show the local geometry around a saddle point.

(2.2) Theorem. *For a triple (x^*, λ^*, μ^*), with $\lambda^* \geq 0$, to be a saddle point of the Lagrangian function it is necessary and sufficient that:*

1. The point x^* is a minimum point of the function $x \mapsto L(x, \lambda^*, \mu^*)$.

2. $g(x^*) \leq 0, \qquad h(x^*) = 0, \qquad \langle \lambda^*, g(x^*) \rangle = 0.$

Furthermore a triple (x^*, λ^*, μ^*) is a saddle point of the Lagrangian function if and only if x^* is the optimal solution of $(\mathcal{P})$, (λ^*, μ^*) is the optimal solution of $(\mathcal{D})$ and there is no duality gap.

Proof. Let us start by showing that the condition is *necessary*, that is to say, let us suppose that (x^*, λ^*, μ^*) is a saddle point for the Lagrangian function L. By means of Definition (2.1), then, x^* must be a minimum of the function $x \mapsto L(x, \lambda^*, \mu^*)$. Furthermore, still from Definition (2.1), we have

$$(2.3) \qquad L(x^*, \lambda^*, \mu^*) \geq L(x^*, \lambda, \mu) = f(x^*) + \langle \lambda, g(x^*) \rangle + \langle \mu, h(x^*) \rangle$$

for all $\lambda \in \mathbb{R}^m$ with $\lambda \geq 0$ and $\mu \in \mathbb{R}^m$. Clearly, this implies that we must have $g(x^*) \leq 0$ and $h(x^*) = 0$, otherwise the previous inequality would fail for a suitable sufficiently large in magnitude choice of t he vector coefficients λ^* and μ^* in (2.3). Now, taking $\lambda = 0$ in (2.3), we obtain that $\langle \lambda^*, g(x^*) \rangle \geq 0$. Noting that $\lambda^* \geq 0$ and that $g(x^*) \leq 0$ imply that $\langle \lambda^*, g(x^*) \rangle \leq 0$, we must have $\langle \lambda^*, g(x^*) \rangle = 0$, and this proves that the condition is necessary.

Let us now prove that the condition is also *sufficient*. To this end, suppose that we are given a triple (x^*, λ^*, μ^*), with $\lambda^* \geq 0$, such that x^* is a minimum of the real function $x \mapsto L(x, \lambda^*, \mu^*)$, subject to the constraints $g(x^*) \leq 0$, $h(x^*) = 0$, and $\langle \lambda^*, g(x^*) \rangle = 0$. Then, in particular,

$$L(x^*, \lambda^*, \mu^*) \leq L(x, \lambda^*, \mu^*).$$

Furthermore,

$$\begin{aligned} L(x^*, \lambda^*, \mu^*) &= f(x^*) + \langle \lambda^*, g(x^*) \rangle + \langle \mu^*, h(x^*) \rangle \\ &\geq f(x^*) + \langle \lambda, g(x^*) \rangle + \langle \mu, h(x^*) \rangle \\ &= L(x^*, \lambda, \mu) \end{aligned}$$

for all $\lambda \in \mathbb{R}^m$ with $\lambda \geq 0$, and $\mu \in \mathbb{R}^p$. Hence, (x^*, λ^*, μ^*) is a saddle point. This proves the first part of the theorem.

To prove the second part, suppose again that (x^*, λ^*, μ^*) is a saddle point. Then, by means of the sufficient condition, x^* is a feasible solution for $(\mathcal{P})$. Since $\lambda^* \geq 0$, then (λ^*, μ^*) is also a feasible solution for $(\mathcal{D})$. Furthermore, by means of the sufficient optimality condition,

$$\begin{aligned} \varphi(\lambda^*, \mu^*) &= L(x^*, \lambda^*, \mu^*) \\ &= f(x^*) + \langle \lambda^*, g(x^*) \rangle + \langle \mu^*, h(x^*) \rangle \\ &= f(x^*). \end{aligned}$$

Thus, x^* and (λ^*, μ^*) solve problems $(\mathcal{P})$ and $(\mathcal{D})$, respectively, and we also have $f(x^*) = \varphi(\lambda^*, \mu^*)$. Hence, we have $g(x^*) \leq 0$, $h(x^*) = 0$ and $\lambda^* \geq 0$.

Moreover, we have by the primal and the dual feasibility that

$$\varphi(\lambda^*, \mu^*) = \min\Big\{ f(x) + \langle \lambda^*, g(x) \rangle + \langle \mu^*, h(x) \rangle : x \in D \Big\}$$
$$\leq f(x^*) + \langle \lambda^*, g(x^*) \rangle + \langle \mu^*, h(x^*) \rangle$$
$$= f(x^*) + \langle \lambda^*, g(x^*) \rangle$$
$$\leq f(x^*)$$

But $\varphi(\lambda^*, \mu^*) = f(x^*)$, by hypothesis. Hence, equality holds, true throughout the chain of inequalities above. In particular, $\langle \lambda^*, g(x^*) \rangle = 0$, so

$$L(x^*, \lambda^*, \mu^*) = f(x^*) = \varphi(\lambda^*, \mu^*) = \min\big\{ L(x, \lambda^*, \mu^*) : x \in \mathbb{R}^n \big\}.$$

Hence, the sufficient condition holds true, so (x^*, λ^*, μ^*) is a saddle point. This completes the proof. $\qquad\square$

The previous result can be further detailed with simple additional constraint qualifications, as the following result shows.

(2.4) Theorem. *Let us suppose that f and $g_1, \ldots, g_m$ are convex and continuously differentiable, $h_1, \ldots, h_p$ are linear and there exists at least one point $\bar{x}$ such that $g_i(\bar{x}) < 0$ for every index $i = 1, \ldots, m$. Thus, if a triple (x^*, λ^*, μ^*) is a solution of the corresponding KKT system, then (x^*, λ^*, μ^*) is a saddle point of the Lagrangian function associated to the non-linear programming problem $(\mathcal{P})$.*

Conversely, if a triple (x^*, λ^*, μ^*) is a saddle point of the Lagrangian function of the primal non-linear programming problem, $(\mathcal{P})$, and $\lambda^* \geq 0$, then x^* is a feasible point for $(\mathcal{P})$, and (x^*, λ^*, μ^*) is a solution of the corresponding KKT system.

Proof. $(\Rightarrow)$: Let (x^*, λ^*, μ^*) be a solution of the KKT system. As we have mentioned in V(4.7), this means that $\nabla_x L(x^*, \lambda^*, \mu^*) = 0$, and, taking into account the convexity hypotheses on f and $g_1, \ldots, g_m$ and the linearity of $h_1, \ldots, h_p$, this implies $L(x^*, \lambda^*, \mu^*) \leq L(x, \lambda^*, \mu^*)$. On the other hand, since (x^*, λ^*, μ^*) is a solution of the KKT system, thanks to Theorem (2.2), we also have $g(x^*) \leq 0$ and $h(x^*) = 0$ and $\langle \lambda^*, g(x^*) \rangle = 0$. Hence,

$$L(x^*, \lambda^*, \mu^*) = f(x^*) \geq f(x^*) + \langle \lambda, g(x^*) \rangle + \langle \mu, h(x^*) \rangle$$

where the inequality follows from the fact that $g(x^*) \leq 0$ and $h(x^*) = 0$. Thus, $L(x^*, \lambda^*, \mu^*) \geq L(x^*, \lambda, \mu)$ and this proves the first part of the theorem.

$(\Leftarrow)$: Let (x^*, λ^*, μ^*) be a saddle point. Then x^* is a minimum of the function $x \mapsto L(x, \lambda^*, \mu^*)$, and therefore $\nabla_x L(x^*, \lambda^*, \mu^*) = 0$. Furthermore, (λ^*, μ^*) is a maximum of the dual function $\varphi(\lambda, \mu) = L(x^*, \lambda, \mu)$, hence

$$\begin{cases} \dfrac{\partial L}{\partial \lambda_i}(x^*, \lambda^*, \mu^*) \leq 0 & \text{if } \lambda_i^* = 0, \\[2ex] \dfrac{\partial L}{\partial \lambda_i}(x^*, \lambda^*, \mu^*) = 0 & \text{if } \lambda_i^* > 0. \end{cases}$$

On the other hand, $(\partial L/\partial \lambda_i)(x^*, \lambda^*, \mu^*) = g_i(x^*)$, for all $i = 1, \ldots, m$, and this gives $g_i(x^*) \leq 0$ and $\lambda_i^* g_i(x^*) = 0$ for every index $i = 1, \ldots, m$. Furthermore, by definition of saddle point we also have $\nabla_\mu L(x^*, \lambda^*, \mu^*) = 0$ and therefore $h(x^*) = 0$, which completes the proof. $\qquad\square$

3. Special cases of duality

In non-linear programming, duality theory offers a robust framework for the analysis of optimisation problems. Among the various special cases, the linear and quadratic scenarios are particularly noteworthy due to their extensive applications and the elegant mathematical properties they possess.

As previously discussed, linear duality emerges in the context of linear programming, where both the objective function and the constraints are linear. In this case, the dual problem also takes the form of a linear program. It will be demonstrated that linear programming duality is a specific instance of the Lagrangian duality addressed in this section.

Another notable special case is that of quadratic programming. Quadratic duality relates to quadratic programming, where the objective function is quadratic, and the constraints remain linear. The dual problem in quadratic programming is likewise a quadratic program. Quadratic duality holds particular significance in fields such as finance, where portfolio optimisation problems frequently involve quadratic objectives, and in machine learning, where support vector machines (SVMs) are formulated as quadratic programs. The duality theory in quadratic programming aids in the development of efficient algorithms and enhances the understanding of the properties of optimal solutions.

Linear programming duality

Let us begin by showing that the Lagrangian duality is consistent with the definition of duality already given in the case of a linear programming problem. To this end, let $A \in \mathfrak{M}(m, n, \mathbb{R})$ be a non-zero matrix, $b \in \mathbb{R}^m$ and $c \in \mathbb{R}^n$ be vectors, and take this problem:

$$(\mathcal{P}): \qquad \begin{cases} \text{minimise} & \langle c, x \rangle \\ \text{subject to} & Ax \geq b \end{cases}$$

which is a linear programming problem in one of the possible standard formats. The Lagrangian function associated to this problem can be written, for $\lambda \in \mathbb{R}^m$, as

$$L(x, \lambda) = \langle c, x \rangle + \langle \lambda, b - Ax \rangle = \langle \lambda, b \rangle + \langle c - A^T \lambda, x \rangle.$$

From this expression for the Lagrangian function, we can now compute the dual function

$$\varphi(\lambda) = \min_{x \in \mathbb{R}^n} L(x, \lambda) = \begin{cases} -\infty & \text{if } c - A^T\lambda \neq 0 \\ \langle \lambda, b \rangle & \text{if } c - A^T\lambda = 0 \end{cases}$$

and this finally gives us the dual problem

$$(\mathcal{D}): \quad \begin{cases} \text{maximise} & \varphi(\lambda) \\ \text{subject to} & \lambda \geq 0 \end{cases} \quad \Rightarrow \quad \begin{cases} \text{maximise} & \langle \lambda, b \rangle \\ \text{subject to} & A^T\lambda = c, \\ & \lambda \geq 0 \end{cases}$$

and observe that this is the dual linear program that we would expect according to the notion of duality introduced in Chapter IV, section 3.

Quadratic programming duality

Another special case of duality that is extremely remarkable for applications is that of the dual of a strictly convex quadratic problem. To introduce this, let $Q \in \mathfrak{M}(n, n, \mathbb{R})$ and $A \in \mathfrak{M}(m, n, \mathbb{R})$ be two non-zero matrices, with Q positive definite, $b \in \mathbb{R}^m$, and $c \in \mathbb{R}^n$ be vectors. Then, the standard form of a strictly convex quadratic programming problem is:

$$(\mathcal{P}): \quad \begin{cases} \text{minimise} & \frac{1}{2}\langle x, Qx \rangle + \langle c, x \rangle \\ \text{subject to} & Ax \leq b \end{cases}$$

The corresponding Lagrangian function is

$$L(x, \lambda) = \tfrac{1}{2}\langle x, Qx \rangle + \langle c, x \rangle + \langle \lambda, Ax - b \rangle.$$

By taking the minimum of the Lagrangian function, we can now compute the dual function as $\varphi(\lambda) = \min_{x \in \mathbb{R}^n} L(x, \lambda)$. To find an analytic expression for this function, let us start by observing that $L(x, \lambda)$ is quadratic and convex function in x, hence its minimum point is, in particular, the stationary point, which can be calculated solving the equation

$$\nabla_x L(x, \lambda) = Qx + c + A^T\lambda = 0,$$

which gives

$$x = -Q^{-1}\left(A^T\lambda + c\right).$$

Thus, the dual function becomes

$$\varphi(\lambda) = -\tfrac{1}{2}\langle \lambda, AQ^{-1}A^T\lambda \rangle - \langle \lambda, b - AQ^{-1}c \rangle - \tfrac{1}{2}\langle c, Q^{-1}c \rangle.$$

After changing the sign to transform the dual into a minimisation problem, the dual problem can be written as

$$(\mathcal{D}): \quad \begin{cases} \text{minimise} & \frac{1}{2}\langle \lambda, P\lambda \rangle + \langle z, \lambda \rangle \\ \text{subject to} & \lambda \geq 0 \end{cases}$$

where $P = AQ^{-1}A^T$ and $z = b - AQ^{-1}c$. Observe that $(\mathcal{D})$ is again a quadratic problem, but it is slightly simpler than the primal problem, because the feasible region is much simpler.

(3.1) Example. In the special uni-dimensional case, let us consider the following quadratic programming problem:

$$(\mathcal{P}) : \begin{cases} \text{minimise} & x^2 \\ \text{subject to} & -x \leq -1. \end{cases}$$

Here, $Q = [2]$, $c = [0]$, $A = [-1]$ and $b = [-1]$. The optimal solution of this problem is $x^* = 1$ and the optimal value is 1. The Lagrangian function is

$$L(x, \lambda) = x^2 + \lambda(1 - x), \qquad \lambda \geq 0.$$

So, the matrix P is equal to $[1/2]$ and the vector z is equal to $[-1]$; the dual problem is

$$(\mathcal{D}) : \begin{cases} \text{maximise} & \frac{1}{4}\lambda^2 - \lambda \\ \text{subject to} & \lambda \geq 0, \end{cases}$$

that is to say, applying an equivalent transformation,

$$(\mathcal{D}) : \begin{cases} \text{minimise} & -\frac{1}{4}\lambda^2 + \lambda \\ \text{subject to} & \lambda \geq 0. \end{cases}$$

The optimal solution of this problem is $\lambda = 2$ which is the KKT conditions associated to $x^* = 1$, and the optimal value is 1.

(3.2) Remark. A special type of quadratic programming problem, that occur every so often in practical settings, is that of finding the minimum norm solution of a linear system of simultaneous equations. More precisely, if $A \in \mathfrak{M}(n, n, \mathbb{R})$ is a given non-singular matrix, and $b \in \mathbb{R}^n$ is a vector, consider the problem

$$(\mathcal{P}) : \begin{cases} \text{minimise} & \frac{1}{2}\langle x, x \rangle \\ \text{subject to} & Ax = b \end{cases}$$

Applying the results developed throughout this section, it is readily shown that its dual problem is

$$(\mathcal{D}) : \begin{cases} \text{maximise} & -\frac{1}{2}\langle A^T\lambda, A^T\lambda \rangle + \langle b, \lambda \rangle \\ \text{subject to} & \lambda \in \mathbb{R}^p \end{cases}$$

which is a concave unconstrained maximisation problem.

Part III

Methods of Optimisation

Chapter VII
Algorithms and Their Convergence

In the previous chapters we have discussed the necessary and sufficient conditions for linear and non-linear programming problems to have an optimal solution. In general, however, the problem of finding this optimal solution is extremely difficult if not impossible. For this reason, it becomes necessary to develop rigorous mathematical procedures that enable us to approximate the optimal solution. Such a procedure constitutes what is called, in technical terms, an 'algorithm', and in this chapter we shall start by introducing the notion of algorithm, discuss the conditions that ensure its convergence towards an optimal solution, and what are the criteria that can be used to 'judge' the goodness of an algorithm. In particular, this chapter will analyse the notion of algorithm and its properties in a general setting, while we shall utilise the general results developed in this chapter to discuss on specific algorithms in the subsequent chapters.

Before moving on, however, it is important to clarify that the word 'algorithm' represents a very general and complex notion, and trying to give a universal definition would be a complex task more suited to Mathematical Logic than to the theory of optimisation, in the same way as it is in the realm of Logic that one can formally define what a 'proof' is. The best we can do, in our limited context of mathematical programming problems, is to introduce a special type of algorithms, which are those utilised in optimisation. They are the so-called iterative algorithms. To treat them in an unified way we can use multi-functions.

A. Carpignani and M. Pappalardo, *Theory and Methods of Optimisation*,
UNITEXT 175, https://doi.org/10.1007/978-3-032-01514-3_7

1. A gentle introduction to the notion of multi-function

A multi-function, or a set-valued mapping, is a special type of mapping that to each point of a domain associates a set. From a purely algebraic point of view, multi-functions are just mappings with values in a power set, but from an applied mathematical point of view, they play a very important role because they describe in a very effective way the notion of iterative algorithm with "variable step". To this end, let us recall that if E is a set, the **power set** of E is the set $\mathcal{P}(E)$ which contains all subsets of E as its elements.

(1.1) Definition. Let X and Y be two sets. A **multi-function** A from X to Y is a mapping $A : X \to \mathcal{P}(Y)$, defined on the set X, with values in the set of all subsets of Y. In other words, a multi-function A from X to Y is a mapping that to each element $x \in X$ associates a subset $A(x)$ of Y.

Let us also observe that to every function $f : X \to Y$ is associated a multi-function $F : X \to \mathcal{P}(Y)$, which is simply defined as $F(x) = \{f(x)\}$. This shall be referred to as the multi-function **associated** to f.

Simple examples of multi-functions are easy to find. For example, if x is a real number, define $A(x)$ as the set of all real numbers that are, in absolute values, smaller than x. In other words, set $A(x) = \{y \in \mathbb{R} : |y| < x\}$. This is a multi-function from $\mathbb{R}$ to $\mathcal{P}(\mathbb{R})$. Observe that, in this example, if x is a negative number, then $A(x)$ is the empty set, because there are no real numbers that are, in absolute value, smaller than a negative number. On the other hand, for $x = 1$, we have $A(1) = (-1, 1)$.

Looking back at Chapter II, Section 5, let us recall that if D is a non-empty convex set in $\mathbb{R}^n$ and $f : D \to \mathbb{R}$ is a convex function, the sub-differential, $\partial f(x)$, is the set of all sub-gradients of f at a point $x \in D$. With the new terminology introduced in this section, we can now also say that the function ∂f that associates to each point x the sub-differential, $\partial f(x)$, of f at x is indeed a multi-function from D to $\mathbb{R}^n$. Let us also see a numerical example of this special type of multi-function.

(1.2) Example. Consider the real function $f(x) = |x|$, defined for every real number x. It is well-known that f is a convex function. Its sub-differential is the multi-function ∂f defined on $\mathbb{R}$ by

$$
\partial f(x) = \begin{cases} \{-1\} & \text{if } x < 0, \\ [-1, 1] & \text{if } x = 0, \\ \{1\} & \text{if } x > 0. \end{cases}
$$

The next definition is a bit more technical, but rather useful. Indeed, as we shall see immediately afterwards, a multi-function is slightly more flexible than a single-valued function (the usual mapping), especially as regard to its domain. Thanks to the fact that the empty set, $\emptyset$, is a (trivial) subset of every set, we shall see shortly that we can always exploit this fact to extend every multi-function to a bigger set in a trivial but effective manner.

(1.3) Definition. Let X and Y be two sets, and let A be a multi-function from X to Y. The **domain** of A is the subset of X of all elements x of X for which $A(x)$ is not empty. We shall denote the domain of A by $\mathrm{dom}(A)$. In symbols, the domain of A is defined as

$$\mathrm{dom}(A) = \Big\{ x \in X : A(x) \neq \emptyset \Big\}.$$

The domain of a multi-function is the part of the 'input' set, X, that gives a non-empty 'output', i.e. that is mapped into 'something'. At first glance, one could be tempted to think that the domain of a multi-function is indeed the part of the input set X 'that matters', but this would be a useless restriction. In fact, the real power of multi-function is actually the fact that they can encode, in this simple way, the fact that sometimes things can go wrong.

(1.4) Remark. If K is a subset of X and $A : K \to \mathcal{P}(Y)$ is a multi-function with domain $\mathrm{dom}(A) \subseteq K$, we can always extend F to a multi-function A^* defined over the set X so that $\mathrm{dom}(A^*) = \mathrm{dom}(A) \subseteq K$. To this end, it suffices to set $A^*(x) = A(x)$ for every $x \in K$ and $A^*(x) = \emptyset$ otherwise. As a result of this, when it looks convenient, we can always assume that a multi-function is defined on the whole set X.

We can now move onto the 'topology' of multi-function, defining what we mean for a multi-function to be 'continuous'. Before doing so, however, let us remark that the next couple of definitions could be given in a much more abstract context, but for the scope of this presentation, and having in mind the applications of multi-functions that we are going to develop in the sequel, we shall limit ourselves to the simple case in which the sets X and Y are subsets of $\mathbb{R}^n$. Hence, unless otherwise specified, we shall make the blanket assumption that in the rest of this section any set is a subset of $\mathbb{R}^n$.

(1.5) Definition. Let X and Y be two sets, A a multi-function from X to Y, and $\bar{x}$ a point in X. We shall say that A is **upper hemi-continuous** at $\bar{x}$ if, for every open set V in Y containing $A(\bar{x})$, there exists a neighbourhood $U \in N(\bar{x})$ such that $A(x) \subseteq V$ for every $x \in U$. In addition, we shall that A is **upper hemi-continuous** if A is upper hemi-continuous at every point x in X.

Roughly speaking, the previous definition formalises the idea that a multi-function A is upper hemi-continuous at $\bar{x}$ if, as the point x approaches $\bar{x}$, the value $A(x)$ of the multi-function does not 'jump out' of a fixed neighbourhood of $A(\bar{x})$. Furthermore, it is worth noticing that this definition turns into the usual definition of continuity if A is the multi-function associated to a function $f : X \to Y$.

(1.6) Definition. Let X and Y be two sets, A a multi-function from X to Y, and $\bar{x}$ a point in X. We shall say that A is **lower hemi-continuous** at $\bar{x}$ if, for every y in $A(x)$ and every neighbourhood $V \in N(y)$, there exists a neighbourhood $U \in N(\bar{x})$ such that $A(x) \cap V \neq \emptyset$ for all $x \in U$. In addition, we shall say that A is **lower hemi-continuous** if A is lower hemi-continuous at every point x in X.

Intuitively, the latter definition assures that, when A is lower hemi-continuous at $\bar{x}$, then at least some of the outputs in $A(x)$ remains near $A(x)$, as x approaches $\bar{x}$. Furthermore, if A is the multi-function associated to a function f, this definition turns into the definition of continuity, and, in this special case, the condition $A(x) \cap V \neq \emptyset$ becomes just $f(x) \in V$.

(1.7) Definition. Let X and Y be two sets, A a multi-function from X to Y, and $\bar{x}$ a point in X. We hall say that A is **continuous** at $\bar{x}$ if it is, at the same time, lower and upper hemi-continuous at $\bar{x}$. In addition, we shall say that A is **continuous** if it is continuous at every point x in X.

(1.8) Example. Let $F, G : \mathbb{R} \to \mathcal{P}(\mathbb{R})$ the multi-functions defined as:

$$F(x) = \begin{cases} [-1, 1] & \text{if } x = 0, \\ \{0\} & \text{if } x \neq 0, \end{cases} \qquad G(x) = \begin{cases} \{0\} & \text{if } x = 0, \\ [-1, 1] & \text{if } x \neq 0. \end{cases}$$

It is not hard to recognise that F is upper hemi-continuous, but not lower hemi-continuous at 0. Similarly, the function G is lower hemi-continuous, but not upper hemi-continuous at 0.

The following definition shows how to construct a multi-function from two multi-functions, through a generalisation of the notion of composition.

(1.9) Definition. Let X, Y and Z be non-empty sets, and $B : X \to \mathcal{P}(Y)$ and $C : Y \to \mathcal{P}(Z)$ two assigned multi-functions. The **composite map** $A = C \circ B$ is defined as the multifunction $A : X \to \mathcal{P}(Z)$ as

$$A(x) = \bigcup_{y \in B(x)} C(y) \qquad \text{for every } x \in X.$$

Let us notice that, in the hypotheses of the previous definition, in the special case where C is the multi-function associated to a mapping $g : Y \to Z$, the multi-function $A = g \circ B$ becomes:

$$A(x) = \Big\{ g(y) : y \in B(x) \Big\} \qquad \text{for every } x \in X.$$

If, in addition, B is the multi-function associated to a mapping $f : X \to Y$, then the multi-function $A = C \circ B$ coincides with the one associated to the composite map $g \circ f$.

The usage of multi-function lends itself well to representing the uncertainty in the value of a function f. Indeed, given a point x, it may happen that the value of $f(x)$ is only partially (or approximately) known. In such cases, for every $x \in X$, one can only assign a set $A(x)$, with the idea that the value $f(x)$ certainly belong to this set. Roughly speaking, a selection, s, of a multi-function, $A : X \to \mathcal{P}(Y)$, represents a way of choosing, for every point x in the set X, a value $s(x)$, among the set of all feasible values, $A(x)$. This idea is formalised in the next definition.

(1.10) Definition. Let X and Y be two sets, and A a multi-function from X to Y. A **selection** of A is a mapping $s : X \to Y$ such that

$$s(x) \in A(x) \qquad \text{for every } x \in X.$$

Loosely speaking, a selection is just a formal procedure to extract a point from a set. An obvious necessary condition for a selection to exist is that $A(x)$ is non-empty, for every $x \in X$; in this case, the existence of a selection is guaranteed by the axiom of choice. In the most general setting, when X and Y are two infinite sets, the existence of such a mapping opens up to incredibly complicated set-theoretical issues, but in the practical examples coming from optimisation, this sort of problems hardly ever occurs, and we shall always be able to assume that, given a multi-function, we can always find a selection. In many cases, in fact, we shall be able to explicitly exhibit such a selection.

2. Algorithms and their properties

An 'algorithm' may be described as a well-defined, mathematically rigorous step-by-step procedure or set of rules designed to solve a specific problem or achieve a desired goal. In its most general sense, an algorithm specifies how to transition from one state to another in a process, ultimately arriving at a solution or at an its approximation. Algorithms are used as specifications for performing calculations and data processing. In particular, in optimisation, an algorithm can be viewed as a procedure that maps a given input (e.g., a problem instance, a point in the domain, or a set of parameters) to a set of outputs, often referred to as 'candidate solutions' or 'iterates'. So, for the purpose of optimisation, an algorithm may be described by assigning a multi-function

$$A : \mathcal{I} \to \mathcal{P}(\mathcal{O})$$

from a set of initial states, $\mathcal{I}$, to a set of final states, $\mathcal{O}$.

At first glance, this idea that an algorithm may be seen as a multi-function may look quite obscure, because the word 'algorithm', in most people's head, generally looks more like a very detailed recipe which is often written in some programming language, or using a more or less mysterious *pseudo-code*.

So how does that detailed recipe that spontaneously appear in the collective imaginary become this rather bizarre representation? To explain this fact, let us start by observing that, to some extent, an algorithm may be thought at least as a mapping, because it takes an initial state and, by means of the step-by-step procedure described in the detailed coded recipe that defines an algorithm, returns an 'output state'. Similarly, a mapping may be described as a black-box that takes input values, or initial states, and returns output values, or final states. So, an algorithm may indeed be thought a special type of black-box, which is somewhat 'transparent', that is to say, which shows the details of the whole internal mechanism within the mapping's black-box. In the end, it looks quite reasonable that an algorithm may be thought as a special type of mapping, but it remains to understand why it is in the language of multi-functions that an algorithm may be better described.

To show this, let us consider a very simple algorithm, that takes a list of n whole numbers, x, and returns those that are even. The recipe, written with a pseudo-code, is described in Algorithm 1.

Algorithm 1 EVEN NUMBERS FINDER

Require: A list x of n whole numbers
Ensure: A list of even numbers
 1: **for** i in x **do**
 2:        **if** i is even **then**
 3:                **return** the number i
 4:        **end if**
 5: **end for**

Looking at Algorithm 1 it is immediately clear that the number of outputs is not prescribed but will depend, ultimately, on the number of even numbers present in the list x. Theoretically, therefore, this algorithm is not different from the multi-function $A : \mathbb{N}^n \to \mathcal{P}(\mathbb{N})$, defined on $\mathbb{N}^n$ by

$$A(x) = \Big\{ y \in x : y \text{ is even} \Big\}.$$

Having clarified how an algorithm can be seen as a multi-function, we now wish to focus on those algorithms that are used in optimisation. In fact, in optimisation, algorithms are typically 'iterative'. Intuitively, an 'iterative' algorithm means that a certain core algorithm is repeated over and over. To understand how an iterative algorithm works, let us consider a general optimisation problem to (say) minimise a function $f(x)$ subject to the constraint $x \in D$. For such a general problem, as we have already mentioned, even with further assumptions to guarantee uniqueness, finding an optimal solution (a global minimum) algebraically is extremely difficult if not impossible. For this reason, the usual approach consists in taking an initial guess of the optimal solution, x^0, and choosing an algorithm, that is to say a multi-function $A : D \to \mathcal{P}(D)$, apply this algorithm to the initial guess x^0, and then picking a new guess $x^1 \in A(x^0)$. With a wise choice of the multi-function A, the new point x^1 will be a better approximation than the initial guess, x^0. Proceeding inductively, then, assuming that a sequence $x^1, \ldots, x^k$ has been generated, we can hope to get an even better approximation of the optimal solution by picking $x^{k+1} \in A(x^k)$.

By means of this procedure, we manage to define a sequence $\{x^k\}$ that, with some luck, and a wise choice of the multi-function A, either converges towards an optimal solution, or at the very least it converges (up to extracting a suitable sub-sequence) towards something that is 'close enough' to be an optimal solution.

This is definitely our ultimate goal. Alas, in most cases, this is also an impossible goal to reach. Indeed, in many practical cases, this is too much of a tight request, and we may have to content ourselves with a less favourable outcome. In fact, if we cannot count on convexity, or if the problem size is too big, an optimisation problem easily becomes unmanageable, and we may wish to limit ourselves to finding an iterative algorithm that generates a sequence $\{x^k\}$ that eventually lies in a specified set that we deem good enough for our purposes. For example, for a non-linear programming problem, we could require that the limit points of the sequence generated by an iterative algorithm exist and lie in one of the following sets:

$$\Omega = \Big\{ x : x \text{ is a global optimal solution of the problem} \Big\},$$

$$\Omega = \Big\{ x : x \text{ is a local optimal solution of the problem} \Big\},$$

$$\Omega = \Big\{ x : x \text{ satisfies the KKT optimality conditions} \Big\}.$$

Generally, convergence of algorithms is made in reference to the set of stationary solutions rather than to the collection of global optimal solutions.

(2.1) Definition. Let D be a non-empty set in $\mathbb{R}^n$, Ω a non-empty subset of D, x a point in D, and $A : D \to \mathcal{P}(D)$ a multi-function. An **optimisation algorithm** consists in defining an iterative sequence $\{x^k\}$ in D such that

$$x^0 = x, \qquad x^{k+1} \in A(x^k) \qquad \text{for every } k \in \mathbb{N}.$$

The set Ω shall be called the **set of solutions** of the optimisation algorithm. We shall also say that the the optimisation algorithm **converges** to Ω if every convergent sub-sequence of $\{x^k\}$ has limit in Ω, that is to say all accumulation points of $\{x^k\}$ belong to Ω.

In the sequel, we shall endeavour to describe conditions that ensure that an optimisation algorithm converges to a given solution set. As we shall see in the next section, this means that either the algorithm terminates in a finite number of steps, achieving a point that lies in the solution set, or the infinite sequence that it generates approaches the solution set, getting closer at each iterate, thereby giving an approximating sequence of a point in the solution set that can be used to approximate a point in the solution set.

3. Convergence of an algorithm

Having established what an iterative algorithm is, and what the meaning of convergence for an algorithm is, we shall now aim, within this section, to state and prove a convergence theorem for multi-functions which is due the American scientist WILLARD I. ZANGWILL. To this end, we shall need to introduce the notion of closed multi-function. Continuity and closure form, together, the two fundamental properties of multi-functions which are needed in optimisation applications.

(3.1) Definition. Let X and Y be two non-empty subsets of $\mathbb{R}^n$ and $\mathbb{R}^m$, respectively. Let $A : X \to \mathcal{P}(Y)$ be a multi-function and let x be a point in X. The map A is said to be **closed** at x if, for any sequence $\{x^k\}$ in X and any sequence $\{y^k\}$ in Y, satisfying the following conditions

$$x^k \to x, \qquad y^k \to y, \qquad y^k \in A(x^k),$$

we have that $y \in A(x)$. The map A is said to be **closed** on $U \subseteq X$ if it is closed at every point in U.

Before moving on, let us see couple of examples, one of a closed multi-function, and one of a multi-function that is not closed.

(3.2) Example. Consider the multi-function $A : \mathbb{R} \to \mathcal{P}(\mathbb{R})$ defined for every real number x by

$$A(x) = \begin{cases} [0, x] & \text{if } x \geq 0, \\ [x, 0] & \text{if } x < 0. \end{cases}$$

Observe that, if $\{x^k\}$ and $\{y^k\}$ are any two sequences of real numbers, the relationship $y^k \in A(x^k)$, for every index k, is equivalent to $0 \leq |y^k| \leq |x^k|$. Thus, if $\{x^k\}$ converges towards 0, the sequence $\{y^k\}$ must also converge towards 0. In addition, $0 \in A(0) = \{0\}$, which shows that A is closed at 0.

(3.3) Example. Consider now the multi-function $A : \mathbb{R} \to \mathcal{P}(\mathbb{R})$ defined for every real number x by

$$A(x) = \begin{cases} [2, x + 1] & \text{for } x \geq 1, \\ \{x\} & \text{for } x < 1. \end{cases}$$

Take the sequences $x^k = 1 - k^{-1}$ and $y^k = 1 - k^{-1}$. Then $x^k \to 1$, $y^k \to 1$, $y^k \in A(x^k)$, but $1 \notin A(1) = \{2\}$. This shows that A is not closed at 1.

In most non-linear programming methods, the multi-function which defines the optimisation algorithm is often composed of several maps. For example, in some algorithms, one first finds a direction d^k to move along, and then one determines the step size λ^k by solving the one-dimensional problem of minimising a function of the form $\varphi(x^k + \lambda d^k)$. In cases such as these, the map A is composed of two multi-functions, $C \circ B$, where B finds the direction d^k, and C finds a step size λ^k.

When the overall multi-function is composed of two or more multi-functions, it is easier to prove that the former map is closed by proving that each of the components are closed. This is the aim of the next result.

(3.4) Proposition. *Let X, Y and Z be non-empty subsets of $\mathbb{R}^n$, $\mathbb{R}^m$ and $\mathbb{R}^p$, respectively. Let $B : X \to \mathcal{P}(Y)$ and $C : Y \to \mathcal{P}(Z)$ be two multi-functions, and consider the composite map $A = C \circ B$. Suppose that B is closed at x, and C is closed on $B(x)$. In addition, suppose that if $\{x^k\}$ is a sequence in X that converges towards x and if $y^k \in B(x^k)$ for every index k, then from the sequence $\{y^k\}$ one can always extract a convergent subsequence. Then A is closed at x.*

Proof. Let $\{x^k\}$ be a sequence in X which converges towards x, $z^k \in A(x^k)$ and $z^k \to z$. We need to show that $z \in A(x)$. By definition of A, for every index k, there is a point $y^k \in B(x^k)$ such that $z^k \in C(y^k)$. By assumption, from the sequence $\{y^k\}$ we can extract a convergent subsequence $\{y^{k_r}\}$ with limit y. Since B is closed at x, then $y \in B(x)$. Moreover, since C is closed on $B(x)$, it is closed at y, and hence, $z \in C(y)$. Thus,

$$z \in C(y) \subseteq (C \circ B)(x) = A(x),$$

whence A is closed at x. $\qquad\square$

(3.5) Corollary. *Let X, Y and Z be non-empty subsets of $\mathbb{R}^n$, $\mathbb{R}^m$ and $\mathbb{R}^p$, respectively. Let $B : X \to \mathcal{P}(Y)$ and $C : Y \to \mathcal{P}(Z)$ be two multi-functions, and consider the composite map $A = C \circ B$. Suppose that B is closed at x, C is closed on $B(x)$, and Y is compact. Then, $A = C \circ B$ is closed at x.*

(3.6) Corollary. *Let X, Y and Z be non-empty subsets of $\mathbb{R}^n$, $\mathbb{R}^m$ and $\mathbb{R}^p$, respectively. Let $B : X \to \mathcal{P}(Y)$ and $C : Y \to \mathcal{P}(Z)$ be two multi-functions, and consider the composite map $A = C \circ B$. Suppose that B is the multi-function associated to a mapping $f : X \to Y$. If f is continuous at x, and C is closed on $f(x)$, then A is closed at x.*

Let us observe that the assumption that from the sequence $\{y^k\}$ one can extract a convergent subsequence in Proposition (3.4) is essential. Without this assumption, in fact, even if the maps B and C are both closed at x and on $B(x)$, respectively, the composition $A = C \circ B$ is not necessarily closed, as shown in the next example due to JAMIE J. GOODE.

(3.7) Example. Let us consider the two multi-functions $B : \mathbb{R} \to \mathcal{P}(\mathbb{R})$ and $C : \mathbb{R} \to \mathcal{P}(\mathbb{R})$ defined as:

$$B(x) = \begin{cases} x^{-1} & \text{for } x \neq 0, \\ 0 & \text{otherwise}; \end{cases} \qquad C(y) = \Big\{ z \in \mathbb{R} : |z| \leq |y| \Big\}.$$

Notice that B and C are closed everywhere. In particular, observe the fact that B is closed at $x = 0$ holds vacuously, for if $x^k \to 0^{\pm}$, the corresponding sequence $y^k \in B(x^k)$ does not have limit points. Now consider the composite map $A = C \circ B$. Then A is given by $A(x) = \{z \in \mathbb{R} : |z| \leq |B(x)|\}$. From the definition of B, one gets

$$A(0) = \{0\}, \qquad A(x) = \left\{z \in \mathbb{R} : |z| \leq |x|^{-1}\right\} \quad \text{for } x \neq 0.$$

Then, the multi-function A is not closed at $x = 0$. To show this, consider the sequence $\{x^k\}$, where $x^k = k^{-1}$, and note that $A(x^k) = \{z \in \mathbb{R} : |z| \leq k\}$, and hence $z^k = 1$ belongs to $A(x^k)$ for every index k. On the other hand, the limit point $z = 1$ does not belong to $A(0) = \{0\}$. Thus, the multi-function A is not closed, even though both B and C are closed. Observe that here, Proposition (3.4) does not apply, because the sequence $\{y^k\}$ such that $y^k \in B(x^k)$ with $x^k = k^{-1}$ does not have any convergent subsequence.

Before we can state Zangwill's Convergence Theorem, we shall also need to introduce another definition, that acts as a sort of 'measure of progress' for an iterative algorithm to seek a solution. Intuitively, using the notion of 'descent function' that is defined below, we can measure if an algorithm is moving towards a 'better' point, that is to say a point that approximates better a solution of an optimisation problem, i.e. a point that reduces the value of this function.

(3.8) Definition. Let D be a non-empty subset of $\mathbb{R}^n$, Ω a non-empty subset of D, and $A : D \to \mathcal{P}(D)$ an multi-function. A continuous real function ψ, defined on D, is said to be a **descent function** for the iterative algorithm generated by A, relative to the solution set Ω, if the following condition holds:

$$\text{if } x \notin \Omega, \text{ then } \psi(y) < \psi(x) \text{ for any } y \in A(x).$$

We are now finally ready to state and prove Zangwill's Convergence Theorem.

(3.9) Theorem (Zangwill's Convergence Theorem). *Let D be a non-empty closed subset of $\mathbb{R}^n$, Ω be a non-empty subset of D, and $A : D \to \mathcal{P}(D)$ be a multi-function on D. Suppose that:*

- *Every sequence generated by A is contained in a compact subset of D.*

- *There exists a descent function ψ for the iterative algorithm generated by A relative to the solution set Ω.*

- *The multi-function A is closed on the complementary of Ω.*

Then either the iterative algorithm stops in a finite number of steps, giving a finite sequence $x^0, x^1, \ldots, x^s$, with the terminal point, x^s, in Ω, or it generates an infinite sequence $\{x^k\}$ such that:

(a) *Every convergent subsequence of $\{x^k\}$ has limit in Ω; that is, all accumulation points of $\{x^k\}$ belong to Ω.*

(b) $\psi(x^k) \to \psi(x)$ *for some* $x \in \Omega$.

Proof. Let $\{x^k\}$ be a iterative sequence generated by the multi-function A. If there existed an index s such that $x^s \in \Omega$, the algorithm would terminate in a finite number of steps, so we may well assume that the sequence generated is infinite and that, for every index k, x^k does not lie in Ω. Since, by hypothesis, this sequence is contained in a compact set, we can always extract from it a convergent subsequence. Let $\{x^{k_\nu}\}$ be any convergent subsequence with limit x. Since ψ is a continuous function, we have

$$\psi(x^{k_\nu}) \to \psi(x).$$

Since ψ is a descent function, then the real sequence $\{\psi(x^k)\}$ is decreasing. Indeed, for every index k, we have $x^{k+1} \in A(x^k)$, whence $\psi(x^{k+1}) < \psi(x^k)$. Now, since $\{\psi(x^k)\}$ is a decreasing sequence of real numbers and one of its subsequences converges, it follows that $\psi(x^k)$ is bounded from below, and hence admits a limit. On the other hand, if a sequence converges, all its subsequences must converge towards the same limit, and since $\{\psi(x^{k_\nu})\}$ converges towards $\psi(x)$, this common limit must be $\psi(x)$. This proves that the sequence $\{\psi(x^k)\}$ converges towards $\psi(x)$.

In order to complete the proof, it suffices to show that $x \in \Omega$. To this end, assume by contradiction that $x \notin \Omega$, and let us extract from $\{x^{k_\nu+1}\}$ a subsequence $\{x^{k_{\nu_r}+1}\}$ that converges towards a point $\bar{x}$ in D. Observe that, on one side, the sequence $\{\psi(x^{k_\nu+1})\}$ converges towards $\psi(\bar{x})$, because ψ is continuous. On the other side, this is also a subsequence of $\{\psi(x^{k_\nu})\}$, hence it converges towards $\psi(x)$. This proves that $\psi(\bar{x}) = \psi(x)$. In addition,

$$x^{k_{\nu_r}} \to x, \qquad x^{k_{\nu_r}+1} \to \bar{x}, \qquad x^{k_{\nu_r}+1} \in A(x^{k_{\nu_r}}).$$

Since the multi-function A is, by hypothesis, closed on the complementary of the set Ω, and since we have assumed, by contradiction, that $x \notin \Omega$, we obtain that A is closed at x, whence $\bar{x} \in A(x)$. Thus, from the fact that ψ is a descent function, it follows that $\psi(\bar{x}) < \psi(x)$, contradicting the fact that $\psi(\bar{x}) = \psi(x)$. Therefore, $x \in \Omega$ and the theorem is fully proved. $\square$

4. Terminating the algorithm

As Zangwill's Convergence Theorem indicates, the algorithm is terminated if one reaches a point in the set of solutions Ω. In most cases, however, a convergence to a point in Ω occurs only in a limited sense, and we must resort to some practical rules for terminating the algorithm. The following rules are frequently used to stop a given algorithm. In the sequel, assume that D is a non-empty set, Ω is a non-empty subset of D, $A : D \to \mathcal{P}(D)$ is a multi-function, $\{x^k\}$ is an iterative sequence generated by the optimisation algorithm associated with D, Ω and A. In this framework, further assume that ε is a positive real number, and k^* is a positive index.

1. 'Control of the time' — this criterion consists in stopping the algorithm if the execution time exceeds a time threshold t^* decided in advance. In other words, one takes only the first part $x^0, \ldots, x^{k^*}$ of the iterative sequence, where k^* is the iteration performed within the time t^* chosen in advance.

2. 'Control of the number of steps of the algorithm' — this criterion consists in stopping the algorithm if the number of points exceeds k^*. In other words, one takes only the first part $x^0, \ldots, x^{k^*}$ of the iterative sequence, where k^* is a number chosen in advance.

3. 'Control of the accuracy' — this criterion consists in stopping the algorithm as soon as the points of the sequence become smaller than a tolerance threshold, ε. In other words, the algorithm terminates when

$$\|x^{k+1} - x^k\| < \varepsilon.$$

4. 'Control of the stationarity' — this criterion, that works only in the special case of unconstrained problems, consists in stopping the algorithm as soon as the norm of the gradient of the objective function on the point of the iterative sequence is smaller than a threshold, ε. In other words, the algorithm terminates when

$$\|\nabla f(x^k)\| < \varepsilon.$$

These four criteria, like many others that we have not mentioned, largely depends on the type of algorithm that is studied, but in the end they all have limitations, and from a theoretical point of view, we can always find counter-examples that show the limits of each of them. Theoretically, in fact, one can always make asymptotic arguments regarding the whole iterative sequence $\{x^k\}$, but from a practical point of view, when an algorithm is implemented to find an approximation of an optimal solution, it becomes necessary to choose a termination criterion, and the choice is eventually left to the person who implements the algorithm.

5. Rates of convergence

One of the critical aspects of solving non-linear programming problems is the convergence speed of the algorithms used. Convergence speed refers to how quickly an algorithm approaches the optimal solution. This is a crucial factor because faster convergence can significantly reduce computational time and resources, making the algorithm more efficient and practical for large-scale problems.

Several factors influence the convergence speed of an algorithm, including the nature of the objective function, the initial starting point, and the specific characteristics of the algorithm itself. Commonly used algorithms in non-linear programming, each have different convergence properties.

Understanding and improving the convergence speed of algorithms is an ongoing area of research in the field of non-linear programming although not central to the development of our treatment.

Roughly speaking, in non-linear programming, convergence rate describes how quickly an algorithm approaches its optimal solution. Here are the formal definitions for different types of convergence rates:

Let's suppose that x^k is the iterate at step k, x^* is the 'target', and x^0 is the initial point.

Linear Convergence. An algorithm has linear convergence if the error decreases at a rate proportional to $\mathcal{O}(\|x^k - x^*\|)$, that is there exists $\rho \in (0, 1)$:

$$\lim_{k \to \infty} \frac{\|x^{k+1} - x^*\|}{\|x^k - x^*\|} = \rho,$$

Super-linear Convergence. An algorithm has super-linear convergence if the error decreases at a rate faster than linear but not necessarily quadratic, that is:

$$\lim_{k \to \infty} \frac{\|x^{k+1} - x^*\|}{\|x^k - x^*\|} = 0,$$

Quadratic Convergence. An algorithm has quadratic convergence if the error decreases at a rate proportional to $\mathcal{O}(\|x^k - x^*\|^2)$, that is there exists $c > 0$:

$$\lim_{k \to \infty} \frac{\|x^{k+1} - x^*\|}{\|x^k - x^*\|^2} = c$$

Moreover, non-linear programming algorithms are also distinguished into algorithms with local or global convergence, depending on whether the convergence of the algorithm depends on the choice of the initial point or not.

Chapter VIII

The Simplex Algorithm

The simplex method, developed by George Dantzig in 1947, is a widely util-ised algorithm for solving linear programming problems. Dantzig's work was driven by the necessity to address large-scale optimisation problems during the Second World War, particularly in the realms of resource allocation and logistics, while he was working for the US Air Force.

This method has heavily transformed the field of operational research and of mathematical optimisation, offering a systematic approach to identifying the optimal solution to linear programming problems. The idea behind the simplex method is indeed very simple: it operates by traversing the edges of the feasible region, as defined by the constraints, moving from one vertex (a 'basic feasible solution') to another in a manner that consistently enhances the objective function until the optimal solution is eventually attained.

Over the decades, the simplex method has been refined and expanded, and it continues to be a milestone in disciplines such as economics, engineering, and management science. Although alternative algorithms, such as interior-point methods, have also been developed, the simplex method remains highly regarded for its practical efficiency and robustness.

In a very concise way, the simplex method is made of three points:

- **Initialisation**: Commencing with an initial feasible solution that lies at a vertex of the feasible region.

- **Iteration**: Progressing to an adjacent vertex that yields a superior ob-jective function value.

- **Termination**: Achieving the optimal solution when no adjacent vertex provides an improvement or determining that the problem is unbounded.

© The Author(s), under exclusive license to Springer Nature Switzerland AG 2026   171
A. Carpignani and M. Pappalardo, *Theory and Methods of Optimisation*,
UNITEXT 175, https://doi.org/10.1007/978-3-032-01514-3_8

1. Sherman-Morrison's formula

In the sequel, we are going to need a tool from Linear Algebra, that provides a computationally fast method to calculate the inverse of a rank-one update to a non-singular matrix. Specifically, if $M \in \mathfrak{M}(r, r, \mathbb{R})$ is a non-singular square matrix, we are going to be interested in computing the inverse of a matrix of the form $M + u \otimes v$ in terms of the inverse of M, where u and v are two column vectors, and where we have denoted by $u \otimes v$ the **outer product** of two vectors, i.e. the linear map from $\mathbb{R}^r$ to $\mathbb{R}^r$ defined by

$$(u \otimes v)(x) = \langle v, x \rangle u \qquad \text{for every } x \in \mathbb{R}^r.$$

It is also useful to recall that the outer product, like the inner product, can always be written in terms of the matrix product. Precisely, if $u, v \in \mathbb{R}^r$, we have $\langle u, v \rangle = u^T v$, and $u \otimes v = uv^T$.

(1.1) Theorem (Sherman-Morrison's formula). *Let $M \in \mathfrak{M}(n, n, \mathbb{R})$ be a non-singular square matrix, and $u, v \in \mathbb{R}^n$ be two column vectors. The matrix $M + u \otimes v$ is not singular if and only if $1 + \langle v, M^{-1}u \rangle \neq 0$. In this case, the inverse of the matrix $M + u \otimes v$ is given by:*

$$(1.2) \qquad \left(M + u \otimes v \right)^{-1} = M^{-1} - \frac{M^{-1}(u \otimes v)M^{-1}}{1 + \langle v, M^{-1}u \rangle}$$

Here, $u \otimes v$ denotes the outer product of vectors u and v.

Before we can prove this formula, we need the following simple result.

(1.3) Lemma (Matrix determinant lemma). *Let $M \in \mathfrak{M}(n, n, \mathbb{R})$ be a non-singular square matrix, and $u, v \in \mathbb{R}^n$ two vectors. Then,*

$$\det(M + u \otimes v) = \left(1 + \langle v, M^{-1}u \rangle \right) \det(M)$$

where $u \otimes v$ denotes the outer product of the two vectors u and v.

Proof. Let us start by observing that, with no loss of generality, we can assume that M is the identity matrix I. Indeed, if the formula is true in this latter special case, then we have

$$\det(M + uv^T) = \det\left(M(I + (M^{-1}u)v^T) \right)$$
$$= \det(M) \det\left(I + v^T(M^{-1}u) \right).$$

Thus, assume that M is indeed the identity matrix. Then,

$$\begin{bmatrix} I & 0 \\ v^T & 1 \end{bmatrix} \begin{bmatrix} I + uv^T & u \\ 0 & 1 \end{bmatrix} \begin{bmatrix} I & 0 \\ -v^T & 1 \end{bmatrix} = \begin{bmatrix} I & u \\ 0 & 1 + v^T u \end{bmatrix}.$$

The determinant of the left-hand side is the product of the determinants of the three matrices. Since the first and the third matrices are triangular matrices, with unit diagonal, their determinants is just 1. The determinant of the middle matrix is instead $\det(I + uv^T)$. On the other hand, the determinant of the right-hand side is $(1 + v^T u)$, so we have

$$\det(I + uv^T) = 1 + v^T u$$

as desired. $\qquad\square$

Proof of Theorem (1.1). ($\Rightarrow$): Assume that $1 + v^T M^{-1} u = 0$. The, by virtue of the matrix determinant lemma (see (1.3)), we obtain

$$\det(M + uv^T) = (1 + v^T M^{-1} u) \det(M) = 0,$$

so the matrix $(M + uv^T)$ is not not singular.

($\Leftarrow$): To prove that $1 + v^T A^{-1} u \neq 0$ implies that $M + uv^T$ is not singular with inverse given by (1.2), let us verify that the matrix Y on the right-hand side of (1.2) is the inverse of $X = M + uv^T$. Since X and Y are square matrices, it suffices to prove that Y is a right inverse, that is to say, it satisfies the equation $XY = I$. Indeed,

$$\begin{aligned}
XY &= \left(M + uv^T\right)\left(M^{-1} - \frac{M^{-1}uv^T M^{-1}}{1 + v^T M^{-1}u}\right) \\
&= MM^{-1} + uv^T M^{-1} - \frac{MM^{-1}uv^T M^{-1} + uv^T M^{-1}uv^T M^{-1}}{1 + v^T M^{-1}u} \\
&= I + uv^T M^{-1} - \frac{uv^T M^{-1} + uv^T M^{-1}uv^T M^{-1}}{1 + v^T M^{-1}u} \\
&= I + uv^T M^{-1} - \frac{u\left(1 + v^T M^{-1}u\right)v^T M^{-1}}{1 + v^T M^{-1}u} \\
&= I + uv^T M^{-1} - uv^T M^{-1} \\
&= I.
\end{aligned}$$

$\qquad\square$

2. Further algebra of linear programming

The idea behind the simplex algorithm is heavily based on the geometrical structure of the feasible region, and utilises the algebraic description of the polyhedron that turns out to be perfectly suited to be automatised into an algorithm. In order to understand the simplex algorithm, it is therefore required to study in more details the links between the 'shape' of a linear programming problem and the geometric properties of its feasible region.

To this end, let us suppose that a non-zero matrix $A \in \mathfrak{M}(m, n, \mathbb{R})$, and two vectors $b \in \mathbb{R}^m$ and $c \in \mathbb{R}^n$ are given, and consider the following pair of primal and dual linear programs in standard form:

$$(\mathcal{P}): \begin{cases} \text{maximise} & \langle c, x \rangle \\ \text{subject to} & Ax \leq b \end{cases} \qquad (\mathcal{D}): \begin{cases} \text{minimise} & \langle b, y \rangle \\ \text{subject to} & A^T y = c, \, y \geq 0. \end{cases}$$

Let us also denote by P and Q the primal and dual feasible regions, respectively. In other words, let us set

$$P = \left\{ x \in \mathbb{R}^n : Ax \leq b \right\}, \qquad D = \left\{ y \in \mathbb{R}^m : A^T y = c, \, y \geq 0 \right\}.$$

(2.1) Remark. Let us quickly recall that Theorem IV(5.6) shows that the primal basic feasible solutions of a linear program are the vertices of the primal polyhedron. If the vertices are non-degenerate, the active constraints at a vertex are exactly n, and they are also referred to as the faces of dimension 0. Similarly, the points of the primal polyhedron which have exactly $n-1$ active constraints are referred to as the **edges**, or faces of dimension 1 of the polyhedron.

The following lemmata are going to be useful to prove the correctness of the simplex algorithm, but also carry a strong geometrical meaning, that is important to grasp solidly to understand the idea of the simplex algorithm.

(2.2) Lemma. *Let $\bar{x}$ and $\bar{y}$ be a pair of complementary basic solutions, relative to a basis B. Assume that $\bar{x}$ is feasible, and set $W = -A_B^{-1}$. Then, the objective function, $\langle c, x \rangle$, increases along the ray $x(\lambda) = \bar{x} + \lambda W^j$, with $\lambda > 0$, if and only if j lies in $J(\bar{y}) = \{ j \in B : \bar{y}_j < 0 \}$.*

Proof. If $J(\bar{y}) = \emptyset$, $\bar{x}$ and $\bar{y}$ are optimal. Otherwise, let us consider the ray leaving the point $\bar{x}$ in the direction W^j. In other words, let us consider:

$$x(\lambda) = \bar{x} + \lambda W^j, \qquad \text{for all } \lambda > 0.$$

Since $\bar{y}_j < 0$, one has

$$\begin{aligned}
\langle c, x(\lambda) \rangle &= \langle c, \bar{x} \rangle + \lambda \langle c, W^j \rangle \\
&= \langle c, \bar{x} \rangle - \lambda \bar{y}_j \\
&> \langle c, \bar{x} \rangle
\end{aligned}$$

for every $\lambda > 0$. This proves that the value of the objective function increases as we move from the point $\bar{x}$ in the direction W^j. $\qquad\square$

Whilst the previous lemma describes the behaviour of the objective functions when moving out-going from the given vertex, the next lemma shows that, moving in the direction defined by the columns of the matrix W, we may either come to the realisation that the polyhedron is unbounded in a certain direction, or we end up into a new vertex. So, while the previous lemma may be referred to as the primal 'out-going lemma', the next one may be referred to as the primal 'in-going lemma'. Clearly, the word 'primal' implies that there are going to be corresponding results for the dual problem, which we are going to discuss immediately afterwards.

(2.3) Lemma. *Let $\bar{x}$ and $\bar{y}$ be a pair of complementary basic solutions, relative to a basis B. Assume that $\bar{x}$ is feasible, and set $W = -A_B^{-1}$. Assume further that the set $J(\bar{y}) = \{j \in B : \bar{y}_j < 0\}$ is non-empty, and let h be an index in the set $J(\bar{y})$. Then, the following conditions hold:*

(a) *If*
$$\langle A_i, W^h \rangle \leq 0 \qquad \text{for every } i \in N,$$

then, the ray $x(\lambda) = \bar{x} + \lambda W^h$ belongs to the primal feasible region, P, for every real number $\lambda \geq 0$.

(b) *If there exists an index i in N such that $\langle A_i, W^h \rangle > 0$, then the ray*

$$x(\lambda) = \bar{x} + \lambda W^h$$

is only included in the primal feasible region for λ ranging over a closed interval of the form $[0, \theta^*]$, with

$$\theta^* = \min\left\{ \frac{b_i - \langle A_i, \bar{x} \rangle}{\langle A_i, W^h \rangle} : i \in N,\ \langle A_i, W^h \rangle > 0 \right\}.$$

Proof. The two statements of the theorem may be proved at the same time. To this end, let λ be a positive real number, and let us analyse when the vector

$$x(\lambda) = \bar{x} + \lambda W^h$$

belongs to the primal polyhedron with equation $Ax \leq b$, which is a compact notation to write the system of simultaneous inequalities $\langle A_i, x \rangle \leq b_i$ with i ranging in $B \cup N$, where B is the given basis, and N the corresponding non-basis. We may reason independently for the basis and the non-basis. Precisely, let us start by assuming that i is an index in the basis B, and observe that

$$\langle A_i, W^h \rangle = \begin{cases} 0 & \text{if } i \in B \setminus \{h\}, \\ -1 & \text{if } i = h, \end{cases}$$

thus

$$\langle A_i, x(\lambda) \rangle = \langle A_i, \bar{x} \rangle + \lambda \langle A_i, W^h \rangle = \begin{cases} b_i & \text{if } i \in B \setminus \{h\}, \\ b_i - \lambda & \text{if } i = h. \end{cases}$$

Since λ is a positive real number, one obviously has $b_i - \lambda \leq b_i$, which ensures that $\langle A_i, x(\lambda) \rangle \leq b_i$ for every positive real number λ.

On the other hand, if i is an index belonging to the the non-basis, N, let us observe that the inequality

$$\langle A_i, x(\lambda) \rangle = \langle A_i, \bar{x} \rangle + \lambda \langle A_i, W^h \rangle \leq b_i \qquad \text{for all } i \in N$$

is only satisfied for every positive real number λ, if $\langle A_i, W^h \rangle \leq 0$. Conversely, if there exists an index i in N such that $\langle A_i, W^h \rangle > 0$, one only has

$$\langle A_i, x(\lambda) \rangle = \langle A_i, \bar{x} \rangle + \lambda \langle A_i, W^h \rangle \leq b_i$$

if and only if

$$0 \leq \lambda \leq \frac{b_i - \langle A_i, \bar{x} \rangle}{\langle A_i, W^h \rangle}.$$

Therefore, by setting

$$\theta^* = \min \left\{ \frac{b_i - \langle A_i, \bar{x} \rangle}{\langle A_i, W^h \rangle} : i \in N, \ \langle A_i, W^h \rangle > 0 \right\},$$

it is ensures that, for every λ in the interval $[0, \theta^*]$, one has $\langle A_i, x(\lambda) \rangle \leq b_i$ for every $i \in N$. Since the same inequality works for every index i in the basis B, this inequality is satisfied for every index i, yielding that $x(\lambda)$ lies in the primal feasible region, for every λ in the interval $[0, \theta^*]$, as required. $\square$

(2.4) Remark. Chosen the outgoing index h, $\theta^* = 0$ means that there exists $i \in N$ and $\langle A_i, W^h \rangle > 0$ such that $b_i - \langle A_i, \bar{x} \rangle = 0$; then we are in a degenerate case.

As announced before, the results discussed above for the primal problem have a corresponding statement for the dual problem. Indeed, even if the dual problem, *per se*, is only an auxiliary problem that is used to help finding the optimal solution of the given one, it is sometimes the case that a practical problem occurs naturally in dual standard form. In such cases, even if that problem is, from our perspective, 'primal', for it is the one upon which we are focused, it would be incredibly onerous, and computationally stupid, to reshape it into primal standard form and then associate to it a dual problem in dual standard form. Conversely, it shall be much more convenient to leave the original problem in dual standard form, and to consider the corresponding primal problem as its dual.

(2.5) Lemma. *Let $\bar{x}$ and $\bar{y}$ be a pair of complementary basic solutions, relative to a basis B. Assume that $\bar{y}$ is feasible, and set $W = -A_B^{-1}$. Then the objective function $\langle b, y \rangle$ decreases along the ray $y(\lambda)$, defined for every positive real number λ, whose components are*

$$y_j(\lambda) = \begin{cases} \bar{y}_j + \lambda \langle A_k, W^j \rangle & \text{if } j \in B, \\ \lambda & \text{if } j = k, \\ 0 & \text{if } j \in N \setminus \{k\} \end{cases}$$

if and only if k lies in $J(\bar{x}) = \{ j \in B : b_j - \langle A_j, \bar{x} \rangle < 0 \}$.

Proof. For each $\lambda > 0$ we have,

$$
\begin{aligned}
\langle b, y(\lambda) \rangle &= \sum_{i \in B} \left(b_i \bar{y}_i + \lambda \langle A_k, W^i \rangle b_i \right) + \lambda b_k \\
&= \sum_{i \in B} b_i \bar{y}_i + \lambda \langle A_k, \sum_{i \in B} W^i b_i \rangle + \lambda b_k \\
&= \langle b, \bar{y}_B \rangle - \lambda \langle A_k, A_B^{-1} b_B \rangle + \lambda b_k \\
&= \langle b, \bar{y} \rangle - \lambda \langle A_k, \bar{x} \rangle + \lambda b_k \\
&= \langle b, \bar{y} \rangle + \lambda (b_k - \langle A_k, \bar{x} \rangle) \\
&< \langle b, \bar{y} \rangle
\end{aligned}
$$

Thus, the value of the objective function decreases along the ray $y(\lambda)$. $\qquad\square$

(2.6) Lemma. *Let $\bar{x}$ and $\bar{y}$ be a pair of complementary basic solutions, relative to a basis B. Assume that $\bar{y}$ is feasible, and set $W = -A_B^{-1}$. Then, the following conditions hold:*

(a) *If*
$$
\langle A_k, W^i \rangle \geq 0 \qquad \text{for every } i \in B,
$$

then, the ray $y(\lambda)$, defined in Lemma (2.5), belongs to the dual feasible region, for every real number $\lambda \geq 0$.

(b) *If there exists at least an index $i \in B$ such that $\langle A_k, W^i \rangle < 0$ then the ray $y(\lambda)$ defined in Lemma (2.5) is only included in the dual feasible region for λ ranging over a closed interval of the form $[0, \theta^*]$, with*

$$
\theta^* = \min \left\{ \frac{\bar{y}_i}{-\langle A_k, W^i \rangle} : i \in B, \langle A_k, W^i \rangle < 0 \right\},
$$

Proof. Let λ be a positive real number, and let us analyse when the element $y(\lambda)$ of the ray defined in Lemma (2.5) lies in the dual feasible region. To this end, let us observe that the constraints $A^T y = c$ are satisfied by every point of the ray. In fact, for every positive real number λ, we have

$$
\begin{aligned}
A^T y(\lambda) &= \sum_{i=1}^{m} A_i^T y_i(\lambda) \\
&= \sum_{i \in B} A_i^T \left(\bar{y}_i + \lambda \langle A_k, W^i \rangle \right) + \lambda A_k \\
&= \sum_{i \in B} A_i^T \bar{y}_i + \lambda A_i^T \sum_{i \in B} \langle A_k, W^i \rangle + \lambda A_k \\
&= A_B^T \bar{y}_B + \lambda A_k \sum_{i \in B} \langle A_i, W^i \rangle + \lambda A_k \\
&= A_B^T \bar{y}_B - \lambda A_k + \lambda A_k \\
&= c
\end{aligned}
$$

Furthermore, the constraints $y_j(\lambda) \geq 0$ are satisfied for every $\lambda \geq 0$, if $j \in N$. On the other hand, if j in B, there are two alternatives: if $\langle A_k, W^j \rangle \geq 0$, then $y_j(\lambda) \geq 0$, but if $\langle A_k, W^j \rangle < 0$, then the constraint $y_j(\lambda) \geq 0$ is satisfied only for

$$
0 \leq \lambda \leq \frac{\bar{y}_i}{-\langle A_k, W^i \rangle}.
$$

Thus, setting

$$\theta^* = \min\left\{ \frac{\bar{y}_i}{-\langle A_k, W^i \rangle} \;:\; i \in B, \ \langle A_k, W^i \rangle < 0 \right\}$$

it is ensured that, for every λ in the interval $[0, \theta^*]$, the vector $y(\lambda)$ lies in the dual feasible region, and the lemma is thus proved. $\qquad\qquad\square$

(2.7) Remark. Chosen the entering index k, $\theta^* = 0$ means that there exists $i \in B$ and $\langle A_k, W^i \rangle < 0$ such that $y_i = 0$; then we are in a degenerate case.

3. Non-degenerate primal simplex algorithm

We have now developed all the necessary tools to finally be able to tackle the simplex algorithm. The aim of this section is therefore to describe the simplex algorithm for solving a linear programming problem in standard primal form, while we shall see in the next section the same application to a linear programming problem in standard dual form. To kick right off, let us suppose that a non-zero matrix $A \in \mathfrak{M}(m, n, \mathbb{R})$, and two vectors $b \in \mathbb{R}^m$ and $c \in \mathbb{R}^n$ are given, and consider the following linear programming problem in primal standard form:

$$(\mathcal{P}) : \quad \begin{cases} \text{maximise} & \langle c, x \rangle \\ \text{subject to} & Ax \le b \end{cases}$$

Let us also assume that the number m of rows in the matrix A be greater than or equal to the number n of columns in the same matrix, and that the rank of the matrix A is equal to n (see IV(5.1)). Furthermore, throughout this section, we shall assume that every basic solution is non-degenerate, delaying the degenerate case to section 5, where we shall develop the rules that ascertain that the algorithm does not cycle in a degenerate vertex.

In the hypotheses described above, let us denote by P the primal feasible region, that is to say, the set $P = \{x \in \mathbb{R}^n : Ax \le b\}$. The simplex algorithm is composed of the following elementary parts:

(a) **Initialisation.** A vertex of the feasible region is given. Algebraically, this is done by assigning a basis, B, and calculating the primal basic feasible solution, $\bar{x} = A_B^{-1} b_B$, and the corresponding dual basic solution, $\bar{y} = (A_B^{-1})^T c$.

(b) **Optimality check.** If $\bar{x}$ and $\bar{y}$ are both feasible for the primal and dual problem, respectively, then the algorithm has found an optimal solution, $\bar{x}$, and it **stops**. Otherwise, the algorithm continues.

(c) **Choice of the out-going direction.** The columns of $W = -A_B^{-1}$ represent the directions outward from the given vertex, and each ray of the form

$$x(\lambda) = \bar{x} + \lambda W^j \qquad \text{for all } \lambda \geq 0,$$

as j runs over the basis B, moves from the current vertex towards an adjacent vertices. The indices of the direction along which the objective function grows are those for which the correspondent component, $\bar{y}_j$, of $\bar{y}$ is negative. The most negative (which gives the greatest increasing of the objective function, of these indices, h, is selected.

(d) **Choice of the in-going direction.** Being in the non-degenerate case $\theta^* > 0$, the vertex is then abandoned following the ray $x(\lambda) = \bar{x} + \lambda W^h$, that is to say by *relaxing* the index h, until a new vertex is reached. To find the vertex reached, one calculates the **step size**, θ_i, for each currently *inactive* constraint, i.e.

$$\theta_i = \frac{b_i - \langle A_i, \bar{x} \rangle}{\langle A_i, W^h \rangle} \qquad \text{for all } i \in N, \text{ with } \langle A_i, W^h \rangle > 0.$$

The index, k, representing the index corresponding to the smallest of these step size, is selected.

(e) **Construction of the new basis.** By removing the index h from the basis, and replacing it with the index k, we obtain a new basis. This new basis can be used to restart the algorithm.

The simplex algorithm with this selection criterion is schematically described in Algorithm 2.

The next result shows that the simplex algorithm, for a non-degenerate linear problem in primal standard form, terminates in a finite number of iterations, giving the optimal solution.

(3.1) Theorem. *The simplex method in primal standard form solves the linear programming problem $(\mathcal{P})$ in finite number of iterations.*

Proof. If $\bar{y}$ is a feasible solution for the dual problem, the algorithm terminates in one step (see Theorem IV(4.7)). Now, by induction, assume that we have already made a certain number of iterations, leading to a certain primal basic solution $\bar{x}$. If the complementary basic solution $\bar{y}$ is feasible for the dual problem, then $\bar{x}$ is optimal for the primal problem, and $\bar{y}$ is optimal for the dual problem, and the algorithm stops.

Thus, let us assume that $\bar{y}$ is not feasible, and assume that the in-going and out-going indices h and k have been calculated as described in step 5 and step 8 of Algorithm 2. Observe that, by construction, $h \neq k$, because these two indices lie in two disjoint sets B and N, respectively. Let us show

---

Algorithm 2 THE SIMPLEX METHOD IN PRIMAL STANDARD FORM

---

1. It is given a basis B that generates a primal feasible solution.

2. Compute the matrix $W = -A_B^{-1}$, inverse of $-A_B$.

3. Calculate $\bar{x} = -Wb_B$, $\bar{y}_B = -W^T c$, $\bar{y}_N = 0$.

4. If $\bar{y}_B = -W^T c \geq 0$, then $\bar{x} = -Wb_B$ is an optimal solution of the primal, $\bar{y} = (\bar{y}_B, 0) = (-W^T c, 0)$ is an optimal solution of the dual and the algorithm **stops**.

5. Set h such that $\bar{y}_h = \min\{\bar{y}_i : i \in B, \bar{y}_i < 0\}$.

6. If $\langle A_i, W^h \rangle \leq 0$ for all $i \in N$, then the problem is unbounded and the algorithm **stops**.

7. Set
$$\theta = \min\left\{ \frac{b_i - \langle A_i, \bar{x} \rangle}{\langle A_i, W^h \rangle} : i \in N, \langle A_i, W^h \rangle > 0 \right\}.$$

8. Set
$$k = \min\left\{ i \in N : \langle A_i, W^h \rangle > 0, \frac{b_i - \langle A_i, \bar{x}_i \rangle}{\langle A_i, W^h \rangle} = \theta \right\}.$$

9. Change B to $B \setminus \{h\} \cup \{k\}$, and N to $N \setminus \{k\} \cup \{h\}$.

10. **Go to** step 3.

---

that $B' = B \setminus \{h\} \cup \{k\}$ is indeed a new basis. To this end, let us start by noting that

$$A_{B'} = A_B - E_h \otimes A_h + E_h \otimes A_k = A_B + E_h \otimes (A_k - A_h),$$

where $E_1, \ldots, E_n$ is the canonical orthonormal basis in $\mathbb{R}^n$, and $A_1, \ldots, A_m$ denote the (column vectors of the) m rows of the matrix A. Thus, by applying Theorem (1.1) yields that the matrix $A_{B'}$ is not singular if and only if

$$1 + \langle A_k - A_h, A_B^{-1} E_h \rangle \neq 0.$$

On the other hand, since $W = -A_B^{-1}$, we have $\langle A_h, W^h \rangle = -1$, and, thanks to the fact that k is chosen such that $\langle A_i, W^h \rangle > 0$ for every index i in N, and that $k \in N$, we also have $\langle A_k, W^h \rangle > 0$. Thus,

$$\begin{aligned}
1 + \langle A_k - A_h, A_B^{-1} E_h \rangle &= 1 - \langle A_k - A_h, W E_h \rangle \\
&= 1 - \langle A_k - A_h, W^h \rangle \\
&= 1 - \langle A_k, W^h \rangle + \langle A_h, W^h \rangle \\
&< 1 - 0 - 1 = 0,
\end{aligned}$$

which proves that $A_{B'}$ is indeed not singular, i.e. B' is a basis. This proves that an iteration of the simplex algorithm, starting from a basis, gives another basis.

Let us now set

$$\theta = \min \left\{ \frac{b_i - \langle A_i, \bar{x} \rangle}{\langle A_i, W^h \rangle} : i \in N, \langle A_i, W^h \rangle > 0 \right\}$$

and observe that the point $x(\theta)$, which is feasible thanks to Lemma $(2.2)(b)$, is nothing but the primal basic feasible solution associated to the new basis $B' = B \setminus \{h\} \cup \{k\}$. To see this, it suffices to note that, if the index i lies in $B \setminus \{h\}$, one has $A_i x(\theta) = b_i$; while, if $i = k$, one has

$$A_k x(\theta) = A_k(\bar{x} + \theta W^h) = A_k \bar{x} + \frac{b_k - A_k \bar{x}}{\langle A_k, W^h \rangle} \langle A_k, W^h \rangle = b_k.$$

Being in the non-degenerate case, the real number θ is strictly positive, we obtain that $x(\theta)$ is necessarily a different vertex of the primal feasible region. Furthermore, by virtue of Lemma (2.3), the value of the objective function on this basic feasible point is strictly higher than the value in the previous vertex. This ensures that the simplex algorithm can never visit the same basis twice. Thus, since the number of feasible bases is finite, the algorithm terminates in a finite number of iterations, either finding a pair of complementary optimal solutions, or by finding that the optimal value of the problem is $+\infty$. $\qquad \square$

(3.2) Example. Let us solve the following linear programming problem in primal standard form

$$
\begin{aligned}
\text{maximise} \quad & x_1 + 3x_2 \\
\text{subject to} \quad & -5x_1 - 2x_2 \leq 0 \\
& -x_1 \leq 1 \\
& x_2 \leq 5 \\
& 3x_1 + 3x_2 \leq 30
\end{aligned}
$$

using the simplex algorithm, starting from the basis $B = \{1, 2\}$.

Iteration 1. The basis matrix and its inverse are:

$$
A_B = \begin{bmatrix} -5 & -2 \\ -1 & 0 \end{bmatrix} \qquad A_B^{-1} = \begin{bmatrix} 0 & -1 \\ -1/2 & 5/2 \end{bmatrix}
$$

The associated basic solution is:

$$
\bar{x} = A_B^{-1} b_B = \begin{bmatrix} -1 \\ 5/2 \end{bmatrix}.
$$

The dual complementary solution is:

$$
\bar{y}_B = (A_B^T)^{-1} c = \left(-\frac{3}{2}, \frac{13}{2} \right), \qquad \bar{y}_N = 0,
$$

therefore $\bar{y} = (-3/2, 13/2, 0, 0)$. The solution $\bar{y}$ is not feasible because $\bar{y}_1 < 0$, thus the out-going index from the basis is 1 and we set W^1 equal to the opposite of the first column of A_B^{-1}, that is

$$
W^1 = \begin{bmatrix} 0 \\ 1/2 \end{bmatrix}.
$$

Let us now compute the scalar products $\langle A_i, W^1 \rangle$, for $i = 3, 4$:

$$
\langle A_3, W^1 \rangle = \frac{1}{2}, \qquad \langle A_4, W^1 \rangle = \frac{3}{2}.
$$

Performing step 7 of the simplex method, the index for which the ratio is minimal among those that follow is determined:

$$
\frac{b_3 - A_3\,\bar{x}}{\langle A_3, W^1 \rangle} = \frac{5/2}{1/2} = 5, \qquad \frac{b_4 - A_4\,\bar{x}}{\langle A_4, W^1 \rangle} = \frac{51/2}{3/2} = \frac{51}{3}.
$$

Iteration 2. The new basis is $B = \{2, 3\}$ and the corresponding basic feasible solution is: $\bar{x} = (-1, 5)$. The basis matrix and its inverse are:

$$
A_B = \begin{bmatrix} -1 & 0 \\ 0 & 1 \end{bmatrix} \qquad A_B^{-1} = \begin{bmatrix} -1 & 0 \\ 0 & 1 \end{bmatrix}.
$$

The complementary dual basic solution is: $\bar{y} = (0, -1, 3, 0)$ that is not feasible. In this case the out-going index is 2. We put:

$$W^2 = \begin{bmatrix} 1 \\ 0 \end{bmatrix}$$

and we compute the scalar products $\langle A_i, W^2 \rangle$, for $i = 1, 4$:

$$\langle A_1, W^2 \rangle = -5, \qquad \langle A_4, W^2 \rangle = 3.$$

Let's compute the quotients:

$$\frac{b_4 - A_4 \bar{x}}{\langle A_4, W^2 \rangle} = \frac{18}{3} = 6.$$

So, the entering index is 4.

Iteration 3. The new basis is therefore $B = \{3, 4\}$, the basis matrix and its inverse are:

$$A_B = \begin{bmatrix} 0 & 1 \\ 3 & 3 \end{bmatrix} \qquad A_B^{-1} = \begin{bmatrix} -1 & 1/3 \\ 1 & 0 \end{bmatrix}.$$

The new primal basic solution is $\bar{x} = (5, 5)$. The dual basic solution is $\bar{y} = \left(0, 0, 2, \frac{1}{3}, 0\right)$ which is feasible, therefore the simplex algorithm terminates providing $\bar{x} = (5, 5)$ as the optimal primal solution and $\bar{y} = \left(0, 0, 2, \frac{1}{3}\right)$ as the optimal dual solution.

Now let us show another example where simplex method detects an unbounded primal linear programming problem.

(3.3) Example. Let us solve the following linear programming problem in primal standard form

$$\begin{array}{rl} \text{maximise} & 5x_1 - x_2 \\ \text{subject to} & -2x_1 + x_2 \leq 4 \\ & x_1 - 2x_2 \leq 2 \\ & x_1 \geq 0 \\ & x_2 \geq 0 \end{array}$$

using the simplex algorithm, starting from the vertex $(0, 0)$.

Iteration 1. The vertex $\bar{x} = (0, 0)$ is generated by the basis $B = \{3, 4\}$ with

$$A_B = A_B^{-1} = \begin{bmatrix} -1 & 0 \\ 0 & -1 \end{bmatrix}$$

Since $\bar{y}_B = (-5, 1)$, the index 3 is the out-going index and $W^3 = (1, 0)$. Furthermore, since

$$\langle A_1, W^3 \rangle = -2, \qquad \langle A_2, W^3 \rangle = 1,$$

the index 2 is the in-going index.

Iteration 2. The new basis is $B = \{2, 4\}$ and

$$A_B = A_B^{-1} = \begin{bmatrix} 1 & -2 \\ 0 & -1 \end{bmatrix}$$

The corresponding basic solution are:

$$\bar{x} = (2, 0), \qquad \bar{y}_B = (5, -9).$$

Therefore 4 is the out-going index and $W^4 = (2, 1)$. Now, since

$$\langle A_1, W^4 \rangle = -3, \qquad \langle A_3, W^4 \rangle = -2,$$

we may conclude that our problem is unbounded (optimum equal to $+\infty$), and the dual polyhedron is empty.

4. Non-degenerate dual simplex algorithm

Sometimes, in practical problems, it may happen that a linear programming problem is naturally formulated in dual standard form. In such cases, one would have two options to solve the problem: one consisting in using equivalent transformations to reshape the problem into primal standard form, or one could leave it in dual standard form. In this case, one would then need a slightly different version of the simplex algorithm, which we shall call the 'dual simplex algorithm'.

To develop this dual algorithm, let us start by considering the following linear programming problem:

$$(\mathcal{D}) : \quad \begin{cases} \text{minimise} & \langle b, y \rangle \\ \text{subject to} & A^T y = c, \, y \geq 0 \end{cases}$$

where $A \in \mathfrak{M}(m, n, \mathbb{R})$ is a matrix, $b \in \mathbb{R}^m$ and $c \in \mathbb{R}^n$ are two vectors.

Algorithm 3 is similar to the primal simplex algorithm, with the difference that at each step, the dual basic solution remains feasible, while the feasibility of the primal solution is checked. If the primal basic solution is feasible, then we have found an optimal pair of complementary solutions; otherwise, the basis needs to be changed. In this section we shall assume that the dual polyhedron has non-degenerate vertices, leaving to section 5 a full discussion of the degenerate case. This assumption that the polyhedron is non-degenerate will ensure that every time the basis is changed, the new dual basic solution is different from the previous one and the value of the objective function improves, i.e. decreases.

In analogy with the primal simplex algorithm, the dual algorithm also finds the optimum in a finite number of steps, as the next result shows.

Algorithm 3 THE SIMPLEX METHOD IN DUAL STANDARD FORM

---

1. It is given a basis B that generates a dual feasible solution.

2. Compute the matrix $W = -A_B^{-1}$, inverse of A_B.

3. Calculate $\bar{x} = -Wb_B$, $\bar{y}_B = -W^T c$, $\bar{y}_N = 0$.

4. If $b_N - A_N \bar{x} \geq 0$, then $\bar{y} = (\bar{y}_B, \bar{y}_N)$ is an optimal solution of the dual problem and the algorithm **stops**.

5. Set k such that $b_k - A_k \bar{x} = \min\{b_i - A_i \bar{x} : i \in N,\ b_i - A_i \bar{x} < 0\}$.

6. If $\langle A_k, W^i \rangle \geq 0$ for each $i \in B$, then the problem is unbounded and the algorithm **stops**.

7. Set
$$\theta = \min \left\{ \frac{\bar{y}_i}{-\langle A_k, W^i \rangle} : i \in B,\ \langle A_k, W^i \rangle < 0 \right\},$$

8. Set
$$h = \min \left\{ i \in B : \langle A_k, W^i \rangle < 0,\ \frac{\bar{y}_i}{-\langle A_k, W^i \rangle} = \theta \right\}.$$

9. Change B to $B \setminus \{h\} \cup \{k\}$, and N to $N \setminus \{k\} \cup \{h\}$.

10. **Go to** step 3.

(4.1) Theorem. *The simplex method in dual standard form solves the linear programming problem $(\mathcal{D})$ in a finite number of iterations.*

Proof. The same proof as in Theorem (3.1) shows that $B' = B \backslash \{h\} \cup \{k\}$ is a basis, because $\langle A_k, W^h \rangle < 0$. Let us argue by induction again, assuming that we have already made a certain number of iterations, leading to a certain dual basic feasible solution $\bar{y}$. If the complementary basic solution $\bar{x}$ is feasible for the primal problem, then the algorithm stops.

Thus, let us assume that $\bar{x}$ is not feasible, and assume the in-going and out-going indices h and k have been calculates as described in Algorithm 3.

Let us set

$$\theta = \min \left\{ \frac{\bar{y}_i}{-\langle A_k, W^i \rangle} : i \in B, \langle A_k, W^i \rangle < 0 \right\},$$

and observe that the point $y(\theta)$, which is feasible thanks to Lemma (2.6)(b), is nothing but the dual basic feasible solution associated to the basis $B' = B \backslash \{h\} \cup \{k\}$. To see this, it suffices to note that, if i is an index in $N \backslash \{k\}$, then $y_i(\theta) = 0$; while, if $i = h$, one has

$$y_h(\theta) = \bar{y}_h + \theta \langle A_k, W^h \rangle = \bar{y}_h - \frac{\bar{y}_h}{\langle A_k, W^h \rangle} \langle A_k, W^h \rangle = 0,$$

that is to say, $y_{N'}(\theta) = 0$. Since the real number θ is strictly positive, we obtain that $y(\theta)$ is necessarily a different vertex of the dual feasible region. Furthermore, by virtue of Lemma (2.5), the value of the objective function on this basic feasible point is strictly lower than the value in the previous point. This ensures that the simplex algorithm can never visit the same basis twice. Thus, since the number of feasible bases is finite, the algorithm terminates after a finite number of iterations, either finding a pair of complementary optimal solutions, or by finding that the optimal value of the problem is $-\infty$.
$\square$

The next example shows how to use the simplex algorithm for a problem in dual standard form.

(4.2) Example. Let us solve the following linear programming problem in dual standard form

$$(\mathcal{D}): \begin{cases} \text{minimise} & 13y_3 + 9y_4 + 7y_5 \\ \text{subject to} & -y_1 + y_3 + y_4 + y_5 = 3 \\ & -y_2 + 2y_3 + y_4 = -4 \\ & y \geq 0 \end{cases}$$

using the simplex algorithm, starting from the basis $B = \{2, 3\}$.

Iteration 1. The basis matrix and its inverse are:

$$A_B = \begin{bmatrix} 0 & -1 \\ 1 & 2 \end{bmatrix} \qquad A_B^{-1} = \begin{bmatrix} 2 & 1 \\ -1 & 0 \end{bmatrix},$$

the dual basic solution is

$$\bar{y}_B = (A_B^{-1})^T (3, -4)^T = (10, 3)^T, \qquad \bar{y} = (0, 10, 3, 0, 0),$$

and the complementary primal is $\bar{x} = (13, 0)$ that is not feasible because

$$b_1 - \langle A_1, \bar{x} \rangle = 13, \qquad b_4 - \langle A_4, \bar{x} \rangle = -4, \qquad b_5 - \langle A_5, \bar{x} \rangle = -6,$$

therefore the in-going index is 5. Let us compute the scalar products

$$\langle A_5, W^2 \rangle = -2, \qquad \langle A_5, W^3 \rangle = -1,$$

and the quotients

$$\frac{\bar{y}_2}{-\langle A_4, W^2 \rangle} = 5, \qquad \frac{\bar{y}_3}{-\langle A_4, W^3 \rangle} = 3,$$

therefore the out-going index is 3.

Iteration 2. The new basis is $B = \{2, 5\}$, the basis matrix and its inverse are:

$$A_B = \begin{bmatrix} 0 & -1 \\ 1 & 0 \end{bmatrix} \qquad A_B^{-1} = \begin{bmatrix} 0 & 1 \\ -1 & 0 \end{bmatrix}$$

the dual basic solution is $\bar{y} = (0, 4, 0, 0, 3)$, and that primal solution is $\bar{x} = (7, 0)$, that is feasible because

$$b_1 - \langle A_1, \bar{x} \rangle = 7, \qquad b_3 - \langle A_3, \bar{x} \rangle = 6, \qquad b_4 - \langle A_4, \bar{x} \rangle = 2,$$

therefore $(0, 4, 0, 0, 3)$ is the optimal solution of the problem $(\mathcal{D})$.

5. Degeneracy and anti-cycling rules

In the previous sections we have introduced the simplex method both for a linear programming problem written in primal and dual standard form, but we have only proved the correctness of the algorithms in the special case where the feasible regions are non-degenerate. When the feasible region is degenerate, in fact, more care must be taken to prevent the algorithm to loop indefinitely onto a degenerate basic solution. This is called, in technical terms, *cycling*. Anti-cycling rules are techniques used in the simplex algorithm to prevent the algorithm from cycling. This is due to the fact that, in the degenerate case, the number θ that appears in Algorithms 2 and 3 could well be 0 and therefore changing the basis may not correspond in geometrically changing a vertex, because the same vertex may be described by several different bases.

The problem of cycling may be solved by introducing some additional rules, which are referred to as *anti-cycling rules*, to limit the choice of the in-going and the out-going indices. One of the most well-known anti-cycling rules is the so-called **Bland's rule**, which ensures that the algorithm makes progress towards an optimal solution by systematically selecting the entering and leaving variables in a particular way.

(5.1) Bland's anti-cycling rule. For the primal simplex algorithm, the out-going index is not the one with the most negative value among the $\bar{y}_i$ but

> *The out-going index should be the smallest index among the ones whose corresponding component of $\bar{y}$ is negative.*

Similarly, in the dual simplex algorithm, the out-going index is not the one with the most negative value of $b_i - A_i\bar{x}$, but

> *The out-going index should be the smallest index among the ones for which the quantity $b_i - A_i\bar{x}$ is negative.*

This simple yet effective rule guarantees that the simplex algorithm shall not revisit the same basis, thus preventing cycling and ensuring convergence to an optimal solution. Anti-cycling rules are crucial for the robustness and reliability of the simplex algorithm, particularly in some practical applications where degeneracy is common.

First of all, we are going to illustrate a linear programming problem in which the simplex algorithm (without using Bland's anti-cycling rules) cycles on the bases of a degenerate vertex when choosing the entering index as the one with the 'most negative' primal slack, that is, the entering index k is the one for which

$$b_k - \langle A_k, \bar{x}\rangle = \min_{i \in N}\Big(b_i - \langle A_i, \bar{x}\rangle\Big).$$

(5.2) Example. Consider the following dual standard problem:

$$
\begin{aligned}
\text{minimise} \quad & -\tfrac{3}{4}y_1 + 150y_2 - \tfrac{1}{50}y_3 + 6y_4 \\
\text{subject to} \quad & \tfrac{1}{4}y_1 - 60y_2 - \tfrac{1}{25}y_3 + 9y_4 + y_5 = 0 \\
& \tfrac{1}{2}y_1 - 90y_2 - \tfrac{1}{50}y_3 + 3y_4 + y_6 = 0 \\
& y_3 + y_7 = 1 \\
& y \geq 0
\end{aligned}
$$

Let us use the dual simplex algorithm, starting from the basis $B = \{5, 6, 7\}$.

Iteration 1. The indexes are divided into two set:

$$B = \{5, 6, 7\}, \qquad N = \{1, 2, 3, 4\}.$$

The basis matrix and its inverse are:

$$A_B = \begin{bmatrix} 1 & 0 & 0 \\ 0 & 1 & 0 \\ 0 & 0 & 1 \end{bmatrix} = A_B^{-1}$$

The primal basic solution and its dual complementary are, respectively:

$$\bar{x} = (0,0,0) \qquad \bar{y} = (0,0,0,0,0,0,1).$$

We have:

$$b_N - A_N\bar{x} = \left(-\tfrac{3}{4}, 150, -\tfrac{1}{50}, 6\right),$$

The entering index, according to the Bland's rule, is $k = 1$. Compute

$$\langle A_1, W^5 \rangle = -\frac{1}{4}, \qquad \langle A_1, W^6 \rangle = -\frac{1}{2}, \qquad \langle A_1, W^7 \rangle = 0,$$

the quotients are:

$$\frac{\bar{y}_5}{-\langle A_1, W^5 \rangle} = 0, \qquad \frac{\bar{y}_6}{-\langle A_1, W^6 \rangle} = 0,$$

and therefore the leaving index is $h = 5$.

Iteration 2. The new partition is:

$$B = \{1,6,7\} \qquad N = \{2,3,4,5\}.$$

The basis matrix and its inverse are:

$$A_B = \begin{bmatrix} 1/4 & 1/2 & 0 \\ 0 & 1 & 0 \\ 0 & 0 & 1 \end{bmatrix} \qquad A_B^{-1} = \begin{bmatrix} 4 & -2 & 0 \\ 0 & 1 & 0 \\ 0 & 0 & 1 \end{bmatrix}$$

The primal basic and dual solutions are::

$$\bar{x} = (-3,0,0) \qquad \bar{y} = (0,0,0,0,0,0,1).$$

We have:

$$b_N - A_N\bar{x} = \left(-30, -\tfrac{7}{50}, 33, 3\right)^T,$$

The entering index is $k = 2$. Compute:

$$\langle A_2, W^1 \rangle = 240, \qquad \langle A_2, W^6 \rangle = -30, \qquad \langle A_2, W^7 \rangle = 0,$$

the only quotient is

$$\frac{\bar{y}_6}{-\langle A_2, W^6 \rangle} = 0,$$

and therefore the leaving index is $h = 6$.

Iteration 3. Now we have

$$B = \{1,2,7\} \qquad N = \{3,4,5,6\}.$$

The basis matrix and its inverse are:

$$A_B = \begin{bmatrix} 1/4 & 1/2 & 0 \\ -60 & -90 & 0 \\ 0 & 0 & 1 \end{bmatrix} \qquad A_B^{-1} = \begin{bmatrix} -12 & -1/15 & 0 \\ 8 & 1/30 & 0 \\ 0 & 0 & 1 \end{bmatrix}$$

The primal and dual basic solutions are:

$$\bar{x} = (-1, -1, 0) \qquad \bar{y} = (0, 0, 0, 0, 0, 0, 1).$$

We have

$$b_N - A_N \bar{x} = \left(-\tfrac{2}{25}, 18, 1, 1\right),$$

the only negative component is the first, therefore the entering index is $k = 3$.
Compute:

$$\langle A_3, W^1 \rangle = -\frac{8}{25}, \qquad \langle A_3, W^2 \rangle = -\frac{1}{500}, \qquad \langle A_3, W^7 \rangle = -1,$$

the quotients are:

$$\frac{\bar{y}_1}{-\langle A_3, W^1 \rangle} = 0, \qquad \frac{\bar{y}_2}{-\langle A_3, W^2 \rangle} = 0, \qquad \frac{\bar{y}_7}{-\langle A_3, W^7 \rangle} = 1$$

and therefore the leaving index is $h = 1$.

Iteration 4. Now we have:

$$B = \{2, 3, 7\} \qquad N = \{1, 4, 5, 6\}.$$

The basis matrix and its inverse are:

$$A_B = \begin{bmatrix} -60 & -90 & 0 \\ -1/25 & -1/50 & 1 \\ 0 & 0 & 1 \end{bmatrix} \qquad A_B^{-1} = \begin{bmatrix} 1/120 & -75/2 & 75/2 \\ -1/60 & 25 & -25 \\ 0 & 0 & 1 \end{bmatrix}$$

The primal and dual basic solutions are

$$\bar{x} = (2, -3, 0) \qquad \bar{y} = (0, 0, 0, 0, 0, 0, 1).$$

We have;

$$b_N - A_N \bar{x} = \left(\tfrac{1}{4}, 3, -2, 3\right),$$

the entering index is $k = 4$. Compute

$$\langle A_4, W^2 \rangle = -\frac{1}{40}, \qquad \langle A_4, W^3 \rangle = \frac{525}{2}, \qquad \langle A_4, W^7 \rangle = -\frac{525}{2},$$

the quotients are

$$\frac{\bar{y}_2}{-\langle A_4, W^2 \rangle} = 0, \qquad \frac{\bar{y}_7}{-\langle A_4, W^7 \rangle} = \frac{2}{525}$$

and the leaving index is $h = 2$.

Iteration 5. Now, we have

$$B = \{3, 4, 7\} \qquad N = \{1, 2, 5, 6\}.$$

The basis matrix and its inverse are:

$$A_B = \begin{bmatrix} -1/25 & -1/50 & 1 \\ 9 & 3 & 0 \\ 0 & 0 & 1 \end{bmatrix} \qquad A_B^{-1} = \begin{bmatrix} 50 & 1/3 & -50 \\ -150 & -2/3 & 150 \\ 0 & 0 & 1 \end{bmatrix}$$

The primal and dual basic solutions are:

$$\bar{x} = (1, -1, 0) \qquad \bar{y} = (0, 0, 0, 0, 0, 0, 1).$$

We have

$$b_N - A_N \bar{x} = \left(-\tfrac{1}{2}, 120, -1, 1 \right),$$

the entering index is $k = 5$. Compute

$$\langle A_5, W^3 \rangle = -50, \qquad \langle A_5, W^4 \rangle = -\frac{1}{3}, \qquad \langle A_5, W^7 \rangle = 50,$$

the quotients are:

$$\frac{\bar{y}_3}{-\langle A_5, W^3 \rangle} = 0, \qquad \frac{\bar{y}_4}{-\langle A_5, W^4 \rangle} = 0,$$

the leaving index is $h = 3$.

Iteration 6. Now we have

$$B = \{4, 5, 7\} \qquad N = \{1, 2, 3, 6\}.$$

The basis matrix and its inverse are:

$$A_B = \begin{bmatrix} 9 & 3 & 0 \\ 1 & 0 & 0 \\ 0 & 0 & 1 \end{bmatrix} \qquad A_B^{-1} = \begin{bmatrix} 0 & 1 & 0 \\ 1/3 & -3 & 0 \\ 0 & 0 & 1 \end{bmatrix}$$

The primal and dual basic solutions are:

$$\bar{x} = (0, 2, 0) \qquad \bar{y} = (0, 0, 0, 0, 0, 0, 1).$$

we have

$$b_N - A_N \bar{x} = \left(-\tfrac{7}{4}, 330, \tfrac{1}{50}, -2 \right),$$

The entering index is $k = 6$. Compute

$$\langle A_6, W^4 \rangle = -\frac{1}{3}, \qquad \langle A_6, W^5 \rangle = 3, \qquad \langle A_6, W^7 \rangle = 0,$$

the only quotient is

$$\frac{\bar{y}_4}{-\langle A_6, W^4 \rangle} = 0,$$

and therefore the leaving index is $h = 4$.

Iteration 7. The new basis is $B = \{5, 6, 7\}$ which coincides with the initial basis. This leads to an infinite loop.

We shall now prove that, adopting Bland's anti-cycling rules, the simplex method does not cycle both in the primal and the dual case.

(5.3) Theorem. *With the rules stated in (5.1), the simplex method in primal standard form does not cycle on a degenerate vertex. In other words, the simplex algorithm either establishes that the degenerate vertex under examination is optimal, and finds the optimal basis, or it leaves the vertex and never returns, if it is not optimal.*

In particular, the primal simplex algorithm performs, in the worst case scenario, a number of iterations equal to the number of bases generating that vertex.

Proof. By contradiction, let us assume that the primal simplex algorithm described in Algorithm 2 produces a sequence of bases

$$B_1, B_2, \ldots, B_\ell = B_1$$

corresponding to the same degenerate primal basic solution. Consider the indices that enter and leave one of the bases in the sequence:

$$\mathcal{I} = \left\{ i : \begin{array}{l} \text{there exists } p \in \{1, \ldots, \ell\} \text{ such that } i \in B_p \\ \text{there exists } q \in \{1, \ldots, \ell\} \text{ such that } i \notin B_q \end{array} \right\}.$$

Let us denote by r the highest index in the set $\mathcal{I}$. In other words, let us set $r = \max\{i : i \in \mathcal{I}\}$. Let B' be a basis in the sequence where the index r is the in-going index, and B'' a basis in the sequence where r is the out-going index. Let us also denote by s the in-going index from the basis B'. Observe that s lies in the set $\mathcal{I}$, and $s < r$.

The bases B' and B'' are associated with the primal basic solutions x', x'', and dual basic solutions y', y'', respectively:

$$x' = A_{B'}^{-1} b_{B'}, \qquad y'_{B'} = (A_{B'}^{-1})^T c, \qquad y'_{N'} = 0,$$
$$x'' = A_{B''}^{-1} b_{B''}, \qquad y''_{B''} = (A_{B''}^{-1})^T c, \qquad y''_{N''} = 0.$$

Since the two bases B' and B'' belong to the sequence $B_1, B_2, \ldots, B_\ell$, they correspond to the same primal basic solution, which yields $x' = x''$. In addition, since the index s leaves the basis B', we have $y'_s < 0$. Denoting by W^s the opposite of the sth column of $A_{B'}^{-1}$, it follows that

$$-y'_s = \langle W^s, c \rangle = y''_{B''} A_{B''} W^s = \sum_{i \in B''} y''_i \langle A_i, W^s \rangle > 0.$$

From this inequality, we get to contradiction by showing that all terms in the previous summation are negative. Let us distinguish three cases:

1. Assume that $i \in B''$ and $i < r$. Since, by means of Bland's anti-cycling rules, r is the out-going index from B'', we have

$$r = \min\Big\{ j \in B'' : y_j'' < 0 \Big\},$$

so $y_i'' \geq 0$. Since, again my means of Bland's anti-cycling rules, r is the in-going index in B', we also have

$$r = \min\Big\{ j \notin B' : \langle A_j, W^s \rangle > 0,\ b_j - \langle A_j, x' \rangle = 0 \Big\}.$$

Moreover, $i \in B''$ so $b_i - \langle A_i, x' \rangle = b_i - \langle A_i, x'' \rangle = 0$. Thus,

- if $i \notin B'$, then $\langle A_i, W_s \rangle \leq 0$ because $i < r$;
- if $i \in B'$, then by construction $\langle A_i, W^s \rangle \leq 0$.

Thus, $y_i'' \langle A_i, W^s \rangle \leq 0$.

2. Now assume that $i \in B''$, and $i = r$. Since r is the out-going index from B'', we have $y_r'' < 0$. On the other hand, r is the index into B', which yields $\langle A_r, W^s \rangle > 0$, whence $y_r'' \langle A_r, W^s \rangle < 0$.

3. Finally, assume that $i \in B''$, and $i > r$. Since r is now the maximum of all the out-going indices, i cannot be the out-going index from any of the bases $B_1, \dots, B_\ell$, and this means that, in particular, $i \neq s$. Moreover, i belongs to all the bases $B_1, \dots, B_\ell$. Thus, $i \in B'$. Therefore, $\langle A_i, W^s \rangle = 0$, whence $y_i'' \langle A_i, W^s \rangle = 0$.

This gives the contradiction that proves the result. $\qquad\square$

(5.4) Theorem. *With the rules stated in (5.1), the simplex method in dual standard form does not cycle on a degenerate vertex. In other words, the simplex algorithm either establishes that the degenerate vertex under examination is optimal, and finds the optimal basis, or it leaves the vertex and never returns, if it is not optimal.*

In particular, the dual simplex algorithm performs, in the worst case scenario, a number of iterations equal to the number of bases generating that vertex.

Proof. By contradiction, let us assume that the dual simplex algorithm described in Algorithm 3 produces a sequence of bases

$$B_1, B_2, \dots, B_\ell = B_1$$

corresponding to the same degenerate dual basic solution. Consider the indices that enter and leave one of the bases in the sequence:

$$\mathcal{I} = \left\{ i : \begin{array}{l} \text{there exists } p \in \{1, \dots, \ell\} \text{ such that } i \in B_p \\ \text{there exists } q \in \{1, \dots, \ell\} \text{ such that } i \notin B_q \end{array} \right\}.$$

Let us denote by r be the highest index in the set $\mathcal{I}$. In other words, let us set $r = \max\{i : i \in \mathcal{I}\}$. Let B' be a basis in the sequence where the index r is the out-going index, and B'' a basis in the sequence where r is the in-going index. Let us also denote by s the out-going index from the basis B'. Observe that s lies in $\mathcal{I}$, and $s < r$.

The bases B' and B'' are associated with the primal basic solutions x', x'', and dual basic solutions y', y'', respectively:

$$x' = A_{B'}^{-1} b_{B'}, \qquad y'_{B'} = (A_{B'}^{-1})^T c, \qquad y'_{N'} = 0,$$

$$x'' = A_{B''}^{-1} b_{B''}, \qquad y''_{B''} = (A_{B''}^{-1})^T c, \qquad y''_{N''} = 0.$$

Since the two bases B' and B'' belong to the sequence $B_1, B_2, \ldots, B_\ell$, they correspond to the same dual basic solution, which yields $y' = y''$. Since s is the in-going index of the basis B', we have $b_s - \langle A_s, x' \rangle < 0$. Denoting by W^i the opposite of the ith column of the matrix $A_{B'}^{-1}$, it follows that

$$
\begin{aligned}
0 &> b_s - \langle A_s, x' \rangle \\
&= b_s - \langle A_s, \textstyle\sum_{i \in B'}(-W^i)b_i \rangle \\
&= b_s + \textstyle\sum_{i \in B'}\langle A_s, W^i \rangle b_i.
\end{aligned}
$$
(5.5)

Moreover, we can write the identity matrix as follows:

$$I = \textstyle\sum_{i \in B'}(-W^i)A_i,$$

thus

$$
\begin{aligned}
\langle A_s, x'' \rangle &= A_s \left(\textstyle\sum_{i \in B'}(-W^i)A_i \right) x'' \\
&= -\textstyle\sum_{i \in B'}\langle A_s, W^i \rangle A_i x''.
\end{aligned}
$$
(5.6)

From relations (5.5) and (5.6), it follows that

$$
\begin{aligned}
0 &> b_s + \textstyle\sum_{i \in B'}\langle A_s, W^i \rangle b_i \\
&= b_s - \langle A_s, x'' \rangle + \textstyle\sum_{i \in B'}\langle A_s, W^i \rangle b_i - \textstyle\sum_{i \in B'} A_s W^i A_i x'' \\
&= b_s - \langle A_s, x'' \rangle + \textstyle\sum_{i \in B'} A_s W^i (b_i - \langle A_i, x'' \rangle).
\end{aligned}
$$

Since r is the entering index in B'', according to Bland's rule, and $s < r$, we have that $b_s - \langle A_s, x'' \rangle \geq 0$. Therefore, we obtain:

$$0 > \textstyle\sum_{i \in B'}\langle A_s, W^i \rangle(b_i - \langle A_i, x'' \rangle).$$

From this inequality, we arrive at a contradiction by showing that all the terms in the previous summation are positive. We distinguish three cases:

1. Assume that $i \in B'$, and $i < r$. If $i \in B''$, then $b_i - \langle A_i, x'' \rangle = 0$ and thus also $A_s W^i (b_i - \langle A_i, x'' \rangle) = 0$. If $i \notin B''$, then $y'_i = y''_i = 0$. Since r enters B'', we have

$$r = \min\Big\{ j \notin B'' : b_j - \langle A_j, x'' \rangle < 0 \Big\}.$$

Thus, $b_i - \langle A_i, x'' \rangle \geq 0$ because $i < r$. Moreover, r leaves B', that is,

$$r = \min\left\{ j \in B' : y'_j = 0, \langle A_s, W^j \rangle < 0 \right\}.$$

Thus, $\langle A_s, W^i \rangle \geq 0$, because $i < r$. So, $A_s W^i (b_i - \langle A_i, x'' \rangle) = 0$.

2. Now assume that $i \in B'$, and $i = r$. Since r is the in-going index in B'', we have $b_r - \langle A_r, x''_r \rangle < 0$. On the other hand, r is the out-going index from B', so $\langle A_s, W^r \rangle < 0$, and thus $A_s W^r (b_r - \langle A_r, x'' \rangle) > 0$.

3. Finally, assume $i \in B'$, and $i > r$. Since r is the maximum of the leaving indices, i cannot be the leaving index from any basis $B_1, \ldots, B_\ell$ and thus belongs to all the bases $B_1, \ldots, B_\ell$, in particular $i \in B''$, so $b_i - \langle A_i, x'' \rangle = 0$ and thus also $A_s W^i (b_i - \langle A_i, x'' \rangle) = 0$.

This gives the contradiction that proves the result. $\qquad\square$

6. Selection criteria for the starting vertex

In order to be able to start the simplex algorithm it is first necessary to find a basis that gives a primal basic feasible solution. This is sometimes very simple to do, but other times it requires the usage of some additional tools to select a basis to start the algorithm with. This is often done employing an auxiliary problem that helps locating a vertex of the polyhedron which represents the feasible region, that is to say a basis that gives a basic feasible solutions. This auxiliary problem is again a linear programming problem, in which a linear objective function is defined over a suitable feasible region. The auxiliary problem is created in such a way that an optimal solution thereof provides a basic feasible solution of the initial problem. To solve the auxiliary problem, of course, one can use the simplex algorithm and one could argue that this makes the problem cyclical as one would need a starting basis for the auxiliary problem as well, but, thanks to the particular structure used to define the auxiliary problem is created, finding such a starting basis is in fact trivial.

Linear programming problems in primal standard form

To describe how the auxiliary problem is created, let us start by taking a linear programming problem in primal standard form, that is to say, consider the problem

$$(\mathcal{P}) : \quad \begin{cases} \text{maximise} & \langle c, x \rangle \\ \text{subject to} & Ax \leq b \end{cases}$$

where $A \in \mathfrak{M}(m, n, \mathbb{R})$ is a matrix, $b \in \mathbb{R}^m$ and $c \in \mathbb{R}^n$ are vectors. Assume that we have a basis B for this problem, whose corresponding basic solution is $\bar{x} = A_B^{-1} b_B$, and assume that this solution is not feasible. Among the non-basic constraints, we may therefore distinguish those satisfied by $\bar{x}$ from those not satisfied:

$$U = \left\{ i \in N : \langle A_i, \bar{x} \rangle \leq b_i \right\}, \qquad V = \left\{ i \in N : \langle A_i, \bar{x} \rangle > b_i \right\}.$$

Using these two sets of indices, we can now take the following auxiliary problem, where the variables ε_i, with $i \in V$, are introduced as suitable positive real numbers to force the constraints that are not feasible to be such:

$$(\mathcal{P}_{\text{aux}}) : \begin{cases} \text{maximise} & -\sum_{i \in V} \varepsilon_i \\ \text{subject to} & \langle A_i, x \rangle \leq b_i & \text{for all } i \in B \cup U \\ & \langle A_i, x \rangle - \varepsilon_i \leq b_i & \text{for all } i \in V \\ & -\varepsilon_i \leq 0 & \text{for all } i \in V \end{cases}$$

In matrix form, the constraints of $(\mathcal{P}_{\text{aux}})$, may be written in a blocks matrix in the form

$$M = \begin{bmatrix} A_B & 0 \\ A_U & 0 \\ A_V & -I \\ 0 & -I \end{bmatrix}$$

where I denotes the identity $k \times k$ matrix, where k is the number of indices in V. Observe that, by construction, the vector $(\bar{x}, \bar{\varepsilon})$ with

$$\bar{\varepsilon} = A_V \bar{x} - b_V \geq 0,$$

is a feasible basic solution of $(\mathcal{P}_{\text{aux}})$ associated to the basis $B \cup V$ and the basis matrix is given by:

$$\begin{bmatrix} A_B & 0 \\ A_V & -I \end{bmatrix}.$$

Starting from this basis, which gives by construction a primal basic feasible solution, we can apply the primal simplex method to solve $(\mathcal{P}_{\text{aux}})$. To this end, let us observe that the optimal value of this auxiliary problem is necessarily a negative number, for it lies between the value $-\sum_{i \in V} \bar{\varepsilon}_i$ and 0. Let us now distinguish two cases:

1. *The optimal value of the auxiliary problem $(\mathcal{P}_{\text{aux}})$ is strictly negative.*

   Then, the original problem $(\mathcal{P})$ has no feasible solution. In fact, by contradiction, let us suppose that $(\mathcal{P})$ has a feasible solution $\bar{x}$. Then, the pair (x, ε), with $x = \bar{x}$ and $\varepsilon = 0$ would be a feasible solution for the problem $(\mathcal{P}_{\text{aux}})$, with the objective function equal to 0, which is a contradiction, because the maximum value of $(\mathcal{P}_{\text{aux}})$ is assumed to be strictly negative.

2. *The optimal value of the auxiliary problem $(\mathcal{P}_{\text{aux}})$ is equal to 0.*

In this case, let us show that we can find a basis for the original problem $(\mathcal{P})$, whose primal basic solution is feasible. To this end, suppose that the optimal basic solution of $(\mathcal{P}_{\mathrm{aux}})$ is (x^*, ε^*) with $\varepsilon^* = 0$. The corresponding sub-matrix of M is obtained by taking a suitable sub-matrix A_{C_1}, where $C_1 \subseteq B$, a suitable sub-matrix A_{C_2}, with $C_2 \subseteq V$ with the property that $C = C_1 \cup C_2$ has overall n indices, that is to say, the matrix A_C is $n \times n$, and the third block is also part of this basis because third groups of constraints are active by hypothesis. In other words, the matrix associated to this basis is:

$$M^* = \begin{bmatrix} A_C & * \\ 0 & -I \end{bmatrix}.$$

The matrix A_C is a $n \times n$ matrix and therefore it suffices to show that this is a non-singular matrix to show that $C = C_1 \cup C_2$ is a basis for $(\mathcal{P})$, which generates a primal basic feasible solution. This is actually trivial, because the upper triangular matrix matrix M^* and therefore $\det(M^*) = \det(A_C)\det(I) = \det(A_C)$ and since the left-hand side of this equality is non-zero, such is the right-hand side.

(6.1) Example. Let us see whether there exists a basis that gives a primal basic feasible solution of the problem:

$$(\mathcal{P}): \begin{cases} \text{maximise} & 2x_1 + x_2 \\ \text{subject to} & x_1 + 2x_2 \le 5 \\ & x_1 - x_2 \le 2 \\ & x_1 + 3x_2 \le 8 \\ & -x_2 \le 0 \end{cases}$$

The basis $\{1, 4\}$ corresponds to the solution $\bar{x} = (5, 0)$ which is not feasible because it violates the second constraint. Let us construct the auxiliary problem corresponding to this basis:

$$(\mathcal{P}_{\mathrm{aux}}): \begin{cases} \text{maximise} & -\varepsilon \\ \text{subject to} & x_1 + 2x_2 \le 5 \\ & x_1 - x_2 - \varepsilon \le 2 \\ & x_1 + 3x_2 \le 8 \\ & -x_2 \le 0 \\ & -\varepsilon \le 0 \end{cases}$$

For this auxiliary problem, the basis $\{1, 2, 4\}$ is feasible and corresponds to the solution $(x_1, x_2, \varepsilon) = (5, 0, 3)$. Starting from this basis, let us apply the primal simplex algorithm. The corresponding dual basic solution is

$$\bar{y}_B = \begin{bmatrix} 1 & 0 & 2 \\ 0 & 0 & -1 \\ 1 & -1 & 3 \end{bmatrix} \begin{bmatrix} 0 \\ 0 \\ -1 \end{bmatrix} = \begin{bmatrix} -1 \\ 1 \\ -3 \end{bmatrix}$$

therefore 1 is the out-going index with $W^1 = (-1, 0, -1)$. Since

$$\langle A_3, W^1 \rangle = -1, \qquad \langle A_5, W^1 \rangle = 1,$$

it is found that 5 is the in-going index. So, the new basis is $\{2, 4, 5\}$. The corresponding basis matrix is

$$\begin{bmatrix} 1 & -1 & -1 \\ 0 & -1 & 0 \\ 0 & 0 & -1 \end{bmatrix}$$

and the corresponding basic solution is $x = (2, 0)$, $\varepsilon = 0$ which is optimal for the auxiliary problem, because the value of the objective function is 0. From the optimal basis matrix for the auxiliary problem, by deleting the third row and the third column, we obtain

$$\begin{bmatrix} 1 & -1 \\ 0 & -1 \end{bmatrix}$$

which is the basis matrix for the problem $(\mathcal{P})$ corresponding to the basis $\{2, 4\}$, which gives the basic feasible solution $x = (2, 0)$.

Linear programming problems in dual standard form

Let us now pass to a linear programming problem that is written in dual standard form and develop a similar method to find a basis that gives a dual basic feasible solution, if it exists, or that establishes that the polyhedron, this time written in a dual standard form, is empty. To this end, consider the following problem in dual standard form:

$$(\mathcal{D}) : \quad \begin{cases} \text{minimise} & \langle b, y \rangle \\ \text{subject to} & A^T y = c \\ & y \geq 0 \end{cases}$$

With no loss of generality, we may assume that $c \geq 0$, by possibly changing the sign of the columns of A corresponding to the negative components of c. Furthermore, we may also assume that there are no redundant equations in the system $A^T y = c$. With these extra hypotheses, let us consider the following auxiliary problem:

$$(\mathcal{D}_{\text{aux}}) : \quad \begin{cases} \text{minimise} & \sum_{i=1}^{n} \varepsilon_i \\ \text{subject to} & A^T y + \varepsilon = c \\ & y, \varepsilon \geq 0 \end{cases}$$

Observe that the corresponding auxiliary matrix is now $M = \begin{bmatrix} A^T & I \end{bmatrix}$, where I is the $n \times n$ identity matrix. The basis formed by selecting the indices corresponding to the auxiliary variables $\varepsilon_1, \ldots, \varepsilon_n$, with the identity matrix as the basis matrix, is a basis for the auxiliary problem, that by construction generates a dual basic feasible solution, and it is $\bar{y} = 0$, $\bar{\varepsilon} = c \geq 0$. Starting from this basis, we can now apply the dual simplex algorithm to solve this auxiliary problem. The optimal value of the auxiliary problem, which is this time a positive value, lying between 0 and $\sum_{i=1}^{n} c_i$, determines whether a feasible basis exists for the problem $(\mathcal{D})$. Let us now distinguish two cases:

1. *The optimal value of the auxiliary problem $(\mathcal{D}_{\mathrm{aux}})$ is strictly positive.*

   Then, the original problem $(\mathcal{D})$ has no feasible solutions. In fact, by contradiction, let us suppose that $(\mathcal{D})$ has a feasible solution $\bar{y}$. Then, $y = \bar{y}$ and $\varepsilon = 0$ would be feasible for $(\mathcal{D}_{\mathrm{aux}})$, with the objective function equal to 0, which is a contradiction, because the minimum value of $(\mathcal{D}_{\mathrm{aux}})$ is assumed to be strictly positive.

2. *The optimal value of the auxiliary problem $\mathcal{D}_{\mathrm{aux}}$ is equal to 0.*

   In this case, let us show that we can find a basis for the original problem $(\mathcal{D})$, whose dual basic solution is feasible. To this end, suppose that the optimal basic solution of $(\mathcal{D}_{\mathrm{aux}})$ is y^*, $\varepsilon^* = 0$. Let us denote by B^* the corresponding optimal basis obtained applying the dual simplex algorithm. Let us recall that B^* is a subset of the set of indices for the auxiliary problem:

$$\{1,\ldots,m+n\} = \{1,\ldots,m\} \cup \{m+1,\ldots,m+n\}.$$

   There are two cases:

   - *The basis B^* is included in $\{1,\ldots,m\}$.*

     In this case, the basis B^* is a basis for the original problem, $(\mathcal{D})$, which generates a dual basic feasible solution, y^*. Utilising this basis one can now start the dual simplex algorithm to solve $(\mathcal{D})$.

   - *The basis B^* can be split into two non-empty sets $C_1 \cup C_2$, with C_1 included in $\{1,\ldots,m\}$ and C_2 included in $\{m+1,\ldots,m+n\}$.*

     In this case, which is necessarily degenerate, because all ε variables are zero and C_2 are the indices relative to the ε variables, this vertex may be described by more than one basis, and we shall need to find a different basis, which is only made of indices in $\{1,\ldots,m\}$. This can be done with a process called **forcing**, which is a variation of the strategy used in the simplex algorithm by swapping two indices, but, unlike the simplex algorithm, in this case the objective function is maintained constant.

It remains to describe the **forcing** process. To this end, let us start by observing that it suffices to describe the process when there is only one index h in C_2. In fact, there were more than one index in C_2, we may repeat the procedure for each index until all are exhausted.

Now, the index h is the out-going index. Then, since the optimal value for the auxiliary problem is 0, if N^* denotes the optimal non-basis in M, and W denotes the matrix $-M_{B^*}^{-1}$, it suffices to take the in-going index in the set

$$(6.2) \qquad \left\{ i \in N^* : i \in \{1, \ldots, m\}, \langle A_i, W^h \rangle \neq 0 \right\}.$$

Observe, in fact, that, for $i \in \{1, \ldots, m\}$, $M_i = A_i$. Let us prove that this set is not empty. To this end, multiplying the matrix A by the matrix W, we obtain a matrix of the form

$$AW = \begin{bmatrix} A_{B^*} \\ A_{N^*} \end{bmatrix} \begin{bmatrix} W^* & W^h \end{bmatrix} = \begin{bmatrix} I & 0 \\ * & w \end{bmatrix}$$

where I is the $(n-1) \times (n-1)$ identity, and the vector w in $\mathbb{R}^{m-n+1}$ cannot be zero, for otherwise, if the last column of the right-hand side of the previous formula is entirely zero, the matrix A would have rank n, against the initial hypothesis that the matrix A is of maximum rank. Since w is the vector with components $\langle A_i, W^h \rangle$ for every $i \in N^*$ such that $i \in \{1, \ldots, m\}$, this proves that the set (6.2) is not empty.

The new basis $B^* \setminus \{h\} \cup \{k\}$ has now one more index in $\{1, \ldots, m\}$ and the procedure may be repeated until the basis B^* is fully formed of indices in $\{1, \ldots, m\}$, whence the basis B^* becomes a basis of the original problem, which generates a dual basic feasible solution y^*.

Let us show two examples which show the method to find a dual feasible basis.

(6.3) Example. Let us see whether there exists a basis that gives a dual basic feasible solution of the problem

$$(\mathcal{D}) : \quad \begin{cases} \text{minimise} & y_1 + y_2 + y_3 \\ \text{subject to} & 2y_1 + y_2 + 2y_3 = 4 \\ & 3y_1 + 3y_2 + y_3 = 3 \\ & y \geq 0 \end{cases}$$

The auxiliary dual is:

$$(\mathcal{D}_{\text{aux}}) : \quad \begin{cases} \text{minimise} & \varepsilon_1 + \varepsilon_2 \\ \text{subject to} & 2y_1 + y_2 + 2y_3 + \varepsilon_1 = 4 \\ & 3y_1 + 3y_2 + y_3 + \varepsilon_2 = 3 \\ & y, \varepsilon \geq 0 \end{cases}$$

For solving the auxiliary problem, let us apply the dual simplex algorithm, starting from the basis $\{4, 5\}$, with corresponding dual basic feasible solution $y = (0, 0, 0)$, $\varepsilon = (4, 3)$. The corresponding primal basic solution is $\bar{x} = (1,1)$ which is not feasible, because the primal is

$$(\mathcal{P}) : \begin{cases} \text{maximise} & 4x_1 + 3x_2 \\ \text{subject to} & 2x_1 + 3x_2 \leq 0 \\ & x_1 + 3x_2 \leq 0 \\ & 2x_1 + x_2 \leq 0 \\ & x_1 \leq 1 \\ & x_2 \leq 1 \end{cases}$$

In fact

$$\langle A_1, \bar{x} \rangle = 5 > 0,$$

thus $k = 1$ is the in-going index. Moreover W is given by:

$$\begin{bmatrix} -1 & 0 \\ 0 & -1 \end{bmatrix}$$

Compute the scalar product of the dual simplex algorithm:

$$\langle A_1, W^4 \rangle = -2 \qquad \langle A_1, W^5 \rangle = -3$$

and the quotients

$$r_4 = -4/-2 = 2 \qquad r_5 = -3/-3 = 1$$

so the out-going index is 5 and the new basis is $\{1,4\}$. The corresponding dual basic feasible solution is

$$(y, \varepsilon) = (1,0,0,2,0);$$

while the primal basic solution is $\bar{x} = (1, -2/3)$ which is not feasible because

$$\langle A_3, x \rangle = 4/3 > 0.$$

The in-going index is 3. Moreover W is given by:

$$\begin{bmatrix} 0 & -1 \\ -1/3 & 2/3 \end{bmatrix}$$

Compute the scalar product of the dual simplex algorithm:

$$\langle A_3, W^1 \rangle = -1/3 \qquad \langle A_3, W^4 \rangle = -4/3$$

and the quotients

$$r_1 = -1/(-1/3) = 3 \qquad r_4 = -2/(-4/3) = 3/2$$

so the out-going index is 4 and the new basis is $\{1,3\}$. The optimal solution of the auxiliary problem is

$$(y, \varepsilon) = (1/2, 0, 3/2, 0, 0).$$

Being now the variables ε out of the basis, they are zero and the objective function of the auxiliary dual has value equal to 0 and therefore the initial polyhedron is not empty. The basis $\{1,3\}$ is the feasible basis of the initial polyhedron.

(6.4) Example. Let us see whether there exists a basis that gives a dual basic feasible solution of the problem

$$(\mathcal{D}): \quad \begin{cases} \text{minimise} & 8y_1 - y_2 - 5y_3 + 8y_4 \\ \text{subject to} & -y_3 - 2y_4 = 0 \\ & y_1 - y_2 + 3y_4 = 1 \\ & y \geq 0 \end{cases}$$

The auxiliary problem is, by definition:

$$(\mathcal{D}_{\text{aux}}): \quad \begin{cases} \text{minimise} & \varepsilon_1 + \varepsilon_2 \\ \text{subject to} & -y_3 - 2y_4 + \varepsilon_1 = 0 \\ & y_1 - y_2 + 3y_4 + \varepsilon_2 = 1 \\ & y \geq 0 \\ & \varepsilon \geq 0 \end{cases}$$

For the auxiliary problem, let us apply the dual simplex algorithm, starting from the basis $\{5, 6\}$, with corresponding dual basic feasible solution $y = (0, 0, 0, 0)$, $\varepsilon = (0, 1)$. The corresponding primal basic solution is $\bar{x} = (1, 1)$ which is not feasible, for

$$\langle A_1, \bar{x} \rangle = 1 > 0.$$

Thus $k = 1$ is the in-going index. Moreover, $\langle A_1, W^5 \rangle = 0$, $\langle A_1, W^6 \rangle = -1$, thus $h = 6$ is the out-going index. The new basis $\{1, 5\}$ corresponds to the dual basic feasible solution $y = (1, 0, 0, 0)$, $\varepsilon = (0, 0)$ which is optimal for the auxiliary problem with value 0. Observe that the solution is degenerate.

To find a dual feasible basis of the initial problem, we need to perform a further step of the dual simplex applied to the auxiliary problem because 5 is an index related to the auxiliary variable ε_1. This may be done even if we are at the optimal solution applying what we have called 'forcing'. It consists in finding another basis which provides the optimal solution.

The inverse of the basis matrix is

$$M = \begin{bmatrix} 0 & 1 \\ 1 & 0 \end{bmatrix}$$

therefore

$$AM = \begin{bmatrix} 1 & 0 \\ -1 & 0 \\ 0 & -1 \\ 3 & -2 \end{bmatrix}$$

Since $\langle A_3, M^5 \rangle = -1$, we can create a new basis by swapping the indices 3 and 5, thereby obtaining $\{1, 3\}$ as a basis for the original problem, whose solution is $y = (1, 0, 0, 0)$. Observe that $\langle A_4, M^5 \rangle = -2$, so index 5 can also be swapped with the index 4, thus another basis for the problem original problem is $\{1, 4\}$, still corresponding to the solution $y = (1, 0, 0, 0)$.

Chapter IX
Methods for Unconstrained Optimisation

Unconstrained optimisation methods are techniques used to find the minimum or maximum of a function with no constraints on the variables. These methods are essential to find numerical solutions and are widely used in various fields such as engineering, economics, and machine learning, where unconstrained optimisation problems often arise.

The goal of unconstrained optimisation is to find the point at which a given objective function reaches its optimal value. As we have aforementioned, this is generally a very complex task from a numerical point of view, and some numerical methods are therefore necessary to achieve at least an approximate solution. Theoretically, however, we know that such points are, in particular, stationary points, i.e. points where the gradient is zero. For this reason, the key point of unconstrained methods involves the search for a point where the gradient (or derivative, in the uni-dimensional case) of the function is zero. This task can be addressed numerically with adequate tools.

There are several popular methods for unconstrained optimisation, and new ones are developed every so often, and it would be impossible to list them all here. A selection of the most classical and popular includes:

Gradient Descent: These methods involve moving in the direction of the negative gradient of the function to find the minimum. They differ from each other in the choice of the step. They are widely used due to its simplicity and effectiveness, especially for large-scale problems.

Conjugate Gradient: These methods involve moving in the direction of the negative gradient linearly combined with the previous direction in order not to forget the already performed steps.

Newton's Method: This method uses second-order derivatives to find the optimal point. It converges faster than gradient descent but requires the computation of the Hessian, which can be computationally expensive

© The Author(s), under exclusive license to Springer Nature Switzerland AG 2026
A. Carpignani and M. Pappalardo, *Theory and Methods of Optimisation*,
UNITEXT 175, https://doi.org/10.1007/978-3-032-01514-3_9

Quasi-Newton Methods: These methods approximate the Hessian matrix to achieve faster convergence without the need to compute second-order derivatives explicitly.

Sub-gradient Methods: These methods involve moving in the direction of a sub-gradient when the function to be minimized is convex and not differentiable.

In the sequel, we shall present these methods to solve an unconstrained problem. To deploy these methods, in each step of the algorithm, it is often going to be necessary to minimise a function of a single variable. For this reason, we are going to start by describing the simplest case, looking at methods to minimise a real function of one real variable, $f : I \to \mathbb{R}$, where I is an open interval. Solving this problem, as we have said, may be extremely complex, and therefore we shall limit ourselves in finding an approximation for a stationary point, that is to say a point x^* such that $f'(x^*) = 0$. Only in special cases, such as when the objective function is convex, will this result in finding an approximation of the global minimum.

1. Methods to find zeros of a single variable function

Given a function of one single variable $f : I \to \mathbb{R}$, defined on an open interval I, that we assume to be at least of class C^1, we wish to find the solutions of the equation $f'(x) = 0$. The simplest method to solve this problem is to work out the first derivative, f', and then look for its zeros, by applying those methods developed to find zeros of a real function. For this reason, in the rest of this section, we are going to discuss two classic and well-known methods for finding the zeros of a continuous function, namely the *bisection method* and *Newton-Raphson's method*.

Bisection method

Given a continuous real function $g : I \to \mathbb{R}$, defined on a suitable open interval I, suppose that we are interested in finding a solution to the equation $g(x) = 0$ with $x \in I$, or, in other words, a zero of the function g. The **bisection method** leverages the properties of continuous real functions, and in particular the so-called 'Intermediate Values Theorem'. Specifically, if the function g is continuous on an interval $[a, b]$, included within the domain I of g, and if the values that g takes at the extremes of this interval have opposite signs — that is, if $g(a)g(b) < 0$, the 'Intermediate Value Theorem' guarantees that there exists at least one zero, x^*, of g within the open interval (a, b). In other words, the Intermediate Value Theorem offers a criterion to establish whether in a given interval included in the domain of a function there is a zero. Before describing how this is going to be implemented into an algorithm, however, it is important to observe that the Intermediate Value Theorem offers a sufficient condition to locate a zero, but that condition may is in general not necessary.

The bisection method approximates x^* as the midpoint of the interval $[a, b]$. Moreover, by determining in which sub-interval the approximated solution lies — through evaluating the sign of $g\left(\frac{a+b}{2}\right)$, the process can be repeated iteratively. This generates a sequence of approximated solutions such that the width of the interval containing the solution is halved at each step.

Now, let us adapt this method to the optimisation problem to find the stationary points of the function f, that is, to solve the equation $f'(x) = 0$. In this case, we may apply the bisection method to the derivative f', provided that f' is continuous on $[a, b]$, and that $f'(a)f'(b) < 0$.

As a stopping criterion for this method, we can choose to repeat the algorithm until the interval $[a^k, b^k]$ is sufficiently small. In other words, the stopping criterion consists in selecting a positive real number ε as a fixed tolerance threshold, and to decide to stop the algorithm when $b^k - a^k < \varepsilon$. Since

$$b^k - a^k = \frac{b - a}{2^k},$$

an approximation for the solution x^* shall be the number x^k, mid-point of the interval $[a^k, b^k]$, with an absolute error less than ε, provided that

$$k > \log_2\left(\frac{b - a}{\varepsilon}\right).$$

Algorithm 4 shows a summary of the bisection method.

Algorithm 4 BISECTION METHOD

---

1. Set $a^0 = a$, $b^0 = b$, and $k = 0$.

2. **while** $|b^k - a^k| > \varepsilon$ **do**

    Set $x^k = (a^k + b^k)/2$.

    if $g(x^k) = 0$ **then stop**

    else if $g(a^k)g(x^k) < 0$ **then**

        Set $a^{k+1} = a^k$ and $b^{k+1} = x^k$.

    else

        Set $a^{k+1} = x^k$ and $b^{k+1} = b^k$.

3. Increase k by 1.

(1.1) Example. Let us suppose we are seeking the stationary points of the function $f(x) = x^4 + 2x^2 - 3x$ in the interval $[0, 1]$. In other words, suppose we wish to find a solution of the equation

$$f'(x) = 4x^3 + 4x - 3 = 0.$$

Observe that $f'(0) = -3$ and $f'(1) = 5$, which ensures that there is a stationary point of the function f in the interval $[0, 1]$. In addition, observe that the function f is convex on $\mathbb{R}$ because its second derivative is

$$f''(x) = 12x^2 + 4,$$

which is positive for every real number x, hence, the stationary point x^* is also the global minimum of f also on the set $\mathbb{R}$.

Applying the bisection method with a threshold $\varepsilon = 0.001$, we obtain:

k	a	b	x	$f(a)$	$f(b)$
1	0.5000	1.0000	0.5000	−0.50000	5.00000
2	0.5000	0.7500	0.7500	−0.50000	1.68750
3	0.5000	0.6250	0.6250	−0.50000	0.47656
4	0.5625	0.6250	0.5625	−0.03809	0.47656
5	0.5625	0.5938	0.5938	−0.03809	0.21228
6	0.5625	0.5781	0.5781	−0.03809	0.08540
7	0.5625	0.5703	0.5703	−0.03809	0.02324
8	0.5664	0.5703	0.5664	−0.00753	0.02324
9	0.5664	0.5684	0.5684	−0.00753	0.00783
10	0.5664	0.5674	0.5674	−0.00753	0.00015

Hence $x^* \simeq 0.5674$.

Newton-Raphson's method

Just like the bisection method, Newton-Raphson's method also seeks a solution to an equation of the form $g(x) = 0$, where $g : I \to \mathbb{R}$ is a real function, defined on an open interval I, which is assumed to be of class $\mathcal{C}^2$. It is worth to mention that Newton-Raphson's method is also called the 'tangents' method' because it starts from a point of the graph of the function g, then takes the tangent to the graph of the function at that point and approximates the zero of the function, taking the x-value where the tangent cuts the x-axis, and utilising this as the new starting point.

More algebraically, starting from a point, x^k, let us approximate the function g with its tangent line at the point $(x^k, g(x^k))$. In other words, assuming that $g'(x_k)$ is different from zero otherwise we have finished, let us take

$$g(x) \simeq g(x^k) + g'(x^k)(x - x^k).$$

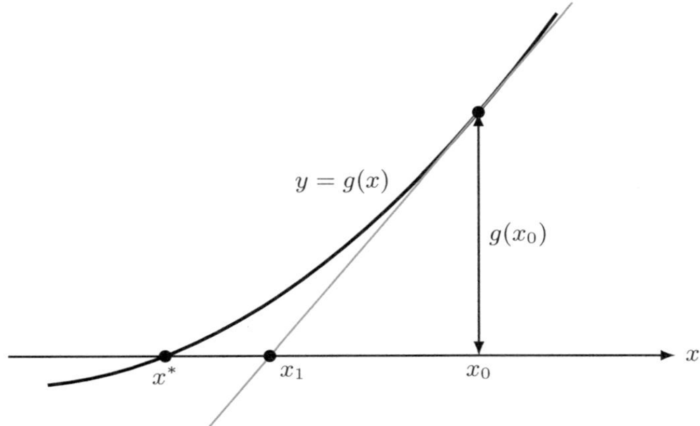

Figure 1.1. The figure shows the first step of Newton-Raphson's method to find the zero, x^*, of a function g. Starting from a random point x_0, in a neighbourhood of x^*, the tangent at $(x_0, g(x_0))$ is calculated and the zero of $g(x)$ is approximated by the zero of this tangent, that is to say x_1. The method can then start back from this new point to get a better approximation of x^*.

The value where this line cuts the x-axis is found setting this expression equal to zero, and rearranging to find x, which is then taken as the new point, x^{k+1}, of the sequence:

$$(1.2) \qquad x^{k+1} = x^k - \frac{g(x^k)}{g'(x^k)}.$$

It is thus defined an iterative sequence $\{x^k\}$. Algorithm 5 shows a summary of the Newton-Raphson's method.

Algorithm 5 NEWTON-RAPHSON'S METHOD

---

1. Given $\varepsilon > 0$, select an initial point x^0 and set $k = 0$.

2. **if** $|g'(x^k)| < \varepsilon$ **then stop**

   else Compute $x^{k+1} = x^k - \dfrac{g(x^k)}{g'(x^k)}$.

3. Increase k by 1 and **go back** to step 2.

To adapt this method to solve an optimisation problem of minimising a function $f(x)$ on an open interval I, we wish to seek the stationary points of the function f, that is, to solve the equation $f'(x) = 0$. In this case, we may apply the Newton-Raphson's method to the derivative f', and this implies that the function f must be of class C^3.

Let us show that Newton-Raphson's method is super-linearly convergent, even though it only locally, that is to say the starting point must be chosen in a suitable neighbourhood of zero.

(1.3) Theorem. *Let $g : I \to \mathbb{R}$ be a real function of class $\mathcal{C}^2$ on an open interval I. Assume that x^* is a zero of g, that is a point in I such that*

$$g(x^*) = 0,$$

and assume that $g'(x^*) \neq 0$. Then there exists a neighbourhood (a,b) of x^* such that, for any initial point x^0 chosen in the interval (a,b), the iterative sequence $\{x^k\}$ defined by (1.2) converges towards x^*. In addition, the convergence is super-linear of order at least 2, and the convergence is quadratic if $g''(x^*) \neq 0$.

Proof. Let us consider the function

$$h(x) = x - \frac{g(x)}{g'(x)}$$

which is defined in a neighbourhood of x^*. Observe that x^* is a fixed point of h, that is to say, $h(x^*) = x^*$. It is not difficult to see that the derivative of this function calculated in x^* is

$$h'(x^*) = \frac{g(x^*)g''(x^*)}{[g'(x^*)]^2} = 0.$$

Since h' is assumed to be continuous at x^*, there exists a neighbourhood (a,b) of the point x^* such that $|h'(x)| \leq \frac{1}{2}$ for every $x \in (a,b)$. No matter how the initial point $x^0 \in (a,b)$ is chosen, if we assume that the all the terms $x^0, x^1, \ldots, x^k$ are in the interval (a,b), then we have:

$$\begin{aligned}
|x^{k+1} - x^*| &= |h(x^k) - h(x^*)| \\
&= |h'(\xi)(x^k - x^*)| \\
&\leq \tfrac{1}{2}|x^k - x^*|.
\end{aligned}$$

Thus, if $x^0 \in (a,b)$, then the iterative sequence $\{x^k\}$ is all contained in (a,b), and we have $|x^k - x^*| \leq 2^{-k}|x^0 - x^*|$. This proves that the iterative sequence $\{x^k\}$ converges towards x^*. Moreover, by definition, we have

$$g'(x^k)(x^{k+1} - x^*) = g'(x^k)(x^k - x^*) - g(x^k),$$

and by means of Taylor's formula

$$0 = g(x^*) = g(x^k) + g'(x^k)(x^* - x^k) + \tfrac{1}{2}(x^k - x^*)^2 g''(\xi^k),$$

where ξ^k is a real number such that $|\xi^k - x^*| < |x^k - x^*|$. Therefore,

$$g'(x^k)(x^{k+1} - x^*) = \tfrac{1}{2}(x^k - x^*)^2 g''(\xi^k),$$

whence

$$\frac{x^{k+1} - x^*}{(x^k - x^*)^2} = \frac{g''(\xi^k)}{2g'(x^k)}.$$

Taking the limit as $k \to \infty$, we obtain

$$\lim_{k \to \infty} \frac{x^{k+1} - x^*}{(x^k - x^*)^2} = \frac{g''(x^*)}{2g'(x^*)}.$$

This proves that the convergence of the method is at least of order 2, and that the convergence is quadratic when $g''(x^*) \neq 0$. $\qquad\square$

Newton-Raphson's method can be computationally expensive, because at each step it is necessary to evaluate in the new point both the function g and its first derivative g'. Moreover, the algorithm is only locally convergent, hence it may not converge, if the starting point is not sufficiently close to the root, as shown by the following example.

(1.4) Example. Consider the function $g(x) = -\frac{1}{4}x^3 + \frac{5}{4}x$. Choosing the initial point $x^0 = 1$, we obtain

$$x^0 = -1, \quad x^1 = 1, \quad x^2 = -1, \quad x^3 = 1, \quad \ldots$$

Thus, the sequence oscillates between 1 and -1, and it clearly does not converge to the zero $x^* = 0$.

Coming back to the initial problem of finding the stationary points of a real function f, we can apply Newton-Raphson's method to its first derivative f'. Let us show another example, applied to f', which emphasises the local convergence of the Newton-Raphson's method in the sense that if x^0 is too far from the minimum x^*, then the generated sequence can be not convergent.

(1.5) Example. Let us consider the real function

$$f(x) = -\frac{1}{16}x^4 + \frac{5}{8}x^2.$$

It is readily shown that

$$f'(x) = -\frac{1}{4}x^3 + \frac{5}{4}x, \qquad f''(x) = -\frac{3}{4}x^2 + \frac{5}{4}.$$

It is easy to see that $x^* = 0$ is a stationary point of f, and since $f''(0) = \frac{5}{4} > 0$, it follows that 0 is a local minimum of f. Nevertheless, if we apply Newton-Raphson's method with starting point $x^0 = 1$, we immediately see that we obtain the oscillating sequence

$$x^0 = -1, \quad x^1 = 1, \quad x^2 = -1, \quad x^3 = 1, \quad \ldots$$

that does not converge to $x^* = 0$.

Let us complete this section by showing an example where Newton-Raphson's method performs well.

(1.6) Example. Assume we wish to find a zero of the function

$$f(x) = x^4 + 2x^2 + 3x - 1$$

using the Newton-Raphson's method starting from the point $x_0 = 1$. The first 4 iterations of the method are shown in the following table:

k	x	$f(x)$	$f'(x)$
1	1.0000	-1.0000	5.0000
2	1.2000	0.3536	8.7120
3	1.1594	0.0172	7.8717
4	1.1572	0.0000	7.8278

2. Gradient descent methods

Gradient descent methods are a backbone of numerical optimisation, offering a robust and intuitive approach to finding optimal solutions for unconstrained non-linear problems. At their heart, these methods rely on the principle of iterative improvement, guided by the local geometry of the objective function. By calculating the gradient of the objective function, the algorithm takes measured steps in the opposite direction, gradually descending towards a local minimum. This elegant interplay between calculus and computation makes gradient descent a versatile tool, widely employed in fields ranging from machine learning to engineering design.

The appeal of gradient descent methods lies in its simplicity and scalability. Unlike methods that require second-order derivatives or extensive matrix computations, gradient descent methods operate efficiently even in high-dimensional spaces, making it particularly well-suited to modern problems with vast datasets. However, its performance hinges on careful tuning of parameters such as the step size, or learning rate, which governs the magnitude of each update. Too large a step may lead to oscillations or divergence, while too small a step risks sluggish convergence.

Thus, while gradient descent is conceptually straightforward, its successful application often requires a nuanced understanding of both the problem at hand and the algorithm's behaviour.

At its core, gradient descent methods seek the stationary points of given real function $f : \mathbb{R}^n \to \mathbb{R}$ using a iterative sequence of the form

$$x^{k+1} = x^k + t_k d^k,$$

where d^k is a given direction in $\mathbb{R}^n$, generally referred to as the **line search**, and t_k is a positive real number, is the so-called **step size** or, especially in machine learning, the **learning rate**.

The search direction is chosen to be a descent direction for the objective function in the sense of Definition V(1.1). Following the aforementioned definition, the aim is to calculate the restriction, φ, of the objective function $f : \mathbb{R}^n \to \mathbb{R}$ to the ray starting at x^k. In other words, let us set

$$\varphi(t) = f(x^k + td^k) \qquad \text{for all } t \in (0, \tau),$$

for some positive real number τ. Differentiating the function φ yields

$$\varphi'(t) = \langle \nabla f(x^k + td^k), d^k \rangle.$$

Thus, evaluating this formula at $t = 0$ gives $\varphi'(0) = \langle \nabla f(x^k), d^k \rangle$, which finally gives that a direction d^k is a descent direction if

$$\langle \nabla f(x^k), d^k \rangle < 0.$$

Among all possible choices, the direction $d^k = -\nabla f(x^k)$, the opposite to the gradient, ensures that this condition is automatically satisfied. In fact,

$$\langle \nabla f(x^k), -\nabla f(x^k) \rangle = -\langle \nabla f(x^k), \nabla f(x^k) \rangle = -\|\nabla f(x^k)\|^2 < 0,$$

so long as $\nabla f(x^k) \neq 0$. Such a direction is called the **steepest descent direction**. So, the class of gradient descent methods consists in taking a more or less random initial point x^0, and by creating an iterative sequence using the equation

$$x^{k+1} = x^k - t_k \nabla f(x^k) \qquad \text{for every index } k.$$

The reason why this is still a class of methods, rather than one single method, lies on the fact that the step size remains to be chosen. As we shall see in the sequel, there is not a single choice for the step size, and different choices lead to different algorithms.

(2.1) Remark. The general definition of a gradient descent method requires that each iteration follows the formula $x^{k+1} = x^k + t_k d^k$, for suitable choices of the positive number t_k and the direction d^k. As regards to the choice of the directions d^k, we have shown that the opposite of the gradient at the point x^k, that is to say the vector $-\nabla f(x^k)$, gives the steepest descent direction, but it should be noted that one could well take any other direction within the convex cone

$$\left\{ d \in \mathbb{R}^n : \langle \nabla f(x^k), d \rangle < 0 \right\}.$$

Within this setting, it ought to be clear that the gradient descent methods follow within the category of the optimisation algorithms as described in abstract in Chapter VII. Indeed, observe that, setting

$$A(x) = \left\{ x + td : \langle \nabla f(x), d \rangle < 0 \text{ and } t \text{ small enough} \right\}$$

this gives a multi-function $A : \mathbb{R}^n \to \mathcal{P}(\mathbb{R}^n)$, and the most general iterative sequence $\{x^k\}$ may be defined as $x^0 \in \mathbb{R}^n$ and $x^{k+1} \in A(x^k)$ for every index k. The choice to take the opposite of the gradient correspond to fixing a continuous selection for the multi-function A.

(2.2) Remark. Let us also observe that, if the objective function is at least of class $\mathcal{C}^2$, and assuming that its Hessian matrix, $Hf(x)$, is positive definite at least in a neighbourhood of the optimal solution, x^*, another option for the direction d^k would be taking

$$d^k = -[Hf(x^k)]^{-1}\nabla f(x^k) \qquad \text{for every index } k.$$

In fact,

$$\langle \nabla f(x^k), d^k \rangle = -\langle \nabla f(x^k), [Hf(x^k)]^{-1}\nabla f(x^k) \rangle < 0.$$

On the other hand, if $Hf(x)$ is not positive definite, one could always choose a positive real number μ greater than the absolute value of the most negative eigenvalue of $Hf(x^k)$, and then take

$$d^k = -[Hf(x^k) + \mu I]^{-1}\nabla f(x^k).$$

Constant rate gradient descent

The first, and easiest, possibility in the choice of the step size consists in picking a suitable positive real constant η. This is called the **constant rate gradient descent**, and a summary of that is shown in Algorithm 6.

Algorithm 6 CONSTANT RATE GRADIENT DESCENT

---

1. Given $\eta > 0$, initialise x^0 to a random initial point and set $k = 0$.

2. Calculate $\nabla f(x^k)$.

3. **if** a STOPPING CRITERION holds, **then stop**

4. Compute $x^{k+1} = x^k - \eta\nabla f(x^k)$.

5. Increase k by 1 and **go back** to step 2.

The following theorem shows the conditions under which the constant rate gradient descent method converges.

(2.3) Theorem. *Let $f : \mathbb{R}^n \to \mathbb{R}$ a real function of class $\mathcal{C}^1$, and η be a positive real number. Assume that:*

(a) *For every real number α, the sublevel set*

$$D_\alpha = \left\{ x \in \mathbb{R}^n : f(x) \leq \alpha \right\}$$

is compact.

(b) *The vector function $\nabla f : \mathbb{R}^n \to \mathbb{R}^n$ is Lipschitz continuous function with constant L, that is to say,*

$$\|\nabla f(x) - \nabla f(y)\| < L\|x - y\| \qquad \text{for all } x, y \in \mathbb{R}^n.$$

(c) *The real number η is smaller than $2/L$, that is to say $0 < \eta < 2/L$.*

Consider the iterative sequence

$$x^0 \in \mathbb{R}^n, \qquad x^{k+1} = x^k - \eta \nabla f(x^k) \qquad \text{for every index } k.$$

Then, either there exists an index k such that $\nabla f(x^k) = 0$ or $\{x^k\}$ has at least one accumulation point and all accumulation points are stationary.

To prove this result we are going to need a preliminary lemma.

(2.4) Lemma. *In the same hypotheses and with the same notations as in Theorem (2.3), the following inequality holds:*

$$f(x + \alpha d) \le f(x) + \alpha \langle \nabla f(x), d \rangle + \tfrac{1}{2} L \alpha^2 \|d\|^2.$$

Proof. Let $y \in \mathbb{R}^n$, and set $z_t = x + t(y - x)$. We have

$$f(y) = f(x) + \int_0^1 \langle \nabla f(z_t), y - x \rangle \, dt$$

and therefore

$$f(y) - f(x) - \langle \nabla f(x), y - x \rangle = \int_0^1 \langle \nabla f(z_t) - \nabla f(x), y - x \rangle \, dt.$$

Thus,

$$\left| f(y) - f(x) - \langle \nabla f(x), y - x \rangle \right| \le \int_0^1 \left| \langle \nabla f(z_t) - \nabla f(x), y - x \rangle \right| \, dt$$

$$\le \int_0^1 \|\nabla f(z_t) - \nabla f(x)\| \|y - x\| \, dt$$

$$\le L \int_0^1 t \|y - x\|^2 \, dt$$

$$= \tfrac{1}{2} L \|y - x\|^2$$

By substituting y with $x + \alpha d$ we have the thesis. $\square$

We are now finally ready to prove the theorem.

Proof of Theorem (2.3). By virtue of Lemma (2.4), with $d = -\nabla f(x^k)$, we have

$$f(x^{k+1}) \le f(x^k) - \eta\|\nabla f(x^k)\|^2 + \tfrac{1}{2}L\eta^2\|\nabla f(x^k)\|^2$$

and therefore

$$f(x^k) - f(x^{k+1}) \ge \eta\big(1 - \tfrac{1}{2}\eta L\big)\|\nabla f(x^k)\|^2 \ge 0$$

Thanks to the assumptions on η, we get $f(x^{k+1}) \le f(x^k)$, hence the whole sequence x^k is included in the sublevel set $D_{f(x_0)}$, which is compact. Thus the thesis holds from the properties of compact sets. $\qquad\qquad\square$

Exact gradient descent

Choosing a constant step size makes it a simple method, but not a particularly efficient one. Other choices may also be taken to choose a different step size at every iteration. The first approach consists in choosing the step size, t_k, by performing an exact line search, which involves finding the minimum of f restricted to the ray starting at x^k in the direction of $-\nabla f(x^k)$, that is to say, solving the problem

$$(2.5) \qquad\qquad \text{minimise } f\big(x^k - t\nabla f(x^k)\big) \text{ subject to } t > 0.$$

To find t_k with rule (2.5) one can utilise one of the methods described in the previous section for functions of one variable. When the step size is chosen solving the optimisation problem (2.5), the method is called **exact gradient descent** method. Note, at step 5, that it entails to call an algorithm for solving the minimisation problem.

Algorithm 7 Exact Gradient Descent

---

1. Initialise x^0 to a random point and set $k = 0$.

2. Calculate $\nabla f(x^k)$.

3. **if** a STOPPING CRITERION holds, **then stop**

4. Calculate $\varphi^k(t) = f\big(x^k - t\nabla f(x^k)\big)$.

5. Set $t_k = $ **minimum of** $\varphi^k(t)$ **subject to** $t > 0$.

6. Compute $x^{k+1} = x^k - t_k\nabla f(x^k)$.

7. Increase k by 1 and **go back** to step 2.

The next result shows the main properties of the exact gradient descent method.

(2.6) Theorem. *Let $f : \mathbb{R}^n \to \mathbb{R}$ a real function of class $\mathcal{C}^1$, x be a point in $\mathbb{R}^n$, and, for every index k, define the iterative sequence $\{x^k\}$ as*

$$x^0 = x, \qquad x^{k+1} = x^k - t_k \nabla f(x^k)$$

where t_k is a solution of the optimisation problem (2.5).

Then, the following conditions hold:

(a) *Two consecutive directions of the exact gradient method are orthogonal.*

(b) *If $\nabla f(x^k) \neq 0$ then $f(x^{k+1}) < f(x^k)$.*

(c) *If the sequence $\{x^k\}$ converges to a point x^*, then $\nabla f(x^*) = 0$.*

Proof. Let us start by proving condition (a), and set $\varphi(t) = f(x^k - t\nabla f(x^k))$. Thanks to the chain rule, we have

$$\varphi'(t_k) = -\langle \nabla f(x^{k+1}), \nabla f(x^k) \rangle.$$

On the other hand, since t_k has been selected to be a minimum point of the function $\varphi(t)$, we also have $\varphi'(t_k) = 0$, which proves condition (a).

Let us move to (b), and recall that $\varphi'(0) = -\|\nabla f(x^k)\|^2 < 0$. Thus, φ is strictly decreasing in a neighbourhood of 0, and therefore

$$f(x^{k+1}) = \varphi(t_k) < \varphi(0) = f(x^k).$$

Finally, let us prove condition (c). To this end, observe that, from (a), for every index k, we have $\langle \nabla f(x^{k+1}), \nabla f(x^k) \rangle = 0$. On the other hand, from the fact that the function ∇f is, by hypothesis, continuous, if the sequence $\{x^k\}$ converges towards a point x^*, taking the limit as $k \to \infty$ gives $\langle \nabla f(x^{k+1}), \nabla f(x^k) \rangle \to \|\nabla f(x^*)\|^2$, and this yields $\nabla f(x^*) = 0$, as requested. $\qquad\square$

The next theorem shows the conditions under which the exact gradient descent method converges.

(2.7) Theorem. *Let $f : \mathbb{R}^n \to \mathbb{R}$ be a real function of class $\mathcal{C}^1$, x a point in $\mathbb{R}^n$, and, for every index k, define the iterative sequence $\{x^k\}$ as*

$$x^0 = x, \qquad x^{k+1} = x^k - t_k \nabla f(x^k)$$

where t_k is a solution of the optimisation problem (2.5). If f has compact sublevels, then, the sequence $\{x^k\}$ has at least one accumulation point, and, in addition to that, every accumulation point is a stationary point of f.

Proof. The assumptions of the theorem ensure that the sequence $\{x^k\}$ is bounded, and therefore, it has at least one accumulation point. The rest follows from the Zangwill's Convergence Theorem, by taking

$$\Omega = \Big\{ x \in \mathbb{R}^n : \nabla f(x) = 0 \Big\}.$$

In fact, the sequence $\{x^k\}$ is bounded because f has compact sublevels. Furthermore, let us choose the function f itself as a descent function. Let us also define the multi-function $A : \mathbb{R}^n \to \mathcal{P}(\mathbb{R}^n)$ by setting, for every x in $\mathbb{R}^n$,

$$A(x) = \Big\{ y = x - t\nabla f(x) : t \in \arg\min\{ f(x - s\nabla f(x)) : s > 0\} \Big\}.$$

It is not difficult to show that this multi-function is closed in $\mathbb{R}^n \setminus \Omega$. Indeed, considering a point $\bar{x} \notin \Omega$, and assuming that $x_i \to \bar{x}$, $y_i \to \bar{y}$ and $y_i \in A(x_i)$, we have

$$(2.8) \qquad\qquad y_i = x_i - t_i \nabla f(x_i)$$

and

$$(2.9) \qquad f(x_i - t_i \nabla f(x_i)) \le f(x_i - s\nabla f(x_i)) \quad \text{for each } s > 0.$$

Then, one has $t_i = \|y_i - x_i\| / \|\nabla f(x_i)\|$, and therefore the sequence $\{t_i\}$ converges towards the number $\bar{t} = \|\bar{y} - \bar{x}\| / \|\nabla f(\bar{x})\|$. Taking the limit in the formula (2.8), we obtain $\bar{y} = \bar{x} - \bar{t}\nabla f(\bar{x})$. Moreover, for every fixed strictly positive real number s, taking the limit in (2.9), we also obtain

$$f(\bar{x} - \bar{t}\nabla f(\bar{x})) \le f(\bar{x} - s\nabla f(\bar{x}))$$

this proves that $\bar{y} \in A(\bar{x})$, that is the multi-function A is closed. This shows that we are in the hypotheses of Zangwill's Convergence Theorem, and this completes the proof. $\qquad\qquad\square$

The main disadvantage of the exact gradient descent method lies in the fact that the 'exact search' makes the whole algorithm very slow, for one needs to find a minimum of a function, even though it is only a single variable function, at every step. Furthermore, the method exhibits a 'zigzag behaviour', where the directions are always one orthogonal to the next, that further contributes to make the whole process slow. It can be shown that, if the function f is of class $\mathcal{C}^2$, and the Hessian matrix, $Hf(x^*)$, is positive definite, then we have

$$\limsup_{k \to +\infty} \frac{\|x^{k+1} - x^*\|}{\|x^k - x^*\|} \le \frac{\lambda_{\max} - \lambda_{\min}}{\lambda_{\max} + \lambda_{\min}},$$

where $\lambda_{\max}$ e $\lambda_{\min}$ are the maximum and the minimum eigenvalue of $Hf(x^*)$, respectively. Hence, in the worst case scenario, the convergence is linear with speed

$$\frac{\lambda_{\max} - \lambda_{\min}}{\lambda_{\max} + \lambda_{\min}},$$

which get closer to 1, when the ratio $(\lambda_{\max} : \lambda_{\min})$ of the eigenvalues of the matrix $Hf(x^*)$, called the **condition number** of the function f, gets higher.

In the end, the convergence of this algorithm can be very slow for functions with a very high condition number, as we are going to show in Example (2.10).

Beforehand, however, let us see what the sequence $\{x^k\}$ looks like in the special case when f is quadratic and convex, that is to say, when f is a function of the form:

$$f(x) = \tfrac{1}{2}\langle x, Qx \rangle + \langle c, x \rangle,$$

where Q is symmetric and positive definite $n \times n$ matrix.

In this special case, we have

$$f\left(x^k - t\nabla f(x^k)\right) = \tfrac{1}{2}\langle g^k, Qg^k \rangle t^2 - \langle g^k, g^k \rangle t + f(x^k),$$

where $g^k = Qx^k + c$. Then,

$$t_k = \frac{\langle g^k, g^k \rangle}{\langle g^k, Qg^k \rangle},$$

and therefore

$$x^{k+1} = x^k - \frac{\langle g^k, g^k \rangle}{\langle g^k, Qg^k \rangle} g^k.$$

The function f is quadratic, hence it does have a unique global minimum at the point $x^* = -Q^{-1}c$. Hence, the condition $x^{k+1} = x^*$ is equivalent to

$$Qg^k = \frac{\langle g^k, g^k \rangle}{\langle g^k, Qg^k \rangle} g^k,$$

that is, if and only if $1/t_k$ is eigenvalue of Q and g^k is the corresponding eigenvector. In fact,

$$-Q^{-1}c = x^k - \frac{\langle g^k, g^k \rangle}{\langle g^k, Qg^k \rangle} g^k,$$

from which

$$-c = Qx^k - \frac{\langle g^k, g^k \rangle}{\langle g^k, Qg^k \rangle} Qg^k,$$

or

$$Qg^k = \frac{\langle g^k, Qg^k \rangle}{\langle g^k, g^k \rangle} g^k.$$

(2.10) Example. Consider the quadratic function $f(x) = x_1^2 + 9x_2^2$, which has global minimum at the point $x^* = (0,0)$. The hessian matrix for this function is

$$Hf(x) = \begin{bmatrix} 2 & 0 \\ 0 & 18 \end{bmatrix} \qquad \text{for every } x \in \mathbb{R}^2,$$

and we have

$$\frac{\lambda_{\max} - \lambda_{\min}}{\lambda_{\max} + \lambda_{\min}} = \frac{18 - 2}{18 + 2} = \frac{4}{5} = 0.8$$

By applying the exact gradient method starting from the point $x^0 = (9, 1)$, and applying the iterative formula, gives

$$x^k = \begin{bmatrix} 9(\frac{4}{5})^k \\ (-\frac{4}{5})^k \end{bmatrix} \qquad \text{for every index } k.$$

Therefore

$$\frac{\|x^{k+1} - x^*\|}{\|x^k - x^*\|} = \frac{4}{5} \qquad \text{for every index } k,$$

or the distance of the sequence from the global minimum x^* is reduced at each step by a factor of 0.8.

3. Inexact gradient descent methods

As we have mentioned before, and we have seen in the two methods of constant rate and exact gradient descent, the choice of the step size may be taken following different approaches, each of which have some advantages and some disadvantages. For example, the exact gradient descent method offers, at each step, the best possible choice of the step size, but at the cost of having to solve an auxiliary optimisation problem, albeit one-dimensional, which may be computationally expensive. On the other hand, a constant rate gradient descent with a very small choice of the learning rate would require many iteration to get close to an optimal solution. In the choice of the step size, therefore, there is a trade-off between the best possible choice of the step size, which reduces the number of steps of the overall algorithm, but that is computationally slow, and a less optimal step size, which may increase the number of steps, but decreases the computational time necessary to perform each of these steps.

There exist many different approaches to handle this trade-off, which are generally described with the phrase **inexact gradient descent** methods. Amid them, an important type of inexact gradient descent methods is the one which requires that the step size satisfies certain limiting conditions that go under the name of **Armijo-Goldstein-Wolfe conditions**:

$$\text{(A)} \quad \langle \nabla f(x^{k+1}), d^k \rangle > \beta \langle \nabla f(x^k), d^k \rangle$$
$$\text{(B)} \quad f(x^{k+1}) < f(x^k) + \alpha t_k \langle \nabla f(x^k), d^k \rangle,$$

where $0 < \alpha < \beta < 1$ are two parameters chosen in advance.

It is worth observing immediately that condition (A) guarantees, to some extent, that the step size t_k is not too small. In fact, if the step size was too small, we would have

$$\langle \nabla f(x^{k+1}), d^k \rangle \simeq \langle \nabla f(x^k), d^k \rangle,$$

and condition (A) could not hold. Conversely, condition (B) ensures that the step size t_k is not too large. In fact, if the step size was too large, the right-hand side of (B) would tend to $-\infty$, and in violation with condition (B).

The following theorem shows under which circumstances the conditions (A) and (B) are satisfied.

(3.1) Theorem. *Let $f : \mathbb{R}^n \to \mathbb{R}$ a real function of class $\mathcal{C}^1$ with compact sublevels. Let x^0 be a point in $\mathbb{R}^n$, and, for every index k, define the iterative sequence $\{x^k\}$ as*

$$x^{k+1} = x^k - t_k d^k,$$

where d^k satisfies the condition $\langle \nabla f(x^k), d^k \rangle < 0$. Then, there exists an interval (a, b), such that every number t_k in the interval (a, b) satisfies the conditions (A) and (B).

Proof. Let us set $\varphi^k(t) = f(x^k + td^k)$, and observe that, since f is by hypothesis differentiable, we have

$$\varphi'_k(0) = \lim_{t \to 0} \frac{f(x^k + td^k) - f(x^k)}{t} = \langle \nabla f(x^k), d^k \rangle < 0.$$

Take a real number α such that $0 < \alpha < 1$. Then, there exists a positive real number ε such that:

$$(3.2) \qquad f(x^k + td^k) - f(x^k) < \alpha t \langle \nabla f(x^k), d^k \rangle \qquad \text{for all } t \in (0, \varepsilon).$$

Observe that the positive number ε cannot be $+\infty$, because f is bounded from above. Furthermore, if we denote by t^* the supremum of all ε that satisfy the condition (3.2), thanks to the fact that the function f is continuous, we also have

$$f(x^k + t^* d^k) - f(x^k) = t^* \alpha \langle \nabla f(x^k), d^k \rangle.$$

By virtue of the mean value theorem, we obtain

$$t^* \langle \nabla f(x^k + t^{**} d^k), d^k \rangle = t^* \alpha \langle \nabla f(x^k), d^k \rangle,$$

with $0 < t^{**} < t^*$. By choosing β such that $\alpha < \beta < 1$ we also obtain

$$\beta \langle \nabla f(x^k), d^k \rangle < \langle \nabla f(x^k + t^{**} d^k), d^k \rangle.$$

In addition, since $t^{**} < t^*$, we also have

$$f(x^k + t^{**} d^k) - f(x^k) < \alpha t^{**} \langle \nabla f(x^k), d^k \rangle.$$

These two inequalities are exactly the conditions (A) and (B) and, if they holds for t^{**}, thanks to the continuity of the two functions f and ∇f, they also hold in a neighbourhood t^{**}, which proves the theorem. $\qquad \square$

If the objective function is convex, one can further show that rule (B) is satisfied in an interval of the form $(0, M)$, while rule (A) is satisfied in an unbounded interval $(N, +\infty)$, starting from N, with $N < M$.

To determine a step size t_k that satisfies conditions (A) and (B), it is generally used a **backtracking** technique: starting with t big enough, the value of t is then reduced by a factor $\rho \in (0, 1)$ until condition (B) is satisfied. Then rule (A) is generally not checked, because this backtracking strategy prevents the step size from being too small.

In practice, the value α is chosen to be quite small, for example $\alpha = 10^{-4}$. With such a choice of the parameter α, condition (B) then yields a decrease in the value of $f(x^k)$ at the new point x^{k+1}.

(3.3) Example. Let us consider the function $f(x_1, x_2) = x_1^4 + x_1^2 + x_2^2$ and suppose that, after a certain number of steps, using an inexact gradient descent method, one gets to $x^k = (1, 1)$ with the corresponding descent direction $d^k = -\nabla f(x^k) = (-6, -2)$. Let us choose the parameters $\alpha = 10^{-4}$, $\beta = 0.1$ and $\rho = 0.5$, and let us apply backtracking to find the step size t_k, starting from $t = 1$. Since

$$f(x^k + td^k) = 651 > f(x^k) + \alpha t \langle \nabla f(x^k), d^k \rangle = 2.996,$$

it is clear that condition (B) is not satisfied, hence we need to reduce the step, which now becomes $t = 0.5$. We now have

$$f(x^k + td^k) = 20 > f(x^k) + \alpha t \langle \nabla f(x^k), d^k \rangle = 2.998.$$

Again, condition (B) is not satisfied, so we need to reduce the step size again, testing $t = 0.25$. Now we have

$$f(x^k + td^k) = 0.5625 < f(x^k) + \alpha t \langle \nabla f(x^k), d^k \rangle = 2.999.$$

Thus, $x^{k+1} = x^k + 0.25d^k = (-\frac{1}{2}, \frac{1}{2})$ does not violate condition (B). Let us also observe that this point satisfies condition (A) as well, because:

$$\langle \nabla f(x^{k+1}), d^k \rangle = 7 > \beta \langle \nabla f(x^k), d^k \rangle = -20.$$

Let us now look at a global convergence result for an inexact gradient descent method, where the step size satisfies the Armijo-Goldstein-Wolfe rules. In choosing the descent direction d^k, we will make the following assumption:

$$(3.4) \qquad \text{there exists } \mu > 0 \text{ such that } \theta^k \le \tfrac{1}{2}\pi - \mu \text{ for every index } k,$$

where θ^k denotes the angle between d^k and $-\nabla f(x^k)$. For instance, such a rule is automatically satisfied taking $d^k = -\nabla f(x^k)$.

Then, the following result holds.

(3.5) Theorem. *Let $f : \mathbb{R}^n \to \mathbb{R}$ a real function of class $\mathcal{C}^1$ with compact sublevels, x^0 be a point in $\mathbb{R}^n$, and, for every index k, define the iterative sequence $\{x^k\}$ as*

$$x^{k+1} = x^k - t_k d^k,$$

where d^k satisfies the condition $\langle \nabla f(x^k), d^k \rangle < 0$ and the condition (3.4), and t_k satisfies the conditions of Armijo-Goldstein-Wolfe.

Then, either there exists and index $\bar{k}$ such that $\nabla f(x^{\bar{k}}) = 0$, or the sequence $\{\|\nabla f(x^k)\|\}$ converges towards zero.

Proof. Since the sequence $\{f(x^k)\}$ is a decreasing, bounded sequence, we know that

$$\lim_{k \to +\infty} \left(f(x^k) - f(x^{k+1}) \right) = 0.$$

Thanks to the rule (B), we have

$$\lim_{k \to +\infty} t_k \langle \nabla f(x^k), d^k \rangle = 0.$$

By contradiction, assume that $\|\nabla f(x^k)\|$ does not tend towards the number 0. This means that there exists a strictly positive real number ε, and a subsequence $\{x^{k_r}\}$ such that

$$\begin{cases} \|\nabla f(x^{k_r})\| > \varepsilon & \text{for every index } r, \\ \|t_{k_r} d^{k_r}\| \to 0. \end{cases}$$

Furthermore, thanks to the rule (A), we have

$$-t_{k_r} \langle \nabla f(x^{k_r}), d^{k_r} \rangle < \frac{t_{k_r} \langle \nabla f(x_{k_{r+1}}) - \nabla f(x_{k_r}), d^{k_r} \rangle}{1 - \beta}$$

and therefore

$$-t_{k_r} \nabla f(x^{k_r})^T d^{k_r} < \frac{\|\nabla f(x^{k_{r+1}}) - \nabla f(x^{k_r})\| \cdot \|t_{k_r} d^{k_r}\|}{1 - \beta}.$$

Since $t_{k_r} d^{k_r} = x^{k_{r+1}} - x^{k_r}$ and, in addition to that, $\|t_{k_r} d^{k_r}\| \to 0$, because ∇f is uniformly continuous, we have

$$\|\nabla f(x^{k_{r+1}}) - \nabla f(x^{k_r})\| \to 0.$$

Thus,

$$0 \leq \cos \theta^{k_r} = \frac{-t_{k_r} \langle \nabla f(x^{k_r}), d^{k_r} \rangle}{\|\nabla f(x^{k_r})\| \|t_{k_r} d^{k_r}\|} \leq \frac{\|\nabla f(x^k_{r+1}) - \nabla f(x^{k_r})\|}{\varepsilon(1 - \beta)}$$

and therefore, $\cos \theta^{k_r} \to 0$ which contradicts the initial assumption (3.4) regarding the angle θ^{k_r}. $\qquad\square$

4. Conjugate gradient method

The conjugate gradient method can be viewed as an enhancement of the gradient method, where the selection of the direction d^k incorporates information from directions computed in prior iterations. This refinement further contributes to the method's effectiveness and convergence properties.

First off, let us show the method in the simpler case of a convex quadratic functions, that is to say, when f is a function of the form

$$(4.1) \qquad f(x) = \tfrac{1}{2}\langle x, Qx \rangle + \langle c, x \rangle,$$

where Q is symmetric and positive definite. In this case, there is a unique global minimum of f which coincides with the stationary point $x^* = -Q^{-1}c$.

In order to describe the conjugate gradient method, let us set $g^k = Qx^k + c$. The direction d^k is given by

$$d^k = \begin{cases} -g^0 & \text{if } k = 0, \\ -g^k + \beta_k d^{k-1} & \text{if } k \geq 1, \end{cases}$$

that is to say, for the initial iteration, the direction is opposite to the gradient of the function f, while, in subsequent iterations, it is a linear combination of the vector opposite to the gradient of f and the direction calculated in the previous iteration.

Here, the parameter β_k is chosen so that the directions d^k and d^{k-1} are **conjugate** with respect to the matrix Q, that is to say, in order to have

$$(4.2) \qquad \langle d^k, Qd^{k-1} \rangle = 0 \qquad \text{for every index } k.$$

Substituting the expression for d^k into (4.2), gives

$$\beta_k = \frac{\langle g^k, Qd^{k-1} \rangle}{\langle d^{k-1}, Qd^{k-1} \rangle} \qquad \text{for } k \geq 1.$$

If an exact gradient descent method is performed at each iteration, then the chosen direction remains a descent direction for f. In fact, if we consider the real function $\varphi(t) = f(x^{k-1} + td^{k-1})$, we have

$$(4.3) \qquad \varphi'(t_{k-1}) = \langle g^k, d^{k-1} \rangle = 0,$$

thus

$$(4.4) \qquad \begin{aligned} \nabla f(x^k)d^k &= \langle g^k, d^k \rangle \\ &= -\|g^k\|^2 + \beta_k \langle g^k, d^{k-1} \rangle \\ &= -\|g^k\|^2 < 0. \end{aligned}$$

Moreover, we can easily calculate the step size t_k. In fact, the function φ is a convex quadratic function in the variable t, because

$$f(x^k + td^k) = \tfrac{1}{2}\langle d^k, Qd^k\rangle t^2 + \langle g^k, d^k\rangle t + f(x^k).$$

As a result, the step size is

$$(4.5) \qquad t_k = -\frac{\langle g^k, d^k\rangle}{\langle d^k, Qd^k\rangle}.$$

The whole algorithm is summarised in the box below.

Algorithm 8 Conjugate Gradient Method for Quadratic Functions

---

1. Choose a point $x^0 \in \mathbb{R}^n$ and set $k = 0$.

2. Calculate $g^k = Qx^k + c$.

3. **If** $g^k = 0$ **then stop**.

4. **If** $k = 0$ **then** $d^k = -g^k$

   else calculate $\beta_k = \dfrac{\langle g^k, Qd^{k-1}\rangle}{\langle d^{k-1}, Qd^{k-1}\rangle}$ and set $d^k = -g^k + \beta_k d^{k-1}$.

5. Calculate $t_k = -\dfrac{\langle g^k, d^k\rangle}{\langle d^k, Qd^k\rangle}$.

6. $x^{k+1} = x^k + t_k d^k$.

7. Increase k by 1 and **go back** to step 2.

(4.6) Theorem. *Let us suppose that f is a convex quadratic function. Then the conjugate gradient method finds the global minimum of f in at most n iterations.*

Proof. Let us start by observing that, at the kth iteration, we have

$$(4.7) \qquad \langle g^k, g^i\rangle = 0 \qquad \text{for all } i = 0, \ldots, k - 1.$$

which meant that the vectors $\{g^0, g^1, \ldots, g^k\}$ are orthogonal to one another. Thus, the method must find the solution x^* in at most n iterations, otherwise at the $(n+1)$th iteration there would be $n+1$ orthogonal vectors in $\mathbb{R}^n$, which is impossible.

Hence, it suffices to prove (4.7) for every index k. This may be done by induction on k. For $k = 1$, we have $\langle g^1, g^0\rangle$, because the first iteration coincides with an iteration of the gradient method. Now suppose that (4.7) holds for a certain $k \geq 1$ and prove it for $k+1$. From the choice of direction d^k and (4.3), it follows that

$$\langle g^{k+1}, g^k \rangle = -\langle g^{k+1}, d^k \rangle - \beta_k \langle g^{k+1}, d^{k-1} \rangle = \beta_k \langle g^{k+1}, d^{k-1} \rangle.$$

Moreover,

$$(4.8) \qquad \begin{aligned} g^{k+1} &= Qx^{k+1} + c \\ &= Q(x^k + t_k d^k) + c \\ &= Qx^k + t_k Q d^k + c \\ &= g^k + t_k Q d^k, \end{aligned}$$

thus from (4.3) and the inductive hypothesis, we obtain

$$\langle g^{k+1}, d^{k-1} \rangle = \langle g^k, d^{k-1} \rangle + t_k \langle d^k, Q d^{k-1} \rangle = 0,$$

therefore

$$(4.9) \qquad \langle g^{k+1}, g^k \rangle = 0.$$

Now fix an index $i \in \{0, 1, \ldots, k - 1\}$. From (4.8) and the inductive hypothesis, it follows that

$$\langle g^{k+1}, g^i \rangle = \langle g^k, g^i \rangle + t_k \langle d^k, Q g^i \rangle = t_k \langle d^k, Q g^i \rangle,$$

moreover, from the choice of d_i and the inductive hypothesis, we have

$$\langle d^k, Q g^i \rangle = -\langle d^k, Q d^i \rangle + \beta_i \langle d^k, Q d^{i-1} \rangle = 0.$$

Hence,

$$(4.10) \qquad \langle g^{k+1}, g^i \rangle = 0.$$

Thus, from (4.9) and (4.10), it follows that

$$(4.11) \qquad \langle g^{k+1}, g^i \rangle = 0 \qquad \text{for all } i = 0, \ldots, k. \qquad \square$$

(4.12) Example. Let us apply the conjugate gradient method to the function $f(x) = x_1^2 + 10x_2^2$, starting from the point $x^0 = (10, 1)$.

1. For the first iteration, we have

$$g^0 = (20, 20),$$

$$d^0 = -g^0 = (-20, -20), \qquad t_0 = -\frac{\langle g^0, d^0 \rangle}{\langle d^0, Q d^0 \rangle} = \frac{1}{11},$$

obtaining the next point

$$x^1 = x^0 + t_0 d^0 = \left(\frac{90}{11}, -\frac{9}{11} \right).$$

2. For the second iteration, we have

$$g^1 = \left(\frac{180}{11}, -\frac{180}{11}\right), \qquad \beta_1 = \frac{\langle g^1, Qd^0 \rangle}{\langle d^0, Qd^0 \rangle} = \frac{81}{121},$$

$$d^1 = -g^1 + \beta_1 d^0 = \left(-\frac{3600}{121}, \frac{360}{121}\right), \qquad t_1 = -\frac{\langle g^1, d^1 \rangle}{\langle d^1, Qd^1 \rangle} = \frac{11}{40},$$

Thus, the next point is $x^2 = x^1 + t_1 d^1 = (0,0)$, which is the global minimum of f.

(4.13) Remark. Other formulas can be obtained for both the step size t_k and the parameter β_k. In fact, from (4.4) and (4.5), we obtain

$$t_k = \frac{\|g^k\|^2}{\langle d^k, Qd^k \rangle}.$$

Moreover, from the choice of d^k, (4.7), and (4.4), we also obtain

$$\langle d^k, g^{k-1} \rangle = -\langle g^k, g^{k-1} \rangle + \beta_k \langle d^{k-1}, g^{k-1} \rangle = -\beta \|g^k\|^2.$$

On the other hand, from (4.8), (4.2), and (4.4), we have

$$\langle d^k, g^{k-1} \rangle = \langle d^k, g^k \rangle - t_{k-1} \langle d^k, Qd^{k-1} \rangle = -\|g^k\|^2.$$

In the end, we get

$$\beta_k = \frac{\|g^k\|^2}{\|g^{k-1}\|^2}.$$

In addition to the property (4.7), it can be shown that if at the kth iteration the conjugate gradient method has not yet found the solution x^*, then

$$\langle d^k, Qd^i \rangle = 0 \qquad \text{for every } i = 0, \ldots, k-1,$$

i.e., the vectors $\{d^0, d^1, \ldots, d^k\}$ are conjugate with respect to the matrix Q.

Now that we have fully discussed the special case of a quadratic function, let us now show the same method for an arbitrary function f.

As we have seen before, in the special case of a quadratic function the step size may be obtained exactly. In the general case, we shall see that the method can be reduced to the one of a quadratic function.

If the step size t_k is to be calculated exactly, then the direction d^k is a descent direction for f, in fact:

$$\nabla f(x^k) d^k = -\|\nabla f(x^k)\|^2 + \beta_k \nabla f(x^k) d^{k-1} = -\|\nabla f(x^k)\|^2 < 0,$$

since $\nabla f(x^k) d^{k-1} = 0$.

Algorithm 9 CONJUGATE GRADIENT METHOD

---

1. Choose a point $x_0 \in \mathbb{R}^n$ and set $k = 0$.

2. **If $\nabla f(x^k) = 0$ then stop.**

3. **If $k = 0$ then $d^k = -\nabla f(x^k)$**

   otherwise calculate $\beta_k = \dfrac{\|\nabla f(x^k)\|^2}{\|\nabla f(x^{k-1})\|^2}$

   and set $d^k = -\nabla f(x^k) + \beta_k d^{k-1}$.

4. Calculate the step size t_k.

5. $x^{k+1} = x^k + t_k d^k$.

6. Increase k by 1 and **go back** to step 2.

If the line search is not exact, the term $\langle \nabla f(x^k), d^{k-1} \rangle$ could dominate $-\|\nabla f(x^k)\|^2$ and the direction d^k would not be a descent direction. However, it can be shown that d^k is a descent direction if the step size t_k satisfies the conditions of Armijo-Goldstein-Wolfe:

$$(4.14) \qquad \begin{aligned} &\text{(A)} \quad f(x^k + t_k d^k) < f(x^k) + \alpha t_k \langle \nabla f(x^k), d^k \rangle, \\ &\text{(B)} \quad |\langle \nabla f(x^k + t_k d^k), d^k \rangle| < -\beta \langle \nabla f(x^k), d^k \rangle, \end{aligned}$$

with $0 < \alpha < \beta < 1/2$.

(4.15) Theorem. *If $f : \mathbb{R}^n \to \mathbb{R}$ has compact sublevels, then the conjugate gradient method, with the step size satisfying the conditions (4.14), generates a sequence $\{x^k\}$ such that every accumulation point is stationary.*

The proof of this theorem is similar to that of Theorem (4.6) with the extra technical complication of being in the general framework.

5. Sub-gradient method

All the methods we have discussed thus far all assume that the objective function is at least of class $\mathcal{C}^1$. However, in some practical problems, it may happen that the objective functions is not overall differentiable, that is to say, it may have points in which the function does not have a gradient.

The sub-gradient method, that we shall develop throughout this section, is an optimisation algorithm used to minimise a convex non-differentiable function f. This method generalises the ideas behind the gradient descent method for differentiable functions to the non-differentiable case by exploiting the properties of sub-gradients, which always exist for a convex function. In fact, the sub-gradient method is an iterative algorithm that updates the current solution x^k using a sub-gradient g^k in $\partial f(x^k)$, and a step size α_k:

$$x^{k+1} = x^k - \alpha_k g^k.$$

In other words, from a generic point x^k, this method seeks the next point x^{k+1} along the direction of a sub-gradient calculated at x^k.

Before going any further with the development of this method, however, is is now paramount to note that while the opposite of gradient of a differentiable function is always an descent direction, a direction opposite to a sub-gradient, in general, may not. Thus, sub-gradient method cannot be included amid the so-called descent methods.

(5.1) Example. Consider the function $f(x) = |x|$ and observe that at the point $x^* = 0$, any sub-gradient different from 0 is an ascent direction for f.

Let us introduce the following notation that should help out with the proof of correctness of the algorithm. To this end, let $f : \mathbb{R}^n \to \mathbb{R}$ be a real function.

For every index k, let us set

$$f_{\text{best}}^k = \min\left\{ f_{\text{best}}^{k-1}, f(x^k) \right\}$$

where x_{best}^k is a point such that $f(x_{\text{best}}^k) = f_{\text{best}}^k$. In other words, let us set

$$f_{\text{best}}^k = \min\left\{ f(x^1), f(x^2), \ldots, f(x^k) \right\}$$

The choice of step size α_k is now crucial for the convergence of the method. Common strategies include the techniques of *constant* step size, *diminishing* step size, and *adaptive* step size.

We are now going to describe a basic version of the sub-gradient method to minimise a convex function $f : \mathbb{R}^n \to \mathbb{R}$. The structure of this method, with a suitable choice of the step-size, α_k, is described in Algorithm 10.
For the sub-gradient method, as described in Algorithm 10, the following convergence result holds.

(5.2) Theorem. *Let $f : \mathbb{R}^n \to \mathbb{R}$ be a convex function. Let us suppose that f admits global minimum x^* and that there exists a real number G such that, for every point x in $\mathbb{R}^n$ and every vector g in $\partial f(x)$, one has $\|g\|^2 \leq G$. Let $\{\alpha_k\}$ be an infinitesimal sequence of positive real numbers such that the series $\sum_{k=0}^{\infty} \alpha_k$ is divergent and the series $\sum_{k=0}^{\infty} \alpha_k^2$ is convergent. Let us take $x^0 \in \mathbb{R}^n$ and define*

$$x^{k+1} = x^k - \alpha_k g^k, \qquad \text{where } g^k \in \partial f(x^k).$$

Algorithm 10 SUB-GRADIENT METHOD

1. Choose $x^0 \in \mathbb{R}^n$ and set $k = 0$.

2. **if** a STOPPING CRITERION holds, **then stop**

3. Calculate a sub-gradient $g^k \in \partial f(x^k)$ and set

$$x^{k+1} = x^k - \alpha_k g^k.$$

4. **if** $0 \in \partial f(x^{k+1})$

       then stop (x^{k+1} is a minimum point of f).

       else increase k by 1 and **go back** to step 2.

Then, the sequence $\{x^k\}$ is convergent and

$$\lim_{k \to +\infty} f(x^k) = f(x^*).$$

Proof. Let us start by observing that

$$
\begin{aligned}
\|x^{k+1} - x^*\|^2 &= \|x^k - a_k g^k - x^*\|^2 \\
&= \|x^k - x^*\|^2 - 2\alpha_k \langle g^k, x^k - x^* \rangle + \alpha_k^2 \|g^k\|^2 \\
&\leq \|x^k - x^*\|^2 - 2\alpha_k [f(x^*) - f^*] + \alpha_k^2 \|g^k\|^2
\end{aligned}
$$

where the last inequality follows from the definition of sub-gradient, and where we have denoted by f^* the value $f(x^*)$. Applying this inequality inductively, we get

$$\|x^{k+1} - x^*\|^2 \leq \|x^0 - x^*\| - 2\sum_{i=0}^{k} \alpha_i [f(x^i) - f^*] + \sum_{i=0}^{k} \alpha_i^2 \|g^i\|^2.$$

Set $\|x^0 - x^*\| = R$ and observe that

$$
\begin{aligned}
2\sum_{i=0}^{k} \alpha_i [f(x^i) - f^*] &\geq \left(\sum_{i=0}^{k} \alpha_i \right) \min\{f(x^i) - f^* : i = 0, \ldots, k\} \\
&= \left(\sum_{i=0}^{k} \alpha_i \right) \left(f_{\text{best}}^k - f^* \right).
\end{aligned}
$$

From the fact that $\|g^i\|^2 \leq G$ for every index i, we also get

$$f_{\text{best}}^* - f^* = \min_{i=0,\ldots,k} \left(f(x^i) - f^* \right) \leq R^2 + G^2 \sum_{i=0}^{k} \alpha_i \|g^i\|^2.$$

Thus,

$$0 \leq f_{\text{best}}^k - f^* \leq \frac{R^2 + G^2 \sum_{i=0}^{k} \alpha_i^2}{2 \sum_{i=0}^{k} \alpha_i}.$$

Taking the limit as $k \to \infty$, from the fact that the series $\sum_{i=0}^{\infty} \alpha_i$ is divergent, and $\sum_{i=0}^{\infty} \alpha_i^2$ is convergent, it follows that the fraction on the right-hand side of the previous chain of inequalities tends towards 0, and the thesis follows. $\qquad\square$

Let us observe that, in general, to calculate a sub-gradient of a convex function is not an easy operation. To see this in more detail, let us consider the following example.

(5.3) Example. Let us consider the function

$$f(x_1, x_2) = x_1^2 + x_2^2 + |x_1 - x_2|,$$

where $(x_1, x_2) \in \mathbb{R}^2$. Since f is convex and non-differentiable, let us apply the sub-gradient method.

1. **Initialisation:** Choose an initial point, for example $(x_1^0, x_2^0) = (0, 0)$, and a step size $\alpha = 0.1$.

2. **Sub-gradient Calculation:** The function $f(x_1, x_2)$ has three terms, the first two of which are differentiable, and the third is not.

   - The derivative of x_1^2 with respect to x_1 is $2x_1$.
   - The derivative of x_2^2 with respect to x_2 is $2x_2$.
   - The sub-gradient of $|x_1 - x_2|$ depends on the values of x_1 and x_2:

     - If $x_1 > x_2$, the sub-gradient with respect to x_1 is 1 and with respect to x_2 is -1.
     - If $x_1 < x_2$, the sub-gradient with respect to x_1 is -1 and with respect to x_2 is 1.
     - If $x_1 = x_2$, any value between -1 and 1 is a sub-gradient for both.

3. **Point Update:** Use the sub-gradient to update the point.

   If $(x_1^0, x_2^0) = (0, 0)$, the sub-gradient of $f(x_1, x_2)$ is $(2x_1^0 + 1, 2x_2^0 - 1) = (1, -1)$.

$$\begin{aligned}
(x_1^1, x_2^1) &= \left(x_1^0 - \alpha \cdot 1, x_2^0 - \alpha \cdot (-1)\right) \\
&= \left(0 - 0.1 \cdot 1, 0 - 0.1 \cdot (-1)\right) \\
&= (-0.1, 0.1).
\end{aligned}$$

4. **Iteration:** Repeat the process:

$$\begin{aligned}
(x_1^2, x_2^2) &= \left(x_1^1 - \alpha(2x_1^1 + 1), x_2^1 - \alpha(2x_2^1 - 1)\right) = (-0.18, 0.18). \\
(x_1^3, x_2^3) &= \left(x_1^2 - \alpha(2x_1^2 + 1), x_2^2 - \alpha(2x_2^2 - 1)\right) = (-0.244, 0.244).
\end{aligned}$$

The process can now be repeated to get a better approximation of the optimal solution.

6. Newton's methods

Newton's methods are inspired by Newton-Raphson's algorithm for a function of one real variable. The Newton's method for multivariate functions seeks a solution to the system of non-linear equations:

$$(6.1) \qquad\qquad\qquad G(x) = 0,$$

where $G : \mathbb{R}^n \to \mathbb{R}^n$ is a map of class $\mathcal{C}^1$, whose Jacobian matrix $JG(x)$ is invertible when x varies in neighbourhood of a solution of the equation (6.1). This method operates by generating a sequence $\{x^k\}$, starting from a point x^0 in the neighbourhood of the solution of (6.1) where the Jacobian matrix is invertible, and then applying the following iterative equation:

$$x^{k+1} = x^k - JG(x^k)^{-1}G(x^k).$$

To find the stationary points of an objective function $f : \mathbb{R}^n \to \mathbb{R}$, it is sufficient to apply the Newton's method to the function $G = \nabla f$.

(6.2) Theorem. *Let $f : \mathbb{R}^n \to \mathbb{R}$ be a real function of class $\mathcal{C}^2$, and x^* be a local minimum point of f such that the Hessian matrix $Hf(x^*)$ is positive definite. Assume that the Hessian matrix is Lipschitz continuous, that is to say, assume that there exists a real number β such that*

$$\|Hf(x) - Hf(y)\| \le \beta \|x - y\| \quad \text{for every } x, y \in \mathbb{R}^n.$$

Then, there exists a neighbourhood of x^* and a positive real number γ such that, given an initial point x^0 in this neighbourhood of x^*, the sequence $\{x^k\}$ defined by the formula

$$x^{k+1} = x^k - [Hf(x^k)]^{-1}\nabla f(x^k)$$

converges to x^* and such that $\|x^{k+1} - x^*\| \le \gamma \|x^k - x^*\|^2$ (that is to say, the convergence is at least quadratic).

Proof. Let us observe that, if x^* is a local minimum, then $\nabla f(x^*) = 0$, hence

$$\begin{aligned}
x^{k+1} - x^* &= x^k - Hf(x^k)^{-1}\nabla f(x^k) - x^* + Hf(x^k)^{-1}\nabla f(x^*) \\
&= Hf(x^k)^{-1}\big(\nabla f(x^*) - \nabla f(x^k) - Hf(x^k)(x^* - x^k)\big).
\end{aligned}$$

Thus,

$$\|x^{k+1} - x^*\| \le \|Hf(x^k)^{-1}\| \|\nabla f(x^*) - \nabla f(x^k) - H(x^k)(x^* - x^k)\|.$$

Since the Hessian matrix H is continuous and positive definite, there exists a neighbourhood U_1 of x^* and a constant α such that, for each x^* in U_1, one has $\|Hf(x)^{-1}\| \leq \alpha$. Furthermore, since the Hessian matrix is also Lipschitz continuous, there exists another neighbourhood U_2 of x^* such that

$$\|\nabla f(x^*) - \nabla f(x^k) - H(x^* - x^k)\| \leq \tfrac{1}{2}\beta\|x^k - x^*\|^2$$

so long as $x^k \in U_2$. Let $V = U_1 \cap U_2$ and observe that, so long as x^k lies in W one has

$$\|x^{k+1} - x^*\| \leq \tfrac{1}{2}\alpha\beta\|x^k - x^*\|^2,$$

Now, up to further reducing the neighbourhood V, we may well assume that $\|x^k - x^*\| \leq 2(\alpha\beta)^{-1}$ and $\|x^k - x^*\| < \varepsilon$ for a given real number ε, with $0 < \varepsilon < 1$, which proves that the sequence $\{x^k\}$ converges towards x^*, so long as $x^* \in V$. Finally, setting $\gamma = \tfrac{1}{2}\alpha\beta$, we get to

$$\|x^{k+1} - x^*\| \leq \gamma\|x^k - x^*\|^2$$

which proves that the convergence is quadratic. $\square$

(6.3) Remark. When f is quadratic:

$$f(x) = \tfrac{1}{2}\langle x, Qx\rangle + \langle c, x\rangle,$$

where Q is a symmetric and positive definite square matrix, the Newton method provides the global minimum of f in a single step, regardless of the initial point x^0. Indeed, if x^* is the minimum of f, that is $Qx^* + c = 0$, we have:

$$x^1 = x^0 - Q^{-1}(Qx^0 + c) = -Q^{-1}c = x^*.$$

The Newton's method has some disadvantages: at each step k, it is necessary to calculate the gradient $\nabla f(x^k)$, the Hessian matrix $Hf(x^k)$, and the inverse of $Hf(x^k)$; moreover, if x^0 is chosen too far from the point x^* (which is the unknown of the problem), the method may not converge. In fact, in this case, $Hf(x^k)$ may not be invertible, and one could obtain $f(x^{k+1}) > f(x^k)$. These disadvantages are avoided by applying the so called *quasi-Newton methods*, in which the matrix $Hf(x^k)^{-1}$ is approximated by a symmetric positive definite matrix H^k, which is updated at each iteration. The general scheme is now described in details.

1. At the step k, we have:

$$\nabla f(x^k) \simeq \nabla f(x^{k+1}) + Hf(x^{k+1})(x^k - x^{k+1}).$$

If we indicate $\delta^k = x^{k+1} - x^k$ and $\gamma^k = \nabla f(x^{k+1}) - \nabla f(x^k)$, we obtain:

$$Hf(x^{k+1})\delta^k \simeq \gamma^k.$$

2. To approximate $Hf(x^{k+1})^{-1}$ with H^{k+1}, it suffices to require that H^{k+1} satisfies the following relation

$$H^{k+1}\gamma^k = \delta^k.$$

3. By setting $H^{k+1} = H^k + a(u \otimes u) + b(v \otimes v)$, where a and b are real numbers, and the vectors u, and v vary in $\mathbb{R}^n$, we obtain the rank two updates of the H^k.

4. By choosing $u = \delta^k$ and $v = H^k \gamma^k$, we obtain:

$$H^{k+1} = H^k + \frac{\delta^k \otimes \delta^k}{\langle \delta^k, \gamma^k \rangle} - \frac{H^k(\gamma^k \otimes \gamma^k)H^k}{\langle \gamma^k, H^k \gamma^k \rangle},$$

which is the basis of the so called **Davidson-Fletcher-Powell method** (generally abbreviated as the DFP method).

It is shown that all matrices H^k are positive definite (thus d^k are descent directions for f), that the convergence rate is super-linear, and that the method is globally convergent, with exact line searches, if every n steps it is reinitialized by setting $H = I$. Its convergence properties are sensitive to the accuracy used in the line search; therefore, each step requires a high number of evaluations of f, which could negate the advantage due to super-linear convergence. In the case where f is quadratic with a positive definite Hessian, then the DFP method, with exact line search, terminates in at most n steps with $H^{n+1} = Q^{-1}$.

An update formula that is somewhat 'complementary' to the DFP method, can be obtained by considering the inverse of the matrices H^k, rather than these matrices themselves. Let us denote these inverses by B^k. Since we want B^{k+1} to approximate $Hf(x^{k+1})$, we shall require that $B^{k+1}\delta^k = \gamma^k$ and, in addition, that $B^{k+1} = B^k + a(u \otimes u) + b(v \otimes v)$. Using a similar argument as the one adopted for the DFP method, we obtain:

$$B^{k+1} = B^k + \frac{\gamma^k \otimes \gamma^k}{\langle \delta^k, \gamma^k \rangle} - \frac{B^k(\delta^k \otimes \delta^k)B^k}{\langle \delta^k, B^k \delta^k \rangle},$$

and therefore,

$$H^{k+1} = H^k + \left[1 + \frac{\langle \gamma^k, H^k \gamma^k \rangle}{\langle \delta^k, \gamma^k \rangle} \right] \frac{\delta^k \otimes \delta^k}{\langle \delta^k, \gamma^k \rangle} - \frac{(\delta^k \otimes \gamma^k)H^k + H^k(\gamma^k \otimes \delta^k)}{\langle \delta^k, \gamma^k \rangle}.$$

The quasi-Newton method obtained by using the sequence $\{H^k\}$ that is given by this formula is generally known as the **Broyden-Fletcher-Goldfarb-Shanno method** (abbreviated as the BFGS method). It also enjoys the same properties as the DFP method. Moreover, the global convergence properties and the super-linear convergence rate remain valid even if an inexact line search is used at each step, according to the Armijo-Goldstein-Wolfe rules, which allows us to significantly reduce the number of evaluations of f.

Chapter X
Methods for Constrained Optimisation

As we have discussed in details in Chapter V, non-linear programming deals with problems where the objective function or the constraints, or both, are non-linear. These problems are more complex to handle than those of linear programming, due to the fact that, for example, non-linear functions may lead to multiple local optima. This difficulty is also and perhaps even more so present when searching numerically for global optimum, making the development of methods even more challenging.

Amid the many non-linear methods developed by researchers, the ones which we going to develop in the sequel are the following:

Gradient-Based Methods: These methods share the same logic with those of the unconstrained case because they use the gradient (and sometimes the Hessian) of the objective function to construct directions that are both descent and feasible directions.

Penalty and Barrier Methods: These methods transform a constrained problem into a series of unconstrained problems by adding penalty terms to the objective function for violating constraints. Exterior penalty methods are simpler to implement but can suffer from numerical instability, while interior penalty (also called barrier) methods are more stable but require careful tuning of parameters.

As we have mentioned, however, this is only a small sample of the wealth of methods developed to find approximated solutions to non-linear programming problems. Indeed, among the most famous ones, we may cite, for example, *sequential quadratic programming* and *trust-region methods*. The reader who is curious about the development of these methods may check the bibliography. It is also worthwhile noting that other classes of methods have been developed in the recent years, such as the so called *heuristic* and *genetic* methods, which are usually utilised when traditional methods fail. If, on one

side, these methods are flexible and can handle a wide variety of problems
that classical method cannot handle, they, on the other hand, do not guarantee, in general, the convergence of the algorithm towards an optimal solution.
For this reason, they are most likely utilised by users of mathematics, rather
than studied by mathematicians, the latter being usually more interested in
theoretically robust results.

1. Frank-Wolfe method

The Frank-Wolfe method, also known as the *conditional gradient method*,
is an iterative optimisation method used to find an approximation of the
solution of a constrained optimisation problems. In its more general form,
Frank-Wolfe method can be applied to a non-linear programming problem of
the form

$$(\mathcal{P}): \quad \begin{cases} \text{minimise} & f(x) \\ \text{subject to} & x \in D, \end{cases}$$

where the function f is of class $\mathcal{C}^1$, and D is a bounded, convex and closed
set included in the domain of f.

At its core, in each step k, the algorithm starts by considering the linearised
problem $(\mathcal{P}_\mathcal{L})$ at the current point x^k, which is obtained by approximating
the objective function with its linear part, $\langle \nabla f(x^k), x \rangle$, at x^k (which exists
by virtue of the fact that f is of class $\mathcal{C}^1$). Then, the method finds the
minimum, y_k, of this corresponding linearised problem, $(\mathcal{P}_\mathcal{L})$, and finally it
minimises the function f over the line segment $[x^k, y^k]$. If the same point x^k
is an optimal solution of the linearised problem, $(\mathcal{P}_\mathcal{L})$, at the point x^k, then
it can be shown that x^k satisfies the KKT conditions for the problem $(\mathcal{P}_\mathcal{L})$.
The formal description of Frank-Wolfe method is given in Algorithm 11.

Then the following result holds.

(1.1) Theorem. *Let E be an open set in $\mathbb{R}^n$ containing the feasible set D,,
which is convex and compact, and f a real function, defined on E, at least
of class $\mathcal{C}^1$. Then for every x^0 in D, the sequence $\{x^k\}$ generated applying
Algorithm 11 has at least one accumulation point, and, in addition, every
accumulation point of the sequence $\{x^k\}$ is a solution of the KKT system.*

Proof. The proof follows from Zangwill's Convergence Theorem. To see
this, let Ω be the set of points that satisfy the KKT conditions for the given
problem $(\mathcal{P})$. Let us observe that the sequence $\{x^k\}$ is contained in the set
D which is, by hypothesis, compact. Now, if x^k does not belong to Ω, then

$$\langle \nabla f(x^k), y^k \rangle < \langle \nabla f(x^k), x^k \rangle,$$

thus, $y^k - x^k$ is a descent direction for the function f, and $f(x^{k+1}) < f(x^k)$,
having chosen the function f as the descent function. In order to be able

---

Algorithm 11 Frank-Wolfe method

---

1. Set $k = 0$ and choose a feasible point x^0.

2. Compute $\nabla f(x^k)$.

3. Compute y^k as a solution of linearised problem

$$(\mathcal{P}_\mathcal{L}): \quad \begin{cases} \text{minimise} & \langle \nabla f(x^k), x \rangle \\ \text{subject to} & x \in D, \end{cases}$$

4. **If** a STOPPING CRITERION holds, **then stop**

5. Let $\varphi^k(t) = f(x^k + t(y^k - x^k))$.

6. Set $t_k = $ **minimum of** $\varphi^k(t)$ **subject to** $t \in [0, 1]$.

7. Compute $x^{k+1} = x^k + t_k(y^k - x^k)$.

8. Increase k by 1 and **go back** to step 2.

---

to apply the Zangwill's Convergence Theorem, it remains to show that the algorithm is described by a closed multi-function. In fact, a step of the algorithm map may be identified with a multi-function of the form $C \circ B$, where, for every x in $\mathbb{R}^n$, $B : \mathbb{R}^n \to \mathcal{P}(\mathbb{R}^n)$ is defined by

$$B(x) = \Big\{ y - x : y \text{ is the optimal solution of } \mathcal{P}_\mathcal{L} \Big\},$$

and $C : \mathbb{R}^n \to \mathcal{P}(\mathbb{R})$ is defined by

$$C(x) = \Big\{ t : t \text{ is the global minimum of } \varphi \text{ on } [0,1] \Big\}.$$

In other words, B is the multi-function defined for every x^k by $B(x^k)$ i.e. the set of directions $y^k - x^k$, where y^k is an optimal solution of the linearised problem. Let us now show that B is a closed multi-function.

To this end, assume that $x^k \to \bar{x}$, and let $\{y^k\}$ be a sequence of points such that $y^k - x^k \in B(x^k)$ and $y^k \to \bar{y}$. Then,

$$\langle \nabla f(x^k), y^k \rangle \leq \langle \nabla f(x^k), x \rangle \quad \text{for every } x \in D,$$

and, since f is of class C^1, we have

$$\langle \nabla f(\bar{x}), \bar{y} \rangle \leq \langle \nabla f(\bar{x}), x \rangle \quad \text{for every } x \in D.$$

Then $\bar{y}$ is optimal for

$$\begin{array}{ll} \text{minimise} & \langle \nabla f(\bar{x}), x \rangle \\ \text{subject to} & x \in D. \end{array}$$

This proves that $\bar{y} - \bar{x} \in B(\bar{x})$, which means that the multi-function B is closed. Moreover, the multi-function C is also closed, thanks to the fact that the functions f and ∇f are both continuous. Hence, the conclusion follows from Zangwill's Convergence Theorem (see VII(3.9)), observing that the composition of closed multi-functions is closed (see VII(3.4)). $\square$

(1.2) Remark. Reading the proof of the convergence of the algorithm 11, we can argue an additional stopping criterion:

$$\langle \nabla f(x^k), y^k - x^k \rangle < \varepsilon,$$

where the strictly positive real number ε is an assigned tolerance threshold.

Let us observe that, in the special case when the feasible region D is a bounded polyhedron, that is to say, when D is of the form

$$D = \left\{ x \in \mathbb{R}^n : Ax \leq b \right\}$$

where $A \in \mathfrak{M}(m, n, \mathbb{R})$ is a non-zero matrix, and $b \in \mathbb{R}^m$, then the linearised problem $(\mathcal{P}_\mathcal{L})$ that appears in the Frank-Wolfe method turns out to be, in fact, a linear programming problem, and therefore the point y_k can be found simply applying the simplex method.

(1.3) Example. Let us consider the following quadratic problem:

$$\begin{aligned}
\text{minimise} \quad & \tfrac{1}{2}x_1^2 + x_2^2 - 2x_1x_2 + x_1 \\
\text{subject to} \quad & -3x_1 + 2x_2 \leq 7 \\
& x_1 + 2x_2 \leq 4 \\
& 0 \leq x_2 \leq 2.
\end{aligned}$$

Starting from the initial point $x^0 = (0, 1)$, let us compute x^1 using Frank-Wolfe algorithm. We have $\nabla f(x^0) = (-1, 2)$; the optimal solution of

$$\begin{aligned}
\text{minimise} \quad & -x_1 + 2x_2 \\
\text{subject to} \quad & -3x_1 + 2x_2 \leq 7 \\
& x_1 + 2x_2 \leq 4 \\
& 0 \leq x_2 \leq 2
\end{aligned}$$

is achieved at the vertex $y^0 = (4, 0)$. To determine the step-size we shall need to know the minimum point of the restriction of f over the segment whose extrema are $x^0 = (0, 1)$ and $y^0 = (4, 0)$. Let us take the following parametrisation of this segment:

$$\begin{cases} x_1(t) = 4t \\ x_2(t) = 1 - t \end{cases} \qquad \text{for } t \in [0, 1].$$

Let us also denote by φ the restriction of f to the segment, whence

$$\varphi(t) = \tfrac{1}{2}(4t)^2 + (1 - t)^2 - 8t(1 - t) + 4t = 17t^2 - 6t + 1,$$

which reaches the point minimum at $t = \frac{3}{17}$, hence

$$x^1 = \left(\tfrac{12}{17}, \tfrac{14}{17} \right).$$

2. Projected gradient method

When the feasible region is a bounded polyhedron, there exists another method to find an approximated solution of a non-linear programming problem, which profoundly exploits the geometric characteristics of polyhedra, which is called the projected gradient method. Precisely, the projected gradient method operates as follows: At each step k, one chooses a descent direction and performs a one-dimensional search along this direction. If the current point, x^k, falls in the interior of the feasible region, one may take the direction $-\nabla f(x^k)$ as the descent direction for this method. On the other hand, if the point x^k lies on the boundary of the feasible region, one will take the projection of the direction $-\nabla f(x^k)$ onto the vector subspace defined by the set of active constraints at the point x^k.

More precisely, let us consider the following non-linear programming problem

$$(\mathcal{P}) : \quad \begin{cases} \text{minimise} & f(x) \\ \text{subject to} & Ax \leq b, \end{cases}$$

where $A \in \mathfrak{M}(m, n, \mathbb{R})$ is a non-zero matrix, $b \in \mathbb{R}^m$, the polyhedron $Ax \leq b$ is bounded, and the function f is of class $\mathcal{C}^1$ on an open set which we shall assume contains the polyhedron $D = \{x : Ax \leq b\}$.

Let us recall, from Chapter V, that we have denoted by $I(x)$ the set of all indices representing the 'active constraints' at the point x, that is to say, the set $I(x) = \{i : A_i x = b_i\}$.

Before describing the method more formally, and in order to help proving the correctness of the algorithm, we are going to need the following preliminary result on the orthogonal projection of a point onto a polyhedron.

(2.1) Lemma. *Let $M \in \mathfrak{M}(r, n, \mathbb{R})$ be matrix, and let us set*

$$H = \begin{cases} I & \text{if } I(x^k) = \emptyset \\ I - M^T(MM^T)^{-1}M & \text{if } I(x^k) \neq \emptyset \end{cases}$$

where $I \in \mathfrak{M}(n, n, \mathbb{R})$ is the identity. Then, for each direction y in $\mathbb{R}^n$, Hy is the orthogonal projection of y onto vector subspace $S = \{x \in \mathbb{R}^n : Mx = 0\}$.

Proof. We need to prove that, for any vector y, one has $Hy \in S$, and that the vector $y - Hy$ is always orthogonal to S, that is to say,

$$\langle y - Hy, x \rangle = 0 \qquad \text{for every } x \in S.$$

We prove easily that $Hy \in S$, because

$$MHy = My - MM^T(MM^T)^{-1}My = My - My = 0.$$

Furthermore, we have

$$
\begin{aligned}
y - Hy &= (I - H)y \\
&= \Big(I - \big(I - M^T(MM^T)^{-1}M\big)\Big)y \\
&= M^T(MM^T)^{-1}My.
\end{aligned}
$$

Then, for each $x \in S$ we have

$$
\langle y - Hy, x \rangle = \langle y - Hy, x \rangle = \langle M^T(MM^T)^{-1}My, x \rangle = 0. \qquad \square
$$

We can now describe in detail the projected gradient method. This is done in Algorithm 12.

(2.2) Theorem. *Let E be an open set in $\mathbb{R}^n$ and f a real function, defined on E, which is assumed to be at least of class $\mathcal{C}^1$. Let $A \in \mathfrak{M}(m, n, \mathbb{R})$ be a non-zero matrix and $b \in \mathbb{R}^m$ and suppose that the polyhedron $Ax \leq b$ is bounded. Then, for every x^0 feasible, the sequence $\{x^k\}$ generated applying Algorithm 12 falls into one of the following two cases:*

1. *If $d^k \neq 0$, then f is decreasing along the direction d^k and the sequence x^k has at least one accumulation point, and, in addition, every accumulation point of the sequence $\{x^k\}$ is a solution of the KKT system.*

2. *If $d^k = 0$ and $u \geq 0$, then x^k satisfies the KKT conditions of the problem $(\mathcal{P})$ and u is the vector of the KKT multipliers.*

Proof. Let us start by proving assertion 1. To do so, let us apply the Zangwill's Convergence Theorem. To this end, let Ω be the set of all points that satisfy the KKT conditions. The sequence $\{x^k\}$ is contained in the polyhedron $Ax \leq b$ which is compact by hypothesis. Having chosen the objective function f as the descent function we prove that the direction chosen by the algorithm is a descent direction. To this end, let us recall that, for d^k to be a descent direction of f, it is necessary that $-\langle \nabla f(x^k), d^k \rangle > 0$. Let us set $g^k = -\nabla f(x^k)$, so that $d^k = Hg^k$, then we have

$$
\begin{aligned}
\langle g^k, d^k \rangle &= \langle g^k, Hg^k \rangle \\
&= \langle g^k - Hg^k + Hg^k, Hg^k \rangle \\
&= \langle g^k - Hg^k, Hg^k \rangle + \|Hg^k\|^2,
\end{aligned}
$$

but, thanks to Theorem (2.1), the vector $g^k - Hg^k$ is orthogonal to Hg^k, therefore

$$
\langle g^k, d^k \rangle = \|Hg^k\|^2 = \|d^k\|^2 > 0,
$$

because $d^k \neq 0$. To apply the Zangwill's Convergence Theorem, it remains to show that the algorithm is described by a closed multi-function. This is

---

Algorithm 12 PROJECTED GRADIENT METHOD

---

1. Set $k = 0$ and choose a feasible point x^0.

2. Let $I(x^k)$ the set of active constraints and M the matrix having as rows the rows A_i of the matrix A, for each $i \in I(x^k)$.

3. Compute the projection matrix:

$$H = \begin{cases} I & \text{if } I(x^k) = \emptyset \\ I - M^T(MM^T)^{-1}M & \text{if } I(x^k) \neq \emptyset \end{cases}$$

   where $I \in \mathfrak{M}(n, n, \mathbb{R})$ is the identity.

4. Compute $\nabla f(x^k)$.

5. Compute the direction $d^k = -H\nabla f(x^k)$.

6. **if** $d^k = 0$ then **go to** step 12 **otherwise go to** step 7.

7. Set $s_k = $ **maximum of** t **subject to** $A(x^k + td^k) \leq b$.

8. Let $\varphi^k(t) = f(x^k + td^k)$

9. Set $t_k = $ **minimum of** $\varphi^k(t)$ **subject to** $[0, s_k]$.

10. Set $x^{k+1} = x^k + t_k d^k$.

11. Increase k by 1, and **go to** step 2.

12. Compute $u = -(MM^T)^{-1}M\nabla f(x^k)$.

13. **if** $u \geq 0$ **then stop** (x^k satisfies the KKT conditions).

    else Choose $u_j = \min\{u_i : i \in I\}$.

    Remove from M the row A_j and **go to** step 3.

---

easily obtained by observing that the desired mapping is of the form $C \circ B$, where B is defined for every point x in $\mathbb{R}^n$ by

$$B(x) = \left\{ t : t \text{ is a solution of problem of step 7} \right\},$$

that is to say, the map that gives the step s_k at each step by solving a linear programming problem and it is obviously a closed function, while C is defined for every point x in $\mathbb{R}^n$ by

$$C(x) = \left\{ t : t \text{ is the global minimum of } \varphi \text{ on } [x^k, x^k + s_k d^k] \right\}.$$

The proof of closure of C is the same as the Frank-Wolfe algorithm. The conclusion follows from Zangwill's Convergence Theorem, observing that the composition of closed multi-functions is closed.

Let us now move onto assertion 2. To this end, observe that, if $d^k = 0$, then

$$
\begin{aligned}
0 &= H\nabla f(x^k) \\
&= (I - M^T(MM^T)^{-1}M)\nabla f(x^k) \\
&= \nabla f(x^k) - M^T(MM^T)^{-1}M\nabla f(x^k) \\
&= \nabla f(x^k) + M^T u.
\end{aligned}
$$

(2.3)

Furthermore, observing that, the columns of the matrix M^T are the vectors row vectors A_i with $i \in I(x^k)$, we also get:

$$ (2.4) \qquad M^T u = \sum_{i \in I} \langle A_i, u_i \rangle $$

Let us rewrite the polyhedron with equation $Ax \leq b$ in the form $g(x) \leq 0$, by setting

$$g_i(x) = \langle A_i, x \rangle - b_i \quad \text{for every } i = 1, \ldots, m.$$

Then the Lagrangian of the problem can be written in the form

$$L(x, \lambda) = f(x) + \sum_{i=1}^{m} \lambda_i g_i(x).$$

Thus, the relations (2.3) and (2.4) yield

$$
\begin{cases}
\nabla f(x^k) + \sum_{i \in I} \langle A_i, u_i \rangle = 0 \\
u_i \geq 0 \qquad \text{for all } i \in I(x^k).
\end{cases}
$$

Setting $u_i = 0$, for each index $i \notin I$, we finally get to

$$
\begin{cases}
\nabla_x L(x^k, u) = \nabla f(x^k) + \sum_{i=1}^{m} u_i \nabla g_i(x^k) = 0 \\
u_i g_i(x^k) = 0 \qquad \text{for all } i = 1, \ldots, m \\
u \geq 0,
\end{cases}
$$

which is the KKT system associated to problem $(\mathcal{P})$. This shows that the point x^k is indeed a solution of the KKT system associated to the problem, and that u is the corresponding vector of the KKT multipliers associated with x^k. $\qquad\qquad\square$

(2.5) Example. Let us consider the following non-linear program:

$$\text{maximise} \quad x_1^2 - x_2^2 + x_1 x_2$$
$$\text{subject to} \quad x \in P.$$

where P is the bounded polyhedron with vertices $(0,0)$, $(1,0)$, $(0,1)$ and $(1,2)$ (see Figure 2.1).

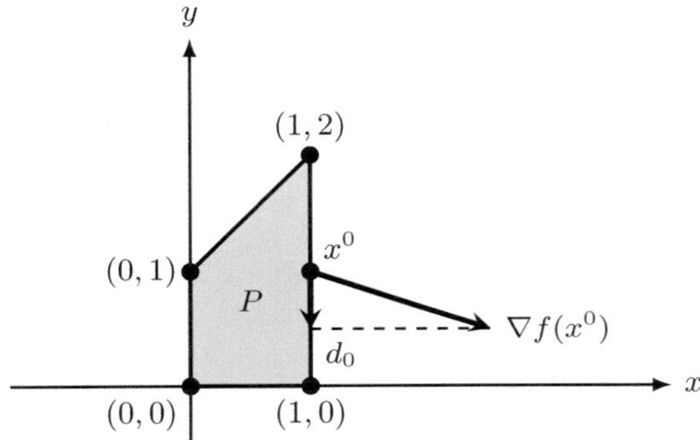

Figure 2.1. The bounded polyhedron that describes the feasible region in Example (2.5).

We shall now illustrate the first step of the projected gradient algorithm starting from the point $x^0 = (1,1)$. Let us now compute the gradient at the point x^0. We have $\nabla f(1,1) = (3,-1)$. Let us observe that the direction points outwards from the polyhedron, so we shall have to project it onto the polyhedron.

The point x^0 has only one active constraint which is $x_1 \leq 1$. Hence, the matrices M and H are:

$$M = (1,0), \qquad H = \begin{bmatrix} 0 & 0 \\ 0 & 1 \end{bmatrix}.$$

Calculating the direction we obtain $d^0 = H\nabla f(1,1) = (0,-1)$.

We can now perform step 4, thereby obtaining $s_0 = 1$; this number is the maximum value of the step-size t before leaving the polyhedron when we move from x^0 along the direction d^0.

Finally we compute the minimum of the function φ which is the restriction of the function f to the segment $[0,1]$ which is $t_0 = \frac{1}{2}$. The obtained new point is $x^1 = (1, \frac{1}{2})$.

(2.6) Example. Let us consider the following non-linear program:

$$\text{minimise} \quad -2x_1^2 - 4x_2^2 - 2x_1 x_2 - 9x_2$$
$$\text{subject to} \quad x \in P.$$

where P is the polyhedron with vertices $(2,0)$, $(5,3)$, $(1,5)$ and $(-1,1)$.

We shall now illustrate the first step of the projected gradient algorithm starting from the point $x^0 = (0, \frac{2}{3})$. Let us now compute the gradient at the point x^0. We have $\nabla f(0, \frac{2}{3}) = (-\frac{4}{3}, -\frac{43}{3})$. Let us observe that the direction points outwards from the polyhedron, so we shall have to project it onto the polyhedron.

The point x^0 has only one active constraint which is $-x_1 - 3x_2 \le -2$. Hence, the matrices M and H are:

$$M = (-1, -3), \qquad H = \begin{pmatrix} 9/10 & -3/10 \\ -3/10 & 1/10 \end{pmatrix}.$$

Calculating the direction we obtain $d^0 = -H\nabla f(0, \frac{2}{3}) = (\frac{31}{10}, -\frac{31}{30})$.

We can now perform step 4, thereby obtaining $s^0 = \frac{20}{31}$; this number is the maximum value of the step-size t before leaving the polyhedron when we move from x^0 along the direction d^0.

Finally we compute the minimum of the function φ^0 onto the segment $[0, \frac{20}{31}]$ which is $t_0 = \frac{5}{16}$. The obtained new point is $x^1 = (\frac{31}{32}, \frac{11}{32})$.

In the next example we shall show a case where we need to pass through step 5 of the algorithm, which happens when the projection of the gradient is the zero vector.

(2.7) Example. Consider the problem:

$$\begin{array}{ll} \text{minimise} & 2x_1^2 + 2x_2^2 - 2x_1 x_2 - 6x_1 \\ \text{subject to} & x_1 + x_2 \le 2 \\ & x_1 \ge 0 \\ & x_2 \ge 0. \end{array}$$

Let us start the algorithm from the point $x^0 = (0,0)$. In x^0 two constraints are active, the second and the third. Therefore

$$M = \begin{bmatrix} -1 & 0 \\ 0 & -1 \end{bmatrix}, \qquad H = \begin{bmatrix} 0 & 0 \\ 0 & 0 \end{bmatrix}.$$

We have $-\nabla f(0,0) = (0,6)$ and its projection onto the polyhedron is given by $d^0 = (0,0)$.

We are ready to perform step 5 of Algorithm 12. We compute the vector u and we obtain $u = (-6, 0)$; then eliminate from the matrix M the first row which is relative to the constraint 2 and we compute again matrix H and the new projection d^0 of the vector $-\nabla f(0,0)$.

$$M = (0, -1), \qquad H = \begin{bmatrix} 1 & 0 \\ 0 & 0 \end{bmatrix}, \qquad d^0 = (6, 0).$$

We can now perform step 4, obtaining $s^0 = \frac{1}{3}$ and $t^0 = \frac{1}{4}$, so $x^1 = (\frac{3}{2}, 0)$.

Let us now perform step 2 with respect the new point x^1.

$$M = (0, -1), \qquad H = \begin{bmatrix} 1 & 0 \\ 0 & 0 \end{bmatrix}, \qquad d^1 = (0, 0).$$

Let us perform again step 5. This gives us

$$u = -3, \qquad M = \varnothing, \qquad H = \begin{bmatrix} 1 & 0 \\ 0 & 1 \end{bmatrix}, \qquad d^1 = (0, 3).$$

So, $s^1 = \frac{1}{6}$ and $t^1 = \frac{1}{6}$, and $x^2 = (\frac{3}{2}, \frac{1}{2})$.

Let us perform step 2 once again.

$$M = (1, 1), \qquad H = \begin{bmatrix} \frac{1}{2} & -\frac{1}{2} \\ -\frac{1}{2} & \frac{1}{2} \end{bmatrix}, \qquad d^2 = (0, 0).$$

Let us perform step 5: this time we have $u = 1$, and therefore x^2 satisfies the KKT conditions. As a result, x^2 is a global minimum, because the function f is convex.

(2.8) Remark. This algorithm of the projected gradient has been generalised to the more general case where the constraint is a closed, bounded and convex. In this case we do not have a closed expression for the projection like that contained in Lemma (2.1). To find the projection of the gradient onto the domain (which exists and is unique because the domain is convex and compact), one must solve a strictly convex quadratic programming problem.

3. Exterior penalty methods

Penalty methods are a class of algorithms used to solve constrained optimisation problems by transforming them into a series of unconstrained problems. The basic idea is to add a penalty term to the objective function that imposes a cost for violating the constraints. This penalty term typically increases as the solution moves further away from the feasible region, thus discouraging constraint violations. Typically they are divided into two classes, namely *exterior penalty methods*, and *interior penalty methods*.

Both methods are particularly useful when dealing with complex constraints that are difficult to handle directly. By converting the constrained problem into a sequence of easier unconstrained problems, penalty methods provide a flexible and powerful approach to solving non-linear programming problems.

Exterior penalty methods start with an infeasible solution and iteratively adjust the penalty term to drive the solution towards feasibility. The penalty term is added to the objective function and imposes a cost for constraint violations. As the penalty parameter increases, the solution converges to the feasible region. A common exterior penalty function is the quadratic penalty, which penalizes the square of the constraint violation. These methods allow the solution to initially violate constraints but gradually reduces violations as the penalty parameter increases. They are useful when starting with an infeasible solution and iterating towards both feasibility and optimality.

Let's describe the so-called exterior penalty methods. To this end, let us consider a non-linear programming problem with only inequality constraints:

$$(3.1) \qquad\qquad (\mathcal{P}) : \quad \begin{cases} \text{minimise} & f(x) \\ \text{subject to} & x \in D \end{cases}$$

where $f : \mathbb{R}^n \to \mathbb{R}$, $g_i : \mathbb{R}^n \to \mathbb{R}$ for $i = 1, \ldots, m$, $g = (g_1, g_2, \ldots, g_m)$, and

$$D = \left\{ x \in \mathbb{R}^n : g(x) \leq 0 \right\}.$$

For any real number a, let us set $a^+ = \max\{a, 0\}$, and let us define on D the function ψ by setting

$$(3.2) \qquad\qquad \psi(x) = \sum_{i=1}^{m} [g_i^+(x)]^2,$$

which shall be called, after the German-American mathematician RICHARD COURANT (1888–1972), the **Courant penalty function** associated to problem $(\mathcal{P})$. Let us also introduce the function Ψ_k, defined on D by

$$\Psi_k(x) = f(x) + k\psi(x).$$

The **unconstrained penalised problem** associated to $(\mathcal{P})$ is defined as

$$(\mathcal{P}_k) : \quad \begin{cases} \text{minimise} & \Psi_k(x) \\ \text{subject to} & x \in \mathbb{R}^n \end{cases}$$

where k is a a positive integer.

Note that $\Psi_k(x)$ coincides with $f(x)$ whenever $x \in D$, and it is strictly greater than $f(x)$ whenever $x \notin D$. Furthermore observe that, if there had been no square in the definition of $\psi(x)$, the function $\psi(x)$ might not be of class $\mathcal{C}^1$ even when g is. Thanks to the presence of the square in the definition of ψ, however, one always has that $g \in \mathcal{C}^1 \Rightarrow \psi \in \mathcal{C}^1$. Let also observe that, in this case, one has

$$\nabla \Psi_k(x) = \nabla f(x) + 2k \sum_{i=1}^{m} g_i^+(x) \nabla g_i(x).$$

The following theorem shows some additional properties of the unconstrained penalised problem.

(3.3) Theorem. *Let us given a problem* $(\mathcal{P})$ *and suppose that* f *has compact sublevels. For every positive integer* k, *assume that* x^k *is an optimal solution of the problem* $(\mathcal{P}_k)$, *and denote by* x^* *an optimal solution of the problem* $(\mathcal{P})$. *Then the following hold:*

1. *If* $f(x)$ *and* $g_i(x)$, *for every* $i \in 1 \cdots m$, *are convex, then* $\Psi_k(x)$ *is convex.*

2. *If* $k_1 \leq k_2$, *then* $\Psi_{k_1}(x^{k_1}) \leq \Psi_{k_2}(x^{k_2}) \leq f(x^*)$,

3. *If* $x^k \in D$ *for some* k, *then* x^k *is optimal for* $(\mathcal{P})$.

4. $\displaystyle \lim_{k \to +\infty} \sum_{i=1}^{m} [g_i^+(x^k)]^2 = 0.$

Proof.

1. The first statement follows from the fact that, if g_i is convex, so is g_i^+, and the conclusion follows from the fact that the sum of convex functions is convex.

2. It suffices to observe that

$$
\begin{aligned}
\Psi_{k_1}(x^{k_1}) &= \min\{\Psi_{k_1}(x) : x \in \mathbb{R}^n\} \\
&\leq \Psi_{k_1}(x^{k_2}) \\
&= f(x^{k_2}) + k_1 \sum_{i=1}^{m}[g_i^+(x^{k_2})]^2 \\
&\leq f(x^{k_2}) + k_2 \sum_{i=1}^{m}[g_i^+(x^{k_2})]^2 \\
&= \Psi_{k_2}(x^{k_2}).
\end{aligned}
$$

Furthermore, for every index i, one has $g_i^+(x^*) = 0$. Hence,

$$
\begin{aligned}
\Psi_{k_2}(x^{k_2}) &\leq \Psi_{k_2}(x^*) \\
&= f(x^*) + k_2 \sum_{i=1}^{m}[g_i^+(x^*)]^2 \\
&= f(x^*).
\end{aligned}
$$

3. Let us assume that $x^k \in D$. Then, for every $x \in D$, we have

$$
f(x^k) = \Psi_k(x^k) \leq \Psi_k(x) = f(x) + k\psi(x) = f(x),
$$

where the last identity follows from the fact that x is an element of D.

4. Since the function f has compact sublevels, there exists a real number c, such that $f(x) \geq c$ for every $x \in \mathbb{R}^n$. Then

$$
\begin{aligned}
c + k \sum_{i=1}^{m}[g_i^+(x^k)]^2 &\leq f(x^k) + k \sum_{i=1}^{m}[g_i^+(x^k)]^2 \\
&= \Psi_k(x^k) \\
&\leq f(x^*),
\end{aligned}
$$

so that

$$0 \leq \sum_{i=1}^{m} [g_i^+(x^k)]^2 \leq kf(x^*) - c$$

for every natural index k. This shows that the sequence with general term $\sum_{i=1}^{m} [g_i^+(x^k)]^2$ converges towards zero. $\qquad\square$

We are now able to prove the convergence theorem for this method.

(3.4) Theorem. *In the same hypotheses as in Theorem (3.3), assume that f has compact sublevels. Every accumulation point of $\{x^k\}$ is a global solution of the problem $(\mathcal{P})$.*

Proof. Let $\bar{x}$ be an accumulation point of $\{x^k\}$. After passing to a subsequence, if necessary, we may assume that $\{x^k\}$ converges towards $\bar{x}$. Then, by continuity, we have

$$\sum_{i=1}^{m} [g_i^+(\bar{x})]^2 = \lim_{h \to +\infty} \sum_{i=1}^{m} [g_i^+(x^k)]^2 = 0.$$

As a result of this, we get that $g_i^+(\bar{x}) = 0$, for every index $i = 1, \ldots, m$. Observe that the latter statement is equivalent to $g(\bar{x}) \leq 0$, that is to say, to the fact that $\bar{x}$ is feasible. Furthermore,

$$\begin{aligned}
f(\bar{x}) &= f(\bar{x}) + \sum_{i=1}^{m} [g_i^+(\bar{x})]^2 \\
&= \lim_{k \to +\infty} \left\{ f(x^k) + \sum_{i=1}^{m} [g_i^+(x^k)]^2 \right\} \\
&\leq \lim_{k \to +\infty} \left\{ f(x^k) + k \sum_{i=1}^{m} [g_i^+(x^k)]^2 \right\} \\
&= \lim_{k \to +\infty} \Psi_k(x^k) \\
&\leq f(x^*).
\end{aligned}$$

Thus, we necessarily have that $f(\bar{x}) = f(x^*)$, that is to say, $\bar{x}$ is a global solution of the problem $(\mathcal{P})$. $\qquad\square$

(3.5) Remark. In the presence of equality constraints, i.e. of constraints the form $h(x) = 0$, one can always reduce the problem to the one with inequality constraints by replacing the equality constraint $h(x) = 0$ with $h(x) \leq 0$ and $h(x) \geq 0$, and observing that

$$\sum_{i=1}^{p} [h_i^+(x)]^2 = \sum_{i=1}^{p} h_i(x)^2.$$

Let us show a simple example.

(3.6) Example. Let $f(x_1, x_2) = x_1^2 + \frac{1}{4}x_2^2$ and $h(x_1, x_2) = x_1 - 2$. We have $\Psi_k(x_1, x_2) = x_1^2 + \frac{1}{4}x_2^2 + k(x_1 - 2)^2$. It is easy to prove that x_k is equal to $(\frac{2k}{1+k}, 0)$ which tends, for k tending to $+\infty$, to $(2, 0)$ which is effectively the solution of the problem.

Even though the previous remark shows that if a non-linear programming problem does have equality constraints, one can always reduce it to a non-linear programming problem with inequality constraints, when a non-linear programming problem does have equality constraints, there is a more direct way to transform the constrained problem into an unconstrained problem, rather than utilising Courant penalty functions. Indeed, one can obtain a similar result by introducing the 'augmented Lagrangian'. To this end, let us consider the following non-linear programming problem:

$$(\mathcal{D}) : \quad \begin{cases} \text{minimise} & f(x) \\ \text{subject to} & h(x) = 0. \end{cases}$$

Let us make a perturbation of the constraints by setting, for every index $j = 1, \ldots, p$, $h_j(x) = \delta_j$, where δ_j is a real number. Applying the penalty function we obtain

$$\begin{aligned} \Psi_k(x) &= f(x) + k \sum_{j=1}^{p} \big(h_j(x) - \delta_j\big)^2 \\ &= f(x) + \sum_{j=1}^{p}(-2k\delta_j)h_j(x) + k \sum_{j=1}^{p} h_j(x)^2 + k \sum_{j=1}^{p} \delta_j^2. \end{aligned}$$

The last term on the right-hand side is a constant, so, after applying an equivalent transformation, we may eliminate it from the problem. Setting now $\mu_j = -2r\delta_j$ the function

$$L_k(x, \mu) = f(x) + \langle \mu, h(x) \rangle + k \sum_{j=1}^{p} h_j(x)^2$$

is the penalty function associated to the perturbation. Observe that $L_k(x, \mu)$ is also the exterior penalty function of the problem

$$\begin{aligned} \text{minimise} \quad & f(x) + \langle \mu, h(x) \rangle \\ \text{subject to} \quad & h(x) = 0. \end{aligned}$$

where μ is a fixed parameter, but also the Lagrangian function of the problem

$$\begin{aligned} \text{minimise} \quad & f(x) + k\|h\|^2 \\ \text{subject to} \quad & h(x) = 0. \end{aligned}$$

This justifies introducing the following function. For any positive real number c, let us set

$$L_c(x, \mu) = f(x) + \langle \mu, h(x) \rangle + \tfrac{1}{2}c\|h(x)\|^2$$

For any $k \in \mathbb{R}$, let us consider the following unconstrained problem:

$$(\mathcal{D}_k) : \quad \begin{cases} \text{minimise} & L_{c^k}(x, \mu^k) \\ \text{subject to} & x \in \mathbb{R}^n \end{cases}$$

with μ^k is a vector in $\mathbb{R}^p$ and c^k is a strictly positive real number. We can then prove the following convergence theorem.

(3.7) Theorem. *In the same hypotheses as in Theorem (3.3), let x^k be a solution of problem $(\mathcal{D}_k)$, $\{\mu^k\}$ is a bounded sequence of vectors in $\mathbb{R}^p$, and $\{c_k\}$ is a divergent increasing sequence of positive real numbers. Then every accumulation point of $\{x^k\}$ is a global solution of $(\mathcal{P})$.*

Proof. Let us denote by $\bar{x}$ be an accumulation point of the sequence $\{x^k\}$. After passing to a subsequence, if necessary, we may assume that $x^k \to \bar{x}$. Then,

$$(3.8) \qquad L_{c^k}(x^k, \mu^k) = \min_{x \in \mathbb{R}^n} L_{c^k}(x, \mu^k) \leq \min_{h(x)=0} L_{c^k}(x, \mu^k) = f(x^*).$$

After passing to a further subsequence, if necessary, we may assume that there exist a vector $\bar{\mu}$ such that $\mu^k \to \bar{\mu}$. Then, evaluating (3.8) at the point x^k gives

$$f(x^k) + \langle \mu^k, h(x^k) \rangle + \tfrac{1}{2} c^k \|h(x^k)\|^2 = L_{c^k}(x^k, \mu^k) \leq f(x^*),$$

and taking the limit superior on both sides, yields

$$(3.9) \qquad f(\bar{x}) + \langle \bar{\mu}, h(\bar{x}) \rangle + \limsup_{r \to +\infty} \tfrac{1}{2} c^k \|h(x^k)\|^2 \leq f(x^*).$$

Since, by hypotheses, $c^k \to \infty$, then $\|h(x^k)\|^2 \to 0$ and, therefore, $h(\bar{x}) = 0$. Moreover, (3.9) entails

$$f(\bar{x}) \leq f(x^*),$$

which shows that $\bar{x}$ is a global solution of problem $(\mathcal{D})$. $\qquad\square$

(3.10) Remark. Often in the solution methods the sequence $\{\mu^k\}$ is not chosen *a priori*, but it is defined iteratively at each step of the algorithm. In particular the most commonly used way to do this is the method known as 'Lagrange multipliers method', which consists in choosing:

$$\begin{cases} \mu^0 \in \mathbb{R}^p \\ \mu^{k+1} = \mu^k + c^k h(x^k). \end{cases}$$

4. Interior penalty methods

Interior penalty methods work by incorporating a penalty term that becomes very large as the solution approaches the boundary of the feasible region. This discourages the solution from violating the constraints. The penalty term is typically designed to be zero within the feasible region and increases rapidly as the constraints are approached. The penalty term is often a barrier function that prevents the solution from leaving the feasible region. For example, a common barrier function is the logarithmic barrier, which becomes infinite at the boundary of the feasible region. These methods are useful when starting with a feasible solution and iterating towards optimality while maintaining feasibility.

Interior penalty methods, which are sometimes referred to as *barrier methods*, are another class of algorithms used to solve constrained optimization problems. Like exterior penalty methods, these methods introduce a penalty function, but in this case the penality function is designed to grow towards infinite, as the solutions approach the boundaries of the constraints, thereby creating a 'barrier' that prevents the solutions from violating the constraints. The basic principle of barrier methods is to transform a constrained optimization problem into an unconstrained one. Roughly speaking, this is done by adding a penalty function, $B(x)$, to the objective function $f(x)$, where x represents the vector of decision variables. As we have said, the penalty function $B(x)$ tends to $+\infty$ when x approaches the 'boundary of the constraints'. Thus, the modified objective function now takes the form

$$\Psi_k(x) = f(x) + k^{-1}B(x)$$

where k is a penalty parameter that controls the severity of the barrier. As k increases, the solution of the unconstrained problem approaches the solution of the original constrained problem. Barrier methods are particularly useful for optimization problems with complex constraints, as they transform the problem into a more manageable form. However, a significant disadvantage is that the penalty function can make the problem numerically unstable, especially when k is very large.

More precisely, assume that E is an open set in $\mathbb{R}^n$ and assume that the real continuous functions $f, g_1, \ldots, g_m$ are all defined in E. Furthermore, assume that f has all compact sub-levels, that is to say, for every real number c, the set $\{x : f(x) \leq c\}$ is compact. Let us set

$$D = \left\{ x \in E : g_1(x) \leq 0, \ldots, g_m(x) \leq 0 \right\}.$$

Let us observe that, the interior $\mathrm{int}(D)$ of D is the set

$$\left\{ x \in E : g_1(x) < 0, \ldots, g_m(x) < 0 \right\}$$

and let us assume that this set is not empty. This crucial assumption makes clear the reason why we assume that the equality constraints $h_j(x) = 0$ here are not present in the description of the feasible domain.

(4.1) Definition. In the framework described above, a real function B, defined on the set $\mathrm{int}(D)$, is called a **barrier function for** D if, for every point $\bar{x}$ in $D \setminus \mathrm{int}(D)$, one has

$$\lim_{x \to \bar{x}} B(x) = +\infty.$$

(4.2) Example. In the framework described above, the real function B_I, defined on $\mathrm{int}(D)$ by

$$B_I(x) = -\sum_{i=1}^{m} g_i(x)^{-1}$$

is clearly a barrier function, which is called the **inverse barrier**. Similarly, the real function B_L, defined on $\mathrm{int}(D)$ by

$$B_L(x) = -\sum_{i=1}^{m} \log\big(-g_i(x)\big)$$

is also a barrier function, called the **log-barrier**.

In the rest of this section, let us also assume that the function $g_1, \ldots, g_m$ are convex and satisfy the Slater condition (see V(5.2)(b)).

(4.3) Proposition. *The following hold:*

(a) *If the functions $g_1(x), \ldots, g_m(x)$ are all of class $\mathcal{C}^1$, such are B_I and B_L.*

(b) *If the functions $g_1(x), \ldots, g_m(x)$ are all convex, such are B_I and B_L.*

Proof. The proof of (a) follows immediately from the the fact that in $\mathrm{int}(D)$, the functions $g_1, \ldots, g_n$ are non-zero, so the composition with the real function $t \mapsto t^{-1}$ and $t \mapsto \log(t)$ maintains differentiability. The proof of (b) is also very simple, for the composition of convex functions in convex. $\square$

We can finally state the convergence theorem, which defines the algorithmic procedure that links to the interior penalty methods.

(4.4) Theorem. *Assume that E is an open set in $\mathbb{R}^n$ and assume that the real function $f, g_1, \ldots, g_m$ are all defined in E and continuous. Assume further that the functions $g_1, \ldots, g_m$ are all convex and satisfy the Slater condition (see V(5.2)(b)) and that the function f has compact sublevels. Let us set*

$$D = \Big\{ x \in E : g_1(x) \le 0, \ldots, g_m(x) \le 0 \Big\}.$$

Let B be a barrier function. For every positive integer k, let us denote by x^k the global minimum of the function $f(x) + k^{-1}B(x)$ on $\mathrm{int}D$. Then, every accumulation point of $\{x^k\}$ is a global minimum of the optimisation problem

$$(\mathcal{P}) : \quad \begin{cases} \text{minimise} & f(x) \\ \text{subject to} & g_i(x) \le 0. \end{cases}$$

Proof. Let $\bar{x}$ be the limit of a subsequence $\{x^k\}$. If $\bar{x}$ belongs to D, we have $\lim_k k^{-1}B(x^k) = 0$. On the other hand, if $\bar{x}$ lies on the boundary of D, by assumption, we have $\lim_k B(x^k) = \infty$. In either case, we obtain that

$$\liminf_{k \to +\infty} k^{-1}B(x^k) \ge 0,$$

which yields

$$(4.5) \qquad \liminf_{k \to +\infty}\Big[f(x^k) + k^{-1}B(x^k)\Big] = f(\bar{x}) + \liminf_{k \to +\infty} k^{-1}B(x^k) \ge f(\bar{x}).$$

The point $\bar{x}$ is a feasible solution of problem $(\mathcal{P})$, since x^k lies in $\text{int}(D)$ and D is a closed set. If $\bar{x}$ were not a global minimum, there would exist a feasible solution x^* such that $f(x^*) < f(\bar{x})$ and therefore also (using the Slater's and convexity assumption, x^* can be approached arbitrarily closely through the interior set D) an interior point $\tilde{x}$ in D such that $f(\tilde{x}) < f(\bar{x})$. We now have, by definition of x^k,

$$f(x^k) + k^{-1}B(x^k) \leq f(\tilde{x}) + k^{-1}B(\tilde{x}) \qquad \text{for every positive integer } k,$$

which, by taking the limit as k tends towards infinity, implies, together with equation (4.5), that $f(\bar{x}) \leq f(\tilde{x})$. This is a contradiction, which entails that $\bar{x}$ is a global minimum for problem $(\mathcal{P})$. $\qquad\square$

(4.6) Example. Let us consider the example give by $f(x_1, x_2) = x_1^2 + x_2^2$ and $g_1(x_1, x_2) = 1 - x_1$. The logarithmic barrier function gives

$$x^k = \tfrac{1}{2}\sqrt{1 + 2/k}.$$

The limit when k approaches to $+\infty$ gives $(1, 0)$ which is the global minimum of problem $\mathcal{P}$.

List of Symbols

$\emptyset$	empty set
∞	infinity (positive infinity)
$\{x : \dots\}$	set of all elements x with the property $\dots$
$a \in A$	a lies in A
$|A|$	the cardinality (number of elements) of the set A
$A \subseteq B$	A is included in B
$A \subset B$	A is included in B, but $A \neq B$
$A \cup B$	union of A and B
$A \cap B$	intersection of A and B
$A \setminus B$	set difference of A and B, i.e. $A \setminus B = \{a \in A : a \notin B\}$
A^c	complementary of A, i.e. $A^c = \{x : x \notin A\}$
$A \times B$	Cartesian product of A and B
A^n	Cartesian product of the set A repeated n times, that is $A^n = \underbrace{A \times A \times \dots \times A}_{n \text{ times}}$
$f : X \to Y$	a map f, defined on the set X, with values in the set Y
$f : X \to \mathcal{P}(Y)$	a multi-function f from X to Y
H^*	polar cone of a set H
H^{**}	bipolar cone of a set H, i.e. $(H^*)^*$
$\mathrm{conv}(A)$	convex hull of the set A
$\mathrm{cone}(A)$	conic hull of the set A
$\mathrm{vert}(P)$	set of vertices of a polyhedral set P
$\mathrm{rec}(P)$	recession cone of a polyhedral set P
$\mathrm{lin}(P)$	lineality space of a polyhedral set P
E_α	sublevel set of E, i.e. $E_\alpha = \{x \in E : f(x) \leq \alpha\}$
$\mathrm{epi}(f)$	epigraph of a function f
$\mathrm{hyp}(f)$	hypograph of a function f

$B(x,r)$	open ball of centre x and radius r
$N(x)$	neighbourhood of the point x
$\text{int}(E), \overline{E}, \partial E$	interior part, closure and boundary of the set E
$\mathcal{C}^0(\Omega)$	vector space of all continuous functions on Ω
$\mathcal{C}^1(\Omega)$	vector space of all differentiable functions on Ω
$\mathcal{C}^2(\Omega)$	vector space of all twice differentiable functions on Ω
$\mathcal{C}^k(\Omega)$	vector space of all function f with $\nabla f \in \mathcal{C}^{k-1}(\Omega)$
$f'(x)$	derivative of the function f at the point x
$f''(x)$	second derivative of the function f at the point x
$g \circ f$	composite function of g and f, i.e. $(g \circ f)(x) = g(f(x))$
$\nabla f(x)$	gradient of the function f at the point x
$Hf(x)$	Hessian matrix of the function f at the point x
$JF(x)$	Jacobian matrix of the function F at the point x
$\nabla_x f(x^*, y^*)$	gradient of the function $f(x,y)$ w.r.t. x
$H_x f(x^*, y^*)$	Hessian of the function $f(x,y)$ at (x^*, y^*) w.r.t. x
$J_x f(x^*, y^*)$	Jacobian of the function $f(x,y)$ at (x^*, y^*) w.r.t. x
$\partial f(x^*)$	sub-differential of f at the point x
$\underset{x \in D}{\arg\min}\, f(x)$	set of all minimum points of f on D
$\displaystyle\int_a^b f(x)\, \mathrm{d}x$	integral from a to b of $f(x)$ in $\mathrm{d}x$
$\mathbb{N}$	the set of natural numbers
$\mathbb{Z}$	the set of integer numbers
$\mathbb{R}$	the set of real numbers
$\mathbb{Z}_+$	the set of positive integer numbers
$\mathbb{R}_+$	the set of positive real numbers
$\mathbb{Z}^n$	the lattice of all n-tuples of integer numbers
$\mathbb{R}^n$	the vector space of all n-tuples of real numbers
$\mathbb{Z}_+^n$	the lattice of all n-tuples of positive integer numbers
$\mathbb{R}_+^n$	the positive orthant, i.e. the set of n-tuples of +ve reals
$\mathfrak{M}(m,n,\mathbb{R})$	set of $m \times n$ matrices with coefficients in $\mathbb{R}$

$[a, b]$	closed interval with extremes a and b
(a, b)	open interval with extremes a and b
$\lfloor x \rfloor$	floor function of x
$\lceil x \rceil$	ceiling function of x
$\{x\}$	fractional part of x

x^+	positive part of x, i.e. $x^+ = \max\{x, 0\}$
x^-	negative part of x, i.e. $x^- = \max\{-x, 0\}$
$|x|$	absolute value (or modulus) x, i.e. $|x| = \max\{x, -x\}$

x^T	transpose vector of x
$\langle x, y \rangle$	scalar product between vectors x and y
$\|x\|$	Euclidean norm of vector x, i.e. $\|x\| = \sqrt{\langle x, x \rangle}$
$x \otimes y$	Outer product of two vector, i.e. $(x \otimes y)(z) = \langle y, z \rangle x$
M^i, M_j	ith column and jth row of a matrix M
$\det(M)$	determinant of a square matrix M
$I(x^*)$	set of indexes of active constraints of x^*

$\mathcal{A}_D(x)$	cone of feasible directions to D in x
$\mathcal{T}_D(x)$	cone of tangents of D at x
$\mathcal{L}_D(x)$	cone of linearised feasible directions to D at x
$\mathcal{S}_D(x)$	cone of strictly linearised feasible directions of D at x
$L(x, \lambda, \mu)$	Lagrangian function, i.e. $L(x, \lambda, \mu) = f(x) + g(x)^T \lambda + h(x)^T \mu$
$C(x^*, \lambda^*, \mu^*)$	critical cone

$\limsup\limits_{x \to x^*} f(x)$	limit superior of a function f as $x \to x^*$
$\liminf\limits_{x \to x^*} f(x)$	limit inferior of a function f as $x \to x^*$
$\lim\limits_{x \to x^*} f(x)$	limit of a function f as $x \to x^*$

$\{x^k\}$	the infinite sequence $x^0, x^1, x^2, \ldots$
$\Psi_k(x)$	Penalty function, i.e $\Psi_k(x) = f(x) + k^{\pm 1} b(x)$

LP	linear programming
NLP	non-linear programming
FJ	Fritz-John conditions
KKT	Karush-Kuhn-Tucker conditions

Bibliography

[1] P. Acquistapace. "Appunti di Analisi Convessa". Appunti delle lezioni di Istituzioni di Analisi Matematica. 2004.

[2] M.S. Bazaraa, H.D. Sherali and C.M. Shetty. *Nonlinear Programming: Theory and Algorithms*. John Wiley & Sons, 2006.

[3] D.P. Bertsekas. *Convex Optimization Theory*. Athena Scientific, 2009.

[4] D.P. Bertsekas. *Nonlinear Programming*. Athena Scientific, 1999.

[5] E.G. Birgin and J.M. Martinez. *Practical Augmented Lagrangian Methods for Constrained Optimization*. SIAM, Philadelphia, 2014.

[6] J.M. Borwein and A.S. Lewis. *Convex Analysis and Nonlinear Opimization*. Springer-Verlag New York, 2000.

[7] D. Braess. *Nonlinear Approximation Theory*. Springer Berlin, 1986.

[8] V. Chvatal. *Linear Programming*. Freeman and Company, 1983.

[9] A.R. Conn, N.I.M. Gould and P.L. Toint. *Trust Region Methods*. SIAM, Series on Optimization, 2000.

[10] G.B. Dantzig and M.N. Thapa. *Linear Programming 1: Introduction*. Springer, 1997.

[11] G.B. Dantzig and M.N. Thapa. *Linear Programming 2: Theory and Extensions*. Linear Programming. Springer, 1997.

[12] G.B. Dantzig and M.N. Thapa. *The Simplex Method*. Springer, 1997.

[13] J.E. Dennis and R.B. Schnabel. *Numerical Methods for Unconstrained Opimization and Nonlinear Equations*. Classics in Applied Mathematics, vol.16, SIAM, 1996.

[14] R. Fletcher. *Practical Methods of Optimization*. Wiley New York, 1987.

[15] L. Grippo and M. Sciandrone. *Introduction to Methods for Nonlinear Opimization*. Springer New York, 2023.

[16] O. Guler. *Foundations of Optimization*. Springer New York, 2010.

[17] J.B. Hiriart Urruty and C. Lemarechal. *Fundamentals of Convex Analysis*. Springer-Verlag Berlin, 2001.

[18] V.G. Karmanov. *Mathematical Programming*. MIR, 1989.

[19] O.L. Mangasarian. *Nonlinear Programming*. McGraw-Hill Series in Systems Science. McGraw-Hill, 1969.

[20] Y.E. Nesterov. *Introductory Lectures on Cnvex Optimization*. Kluwer, 2004.

[21] J. Nocedal and S.J. Wright. *Numerical Opimization*. Springer New York, 2006.

[22] M. Pappalardo and M. Passacantando. *Ricerca Operativa*. Pisa University Press, 2012.

[23] P. Pedregal. *Introduction to Opimization*. Springer, 2003.

[24] E. Polak. *Optimization*. Springer New York, 1997.

[25] B.T. Polyak. *Introduction to Optimization*. Springer New York, 1987.

[26] B.N. Pshenichny and Y.M. Danilin. *Numerical Methods in Extreml Problems*. MIR Publishers, 1978.

[27] R.T. Rockafellar. *Convex Analysis*. Vol. 18. Princeton University Press, 1997.

[28] A. Schrijver. *Theory of Linear and Integer Programming*. Wiley, 1986.

[29] S. Wright and B. Recht. *Optimization for Data Analysis*. Springer New York, 2022.

[30] W.I. Zangwill and B. Mond. *Nonlinear Programming: A Unified Approach*. Prentice-Hall International Series in Management. Prentice-Hall, 1969.

[31] G.M. Ziegler. *Lectures on Polytopes*. Springer-Verlag, 1995.

Index